METHODS IN MOLECULAR BIOLOGY™

For further volumes:
http://www.springer.com/series/7651

For further volumes:
http://www.springer.com/series/[illegible]

Next Generation Microarray Bioinformatics

Methods and Protocols

Edited by

Junbai Wang

Department of Pathology, Oslo University Hospital, Radium Hospital, Montebello, Oslo, Norway

Aik Choon Tan

*Division of Medical Oncology, Department of Medicine, School of Medicine,
University of Colorado Anschutz Medical Campus, Aurora, CO, USA*

Tianhai Tian

School of Mathematical Sciences, Monash University, Melbourne, VIC, Australia

Editors
Junbai Wang, Ph.D.
Department of Pathology
Oslo University Hospital
Radium Hospital
Montebello, Oslo, Norway
junbai.wang@rr-research.no

Aik Choon Tan, Ph.D.
Division of Medical Oncology
Department of Medicine School of Medicine
University of Colorado Anschutz Medical Campus
Aurora, CO, USA
aikchoon.tan@ucdenver.edu

Tianhai Tian, Ph.D.
School of Mathematical Sciences
Monash University
Melbourne, VIC, Australia
tianhai.tian@monash.edu

ISSN 1064-3745 e-ISSN 1940-6029
ISBN 978-1-61779-399-8 e-ISBN 978-1-61779-400-1
DOI 10.1007/978-1-61779-400-1
Springer New York Dordrecht Heidelberg London

Library of Congress Control Number: 2011943561

Printed on acid-free paper

Humana Press is part of Springer Science+Business Media (www.springer.com)

Preface

The twenty-first century is the time of excitement and optimism for biomedical research. Since the completion of the human genome project in 2001, we are entering into the postgenome era where the key research efforts are now interpreting and making sense of these massive genomic data, in order to translate into disease treatment and management. Over the past decade, DNA-based microarrays have been the assays of choice for high-throughput studies of gene expression. Microarray-based expression profiling was provided, for the first time, by means of monitoring genome-wide gene expression changes in a single experiment. Though microarray technology has been widely employed to reveal molecular portraits of gene expression in various cancers' subtypes and correlations with disease progression as well as response to drug treatments, it is not limited to measure gene expression. As the technology became established in early 2000, researchers began to use microarrays to measure other important biological phenomena. For example, (1) Microarrays are being used to genotype single-nucleotide polymorphisms (SNPs) by hybridizing the DNA of individuals to arrays of oligonucleotides representing different polymorphic alleles. The SNP microarray has accelerated genome-wide association studies over the last 5 years, and many loci that are associated with diseases have been discovered and validated. Similarly, another innovative application of the SNP microarray is to interrogate allele-specific expression for identifying disease-associated genes. (2) Array-comparative genomic hybridization (aCGH) is being used to detect genomic structural variations, such as segments of the genome that have varying numbers of copies in different individuals. (3) Epigenetic modifications such as methylation at CpG sites can also be assessed by microarray. (4) Using ChIP-chip assay, genome-wide protein–DNA interactions and chromatin modifications can be profiled by microarrays. (5) More recently, microarray has been used to measure genome-wide microRNA expression patterns to reveal the regulatory role of these noncoding RNAs in disease states. Obviously, the progress of microarray applications is tightly associated with the development of novel computational and statistical methods to analyze and interpret these data sets.

Recent improvements in the efficiency, quality, and cost of genome-wide sequencing have prompted biologists and biomedical researchers to move away from microarray-based technology to ultrahigh-throughput, massively parallel genomic sequencing (Next Generation Sequencing, NGS) technology. NGS technology opens up new research avenues for the investigation of a wide range of biological and medical questions across the entire genome at single base resolution; for example, sequencing of several human genomes, monitoring of genome-wide transcription levels (RNA-seq), understanding of epigenetic phenomena, DNA–protein interactions (ChIP-seq), and de novo sequencing of several genomes. Despite the differences in the underlying sequencing technologies of various NGS machines, the common output from them are the capability to generate tens of millions of short reads (tags) from each experimental run. Thus, NGS technology shifts the bottleneck in sequencing processes from experimental data production to computationally intensive informatics-based data analysis. As in the early days of microarray data analysis, novel computational and statistical methods tailored to NGS are urgently needed for drawing meaningful and accurate conclusions from the massive short reads. Furthermore, it is expected that NGS technology may eventually replace microarray technology in the next decade, which will grow from a pioneering method applied by innovators at the cutting edge research to a ubiquitous technique that will allow researchers to investigate "big-picture" questions in biology at much higher resolution.

This book, Next Generation Microarray Bioinformatics, is our attempt to bring together current computational and statistical methods in analyzing and interpreting both microarray and NGS data. Here, we have compiled and edited 26 chapters that cover a wide range of methodological and application topics in microarray and NGS bioinformatics. These chapters are organized into five thematic sections: (1) Resources for Microarray Bioinformatics; (2) Microarray Data Analysis; (3) Microarray Bioinformatics in Systems Biology; (4) Next Generation Sequencing Data Analysis; and (5) Emerging Applications of Microarray and Next Generation Sequencing. Each chapter is a self-contained review of a specific methodological or application topic. Every chapter typically starts with a brief review of a particular subject, then describes in detail the computational and statistical techniques used to solve the biological questions, and finally discusses the computational results generated by these bioinformatics tools. Therefore, the reader need not read the chapters in a sequential manner. We expect this book would be a valuable methodological resource not only to molecular biologists and computational biologists who are interested in understanding the principle of these methods and designing future research project, but also to computer scientists and statisticians who work in a microarray core facility or other similar organizations that provide service for the high-throughput experiment community.

The first section of this book contains three important resource chapters of microarray and NGS bioinformatics community. The introductory chapter provides an overview on the current state of microarray technologies and is contributed by Kuo and colleagues. The second chapter is contributed by the KEGG group. The KEGG database represents one of the earliest databases to store, manage, integrate, and visualize genomics data. In this chapter, Kotera and colleagues provide the latest developments of the KEGG efforts in analyzing and interpreting omics data. The NCBI Gene Expression Omnibus (GEO) group writes the third chapter in this section, which is one of the major data repositories for high-throughput microarray and next-generation sequencing data. White and Barrett describe various strategies to explore functional genomics data sets in the GEO database.

The second section of this book consists of eight chapters that describe methods to analyze microarray data from the top down approach. The first chapter, contributed by Van Loo and colleagues, that described a novel R-package ASCAT specifically designed to delineate genomic aberration in cancer genomes from SNP microarrays. Then Cheung, Meng, and Huang wrote the following two chapters of advanced machine learning methods in investigating disease classification and time-series microarray data analysis, respectively. Lin and colleagues provide a tutorial on a novel R-package, GeneAnswers, to perform gene-concept network analysis in the next chapter. Nair contributed the next chapter, which emphasizes the utility of R/Bioconductor, an open source software for bioinformatics, in the analysis and interpretation of splice isoforms in microarray. The next three chapters focusing on cross-platform comparisons of microarray data and integrative approaches for microarray data analysis were delivered by Li et al., Hovig et al., and Huttenhower et al., respectively.

The third section of this book concentrates on the bottom-up approaches for establishing different types of models based on microarray expression datasets in which the number of genes is much larger than that of samples. The first chapter written by Yu and colleagues discussed a general profiling method to estimate parameters in the ordinary differential equation models from the time-course gene expression data. To deal with inhomogeneity and nonstationarity in temporal processes, Husmeier and colleagues described the inhomogeneous dynamic Bayesian networks which allow the network structure to change over time in the second chapter. Castelo and Roverato contributed the third chapter that introduced an R package of a graphic approach for inferring regulatory networks from microarray datasets. Wang and Tian contribute the final chapter of this section. They introduced a nonlinear model, which can be used to infer the transcriptional factor activities from the microarray expression data of the target genes as well as to predict the regulatory relationship between transcriptional factors and their target genes.

The fourth section of this book contains six chapters, specifically devoted to NGS data analysis. It starts from an overview of the NGS data analysis by Gogol-Döring and Chen, which includes the basic steps for analyzing NGS such as quality check and mapping to a reference genome. The second chapter is written by Sandber and colleagues, where the authors provide a detailed illustration of how to analyze gene expression using RNA-Sequencing data through several real examples. Lin and colleagues contributed to the third chapter that introduces the low level ChIP-seq data analysis such as preprocessing, normalization, differential identification, and binding pattern characterization. The fourth chapter is contributed by Xu and Sung, in which reader will find how to use Hidden Markov Model to identify differential histone modification sites from ChIP-seq data. The last two chapters describe two software packages (SISSRs developed by Narlikar and Jothi and ChIPMotifs developed by Jin and colleagues) that are designed to study protein–DNA interactions (e.g., peak finder and de novo motif discovery) by analyzing ChIP-based high-throughput experiments.

The final section of this book contains five methodological chapters that cover the emerging applications of microarray and next-generation sequencing in biomedical researchers. In Wei's chapter, it describes Hidden Markov Models for controlling false-discovery rate in genome-wide association analysis. Tan describes Gene Set Top Scoring Pairs (GSTSP), a novel machine learning method in identifying discriminative gene set classifier, based on the relative expression concept. In the next chapter, Wu and Ji focus on JAMIE, a software tool that can perform jointly analysis on multiple ChIP-chip experiments. In the chapter written by Pelligrini and Ferrari, they described an overview on bioinformatics methods in analyzing epigenetic data. The final chapter is a bioinformatics workflow for the analysis and interpretation of genome-wide shRNA synthetic lethal screen based on next-generation sequencing written by Kim and Tan.

We would like to acknowledge the contribution of all authors to the conception and completion of this book. We would like to thank Prof. John M. Walker, the Methods in Molecular Biology series editor, for entrusting and giving us this opportunity to edit this volume. We also like to thank the staff at the Humana Press and Springer publishing company for their professional assistance in preparing this volume. Finally, we would like to thank our families for their love and support.

Oslo, Norway | *Junbai Wang*
Aurora, CO, USA | *Aik Choon Tan*
Melbourne, VIC, Australia | *Tianhai Tian*

Contents

Part III Microarray Bioinformatics in Systems Biology (Bottom-Up Approach)

Part IV Next Generation Sequencing Data Analysis

Part V Emerging Applications of Microarray and Next Generation Sequencing

Contributors

AZZA E. ABDALLA · *Department of Biology, University of South Carolina, Columbia, SC, USA*
TANYA BARRETT · *National Center for Biotechnology Information, National Library of Medicine, National Institutes of Health, Bethesda, MD, USA*
ANNE-LISE BØRRESEN-DALE · *Department of Genetics, Institute for Cancer Research, Oslo University Hospital Radiumhospitalet, Oslo, Norway; Institute for Clinical Medicine, Faculty of Medicine, University of Oslo, Oslo, Norway*
JIGUO CAO · *Department of Statistics and Actuarial Science, Simon Fraser University, Burnaby, BC, Canada*
ROBERT CASTELO · *Research Program on Biomedical Informatics, Department of Experimental and Health Sciences, Universitat Pompeu Fabra, and Institut Municipal d'Investigació Mèdica, Barcelona, Spain*
TAO CHEN · *Division of Genetic and Molecular Toxicology, National Center for Toxicological Research, U.S. Food and Drug Administration, Jefferson, AR, USA*
WEI CHEN · *Berlin Institute for Medical Systems Biology, Max-Delbrück-Center for Molecular Medicine, Berlin, Germany*
LEO WANG-KIT CHEUNG · *Bioinformatics Core, Department of Preventive Medicine and Epidemiology, Stritch School of Medicine, Loyola University Medical Center, Maywood, IL, USA*
HILARY A. COLLER · *Department of Molecular Biology, Princeton University, Princeton, NJ, USA*
FRANK DONDELINGER · *Biomathematics and Statistics Scotland, Scotland, UK School of Informatics, University of Edinburgh, Edinburgh, UK*
PEGGY J. FARNHAM · *Department of Biochemistry & Molecular Biology, Norris Comprehensive Cancer Center, University of Southern California, Los Angeles, CA, USA*
GANG FENG · *Biomedical Informatics Center, Clinical and Translational Sciences Institute, Northwestern University, Chicago, IL, USA*
ROBERTO FERRARI · *Department of Biological Chemistry, University of California, Los Angeles, CA, USA*
ANDREW FRASER · *Department of Allergy and Inflammation, BIDMC, Boston, MA, USA*
ANDREAS GOGOL-DÖRING · *Berlin Institute for Medical Systems Biology, Max-Delbrück-Center for Molecular Medicine, Berlin, Germany*
SUSUMU GOTO · *Bioinformatics Center, Institute for Chemical Research, Kyoto University, Uji, Kyoto, Japan*
GARY HARDIMAN · *Department of Allergy and Inflammation, BIDMC, Boston, MA, USA*
STEVEN Y. HE · *Department of Medicine, University of California San Diego, San Diego, CA, USA*

Mika Hirakawa • *Bioinformatics Center, Institute for Chemical Research, Kyoto University, Uji, Kyoto, Japan*
Eivind Hovig • *Departments of Tumor Biology and Medical Informatics, Institute for Cancer Research, Norwegian Radium Hospital, Montebello, Oslo, Norway*
Kun Huang • *Department of Biomedical Informatics, The Ohio State University, Columbus, OH, USA*
Tim H.-M. Huang • *Department of Molecular Virology, Immunology & Medical Genetics, The Ohio State University, Columbus, OH, USA*
Yufei Huang • *Department of Electrical and Computer Engineering, University of Texas at San Antonio, San Antonio, TX, USA; Greehey Children's Cancer Research Institute, University of Texas Health Science Center at San Antonio, San Antonio, TX, USA*
Dirk Husmeier • *Biomathematics and Statistics Scotland, Scotland, UK*
Curtis Huttenhower • *Department of Biostatistics, Harvard School of Public Health, Boston, MA, USA*
Tor-Kristian Jenssen • *PubGene AS, Vinderen, Oslo, Norway*
Hongkai Ji • *Department of Biostatistics, The Johns Hopkins Bloomberg School of Public Health, Baltimore, MD 21205, USA*
Victor X. Jin • *Department of Biomedical Informatics, The Ohio State University, Columbus, OH, USA*
Raja Jothi • *National Institutes of Environmental Health Sciences, National Institutes of Health, Research Triangle Park, NC, USA*
Minoru Kanehisa • *Bioinformatics Center, Institute for Chemical Research, Kyoto University, Uji, Kyoto, Japan*
Ersen Kavak • *Department of Cell and Molecular Biology, Karolinska Institutet and Ludwig Institute for Cancer Research, Stockholm, Sweden*
Brian A. Kennedy • *Department of Biomedical Informatics, The Ohio State University, Columbus, OH, USA*
Warren A. Kibbe • *Biomedical Informatics Center, Clinical and Translational Sciences Institute, Northwestern University, Chicago, IL, USA*
Jihye Kim • *Division of Medical Oncology, Department of Medicine, School of Medicine, University of Colorado Anschutz Medical Campus, Aurora, CO, USA*
Masaaki Kotera • *Bioinformatics Center, Institute for Chemical Research, Kyoto University, Uji, Kyoto, Japan*
Vessela N. Kristensen • *Department of Genetics, Institute for Cancer Research, Oslo University Hospital Radiumhospitalet, Oslo, Norway; Institute for Clinical Medicine, Institute for Clinical Epidemiology and Molecular Biology (EpiGen), Akershus University Hospital, Faculty of Medicine, University of Oslo, Nordbyhagen, Norway*
Winston Patrick Kuo • *Harvard Catalyst – Laboratory for Innovative Translational Technologies, Harvard Medical School, Boston, MA, USA; Department of Developmental Biology, Harvard School of Dental Medicine, Boston, MA, USA*
Joshua C. Kwekel • *Division of System Biology, National Center for Toxicological Research, U.S. Food and Drug Administration, Jefferson, AR, USA*

JESSICA N. LACY · *Harvard Catalyst – Laboratory for Innovative Translational Technologies, Harvard Medical School, Boston, MA, USA*
XUN LAN · *Department of Biomedical Informatics, The Ohio State University, Columbus, OH, USA*
SOPHIE LÈBRE · *Université de Strasbourg, LSIIT – UMR 7005, Strasbourg, France*
ZHIGUANG LI · *Division of Genetic and Molecular Toxicology, National Center for Toxicological Research, U.S. Food and Drug Administration, Jefferson, AR, USA*
SHILI LIN · *Department of Statistics, The Ohio State University, Columbus, OH, USA*
SIMON M. LIN · *Biomedical Informatics Center, Clinical and Translational Sciences Institute, Northwestern University, Chicago, IL, USA*
OLE CHRISTIAN LINGJÆRDE · *Biomedical Research Group, Department of Informatics, Centre for Cancer Biomedicine, University of Oslo, Oslo, Norway*
FANG LIU · *Department of Tumor Biology, Institute for Cancer Research, Norwegian Radium Hospital, Montebello, Oslo, Norway; PubGene AS, Vinderen, Oslo, Norway*
JIA MENG · *Department of Electrical and Computer Engineering, University of Texas at San Antonio, San Antonio, TX, USA*
T. MURLIDHARAN NAIR · *Departments of Biological Sciences, Computer Science/Informatics, Indiana University South Bend, Bloomington, IN, USA*
LEELAVATI NARLIKAR · *National Institutes of Environmental Health Sciences, National Institutes of Health, Research Triangle Park, NC, USA; Centre for Modeling and Simulation, University of Pune, Pune, Maharashtra, India*
GRO NILSEN · *Biomedical Research Group, Department of Informatics, Centre for Cancer Biomedicine, University of Oslo, Oslo, Norway*
SILJE H. NORDGARD · *Department of Genetics, Institute for Cancer Research, Oslo University Hospital Radiumhospitalet, Oslo, Norway*
LUCILA OHNO-MACHADO · *Division of Biomedical Informatics, University of California San Diego, San Diego, CA, USA*
MATTEO PELLEGRINI · *Department of Molecular, Cell and Developmental, University of California, Los Angeles, CA, USA*
XIN QI · *School of Public Health, Yale University, New Haven, CT, USA*
DANIEL RAMSKÖLD · *Department of Cell and Molecular Biology, Karolinska Institutet and Ludwig Institute for Cancer Research, Stockholm, Sweden*
CHIARA RIVAS-MORELLO · *Harvard Catalyst – Laboratory for Innovative Translational Technologies, Harvard Medical School, Boston, MA, USA*
JAE-HYUNG ROBERT · *Department of Developmental Biology, Harvard School of Dental Medicine, Boston, MA, USA*
STEVEN T. ROSEN · *Robert H. Lurie Comprehensive Cancer Center, Northwestern University, Chicago, IL, USA*
ALBERTO ROVERATO · *Department of Statistical Science, Università di Bologna, Bologna, Italy*
RICKARD SANDBERG · *Department of Cell and Molecular Biology, Karolinska Institutet and Ludwig Institute for Cancer Research, Stockholm, Sweden*

PAMELA SHAW • *Galter Health Sciences Library, Northwestern University, Chicago, IL, USA*
WING-KIN SUNG • *Department of Computational and Mathematical Biology, Genome Institute of Singapore, Singapore, Singapore; School of Computing, National University of Singapore, Singapore, Singapore*
AIK CHOON TAN • *Division of Medical Oncology, Department of Medicine, School of Medicine, University of Colorado Anschutz Medical Campus, Aurora, CO, USA*
CENNY TASLIM • *Department of Molecular Virology, Immunology & Medical Genetics, The Ohio State University, Columbus, OH, USA; Department of Statistics, The Ohio State University, Columbus, OH, USA*
TIANHAI TIAN • *School of Mathematical Sciences, Monash University, Melbourne, VIC, Australia*
TOSHIAKI TOKIMATSU • *Bioinformatics Center, Institute for Chemical Research, Kyoto University, Uji, Kyoto, Japan*
ALEXANDER J. TRACHTENBERG • *Harvard Catalyst – Laboratory for Innovative Translational Technologies, Harvard Medical School, Boston, MA, USA*
ALLISON TRUONG • *Department of Biology, University of California Los Angeles, Los Angeles, CA, USA*
PETER VAN LOO • *Cancer Genome Project, Wellcome Trust Sanger Institute, Hinxton, Cambridge, UK; Department of Molecular and Developmental Genetics, VIB, Leuven, Belgium; Department of Human Genetics, University of Leuven, Leuven, Belgium*
HANS KRISTIAN MOEN VOLLAN • *Department of Genetics, Institute for Cancer Research, Oslo University Hospital Radiumhospitalet, Oslo, Norway; Institute for Clinical Medicine, Faculty of Medicine, University of Oslo, Oslo, Norway; Division of Surgery and Cancer, Department of Breast and Endocrine Surgery, Oslo University Hospital Ulleval, Oslo, Norway*
LEVI WALDRON • *Department of Biostatistics, Harvard School of Public Health, Boston, MA, USA*
JUNBAI WANG • *Department of Pathology, Oslo University Hospital, Radium Hospital, Montebello, Oslo, Norway*
ZHI WEI • *Department of Computer Science, New Jersey Institute of Technology, Newark, NJ, USA*
STEPHEN E. WILHITE • *National Center for Biotechnology Information, National Library of Medicine, National Institutes of Health, Bethesda, MD, USA*
HAO WU • *Department of Biostatistics and Bioinformatics, Emory University, Atlanta, GA, USA*
HAN XU • *Department of Computational and Mathematical Biology, Genome Institute of Singapore, Singapore, Singapore*
HONGYU ZHAO • *School of Public Health, Yale University, New Haven, CT, USA*

Part I

Introduction and Resources for Microarray Bioinformatics

Chapter 1

A Primer on the Current State of Microarray Technologies

Alexander J. Trachtenberg, Jae-Hyung Robert, Azza E. Abdalla, Andrew Fraser, Steven Y. He, Jessica N. Lacy, Chiara Rivas-Morello, Allison Truong, Gary Hardiman, Lucila Ohno-Machado, Fang Liu, Eivind Hovig, and Winston Patrick Kuo

Abstract

DNA microarray technology has been used for genome-wide gene expression studies that incorporate molecular genetics and computer science analyses on massive levels. The availability of microarrays permit the simultaneous analysis of tens of thousands of genes for the purposes of gene discovery, disease diagnosis, improved drug development, and therapeutics tailored to specific disease processes. In this chapter, we provide an overview on the current state of common microarray technologies and platforms. Since many genes contribute to normal functioning, research efforts are moving from the search for a disease-specific gene to the understanding of the biochemical and molecular functioning of a variety of genes whose disrupted interaction in complicated networks can lead to a disease state. The field of microarrays has evolved over the past decade and is now standardized with a high level of quality control, while providing a relatively inexpensive and reliable alternative to studying various aspects of gene expression.

Key words: Microarrays, Gene expression, One dye, Two dye, High throughput, QRT-PCR, Cross platform

1. Introduction

The term "microarray" refers to the orderly arrangement, "array," of the probes of interest in a grid format used at a small size, "micro." The genomics context for the term "microarray" often refers to the apparatus where single-stranded DNA oligonucleotides (short sequences of nucleotides) or "oligos" are affixed to a solid surface. Single-stranded DNA has a natural affinity, under particular chemistry and conditions, to anneal to its complementary sequence of single-stranded DNA or RNA. Because of its

Junbai Wang et al. (eds.), *Next Generation Microarray Bioinformatics: Methods and Protocols*,
Methods in Molecular Biology, vol. 802, DOI 10.1007/978-1-61779-400-1_1, © Springer Science+Business Media, LLC 2012

affinity to become double stranded, when a sample, in appropriate buffer, is added to the surface of the microarray, the free floating samples will hybridize to the immobilized complementary DNA oligo. Depending on the protocol, a fluorescent dye is either added prior to sample addition and hybridization or after the DNA hybridization to the microarray. Prior to sample addition, one or two fluorescent dyes can be used. In this context, a microarray is a high-throughput DNA or RNA hybridization platform for performing gene expression analysis (although protein arrays are also available, this chapter focuses on DNA/RNA microarrays). Unlike its predecessors in gene expression studies (such as differential/subtractive hybridization and RNase protection assay), microarray allow for gene expression analysis of thousands of genes, capable of covering the whole genome (approximately 25,000 genes for the human genome) from as little as 50–100 ng of total RNA. The technology was revolutionized by the ability to synthesize gene-specific probes onto a silicon surface, as achieved by Affymetrix®. This is in contrast to the early days of microarray technology, where individual laboratories immobilized prefabricated cDNA/oligos onto derivative glass slides using robotic printing instruments. Today, multiple commercial platforms provide microarrays customized to an individual's specific needs (focus/pathway/disease-specific arrays).

2. Materials

2.1. Materials Needed for a Microarray Experiment

1. RNA (isolation from a biological sample).
2. Microarray chip (preferably commercial platforms).
3. In vitro transcription/RNA amplification kit (if starting RNA levels are low).
4. Labeling kit (often specific and optimized to the microarray platform of interest).
5. Hybridization station and chambers (often specific to the microarray platform).
6. Scanner for image capture (see Note 1).
7. Software for data analysis.

2.2. Basic Microarray Menu of Methodologies

2.2.1. Sample Preparation (RNA Isolation)

The first step in running a microarray experiment is the isolation of RNA from a biological sample. Once RNA is extracted, the samples should be processed using an Agilent 2100 Bioanalyzer (Agilent Technologies, Santa Clara, CA) to check for integrity and purity of mRNA – A260/A280 ratio (see Note 2). While protein and DNA contaminations will interfere with proper measurement of RNA being assayed, organic solvent contamination (e.g., ethanol), as measured by A260/A230 ratio, would interfere with labeling this RNA by hindering efficacy of the cDNA synthesis reaction (see Note 3).

2.2.2. Generation of cDNA or aRNA from Isolated mRNA

Once RNA is obtained, mRNA is converted to cDNA using reverse transcription. The conversion of mRNA (or genetic DNA) into cDNA or aRNA may also involve the tagging of nucleic acids for subsequent labeling reaction (following the manufacturer's protocol). Optional in vitro amplification of RNA can be performed using commercial amplification kits when the starting RNA concentration is low. The created cDNA represents one mRNA in the sample.

2.2.3. Labeling of the In Vitro Transcribed Transcripts

cDNA needs to be labeled to provide a fluorescent signal during hybridization. The most common labeling dyes used for microarray detection are Cy3 and Cy5 dyes. These fluorescent dyes are usually conjugated to a secondary complex that stably interacts with the tag that is incorporated into the cDNA. As an example, the secondary complexes can be primers complementary to the tag or streptavidin if biotinylated primers are used for generating transcripts.

2.2.4. Hybridization to Gene-Specific Oligo-Probes

The hybridization step aims in placing the labeled cDNA on the surface of the microarray under stringent conditions to facilitate sequence-specific binding. This is a rate-limiting step in the microarray process that can last as long as 20 h (overnight), although the use of microfluidics has significantly reduced the hybridization time. If microarray chips are on glass slides, it is highly advisable to use closed chambers with slide hybridization stations to limit evaporation and gentle agitation to increase hybridization efficiency.

2.2.5. Scanning/Data Acquisition

After the microarray experiment is completed, the slide/chip is ready for scanning. Laser-based scanners are used to generate an image of the microarray that has the labeled cDNA samples that are bound to the probes. The image is then used to decipher the hybridization efficiency of each feature/spot on the microarray that correlates to the relative abundance of the target gene in the sample of interest.

2.2.6. Data Analysis

Whole genome microarrays contain approximately 25,000 genes; each gene may be represented by multiple probes. Ideally, each experimental condition consists of biological triplicates, thus further burdening data analysis. Several software packages have been commonly used, for example, JMP® Genomics (SAS Institute, Cary, NC), MatLab® (The MathWorks, Natick, MA), and R software environments such as BioConductor (1).

3. Methods

3.1. Gene Expression Profiling

The primary use of microarray technology is gene expression analysis. Gene expression is an intermediate step before the assembly of proteins from their amino acid building blocks. When a gene is expressed, messenger RNA (mRNA) is produced ("transcribed") from the gene's DNA sequence, and it serves as a template to guide the synthesis of a protein, allowing particular amino acids to be systematically incorporated into a protein (Fig. 1). The mRNA transcript is a complement of a corresponding part of the DNA coding region. The purpose of a gene expression microarray is to measure how much mRNA corresponding to a particular gene is present in the cell(s) or tissue of interest. The principle behind microarrays is that complementary sequences will bind to each other under proper conditions, whereas noncomplimentary sequences will not bind. For example, if the DNA sequence on an array is ten nucleotides long, TACCGAACTG, the sequence ATGGCTTGAC will "hybridize" to the

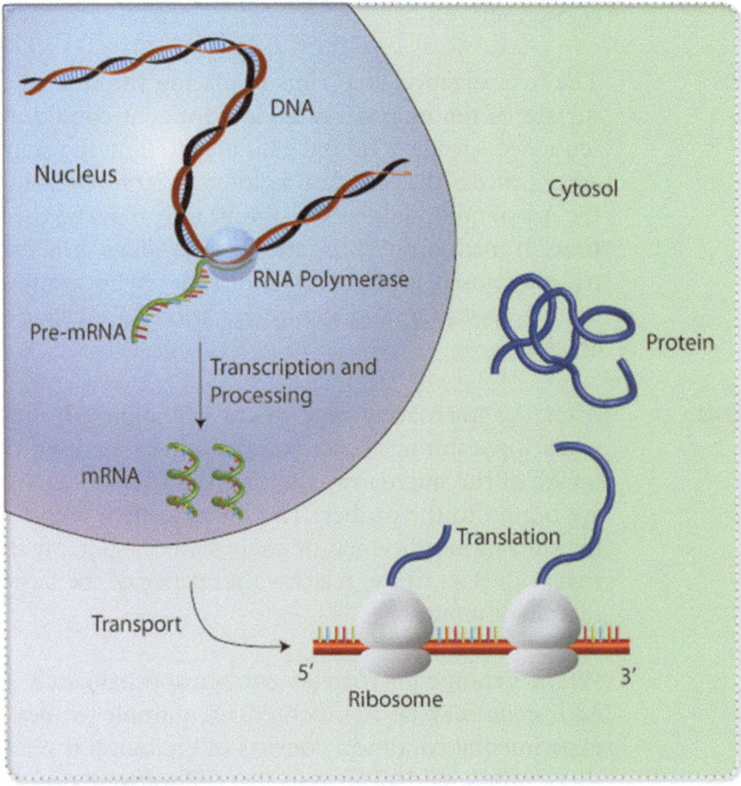

Fig. 1. Transcription of DNA to mRNA and translation of mRNA to protein. Activities of the cell are controlled by instructions contained in the DNA sequences, through mRNA that carries the genetic information (transcription) from the cell to the cytoplasm, where proteins are produced (translation).

probe ("A" nucleotides complement "T" and "C" nucleotides complement "G"). Probes are designed to be specific to a gene that is positioned on the microarray. In general, differential gene expression response to a specific stimulus is compared to untreated samples, thereby distinguishing stimuli-specific gene expression responses. Kulesh et al. first used microarray analysis to identify interferon-induced genes (2). Since this study, tens of thousands of studies using different microarray platforms have been published, with the majority of these involving differential gene expression analyses.

3.2. Microarray Design

A standard microarray consists of gene-specific probes cross-linked onto a solid surface such as glass, plastic, or silicon biochip. Although microarray chips can be produced "in-house," the consistency and quality of commercial arrays more than justifies their cost (3). The probes are generally oligos (ranging from 25 to 85 bp in length), although gene fragments or PCR products have also served as probes in the past. The probes can be deposited onto an array surface either by "spotting" presynthesized oligos or cDNA (otherwise known as Stanford type cDNA array) or by directly synthesizing or "ink-jet printing" the oligos on the array surface. Due to the logistics of synthesizing and cataloging the thousands of presynthesized oligos, "spotting" tends to be much more difficult to do, although it remains a commercially available technology. In contrast, two technologies, photolithographic synthesis (as advanced by Affymetrix®) (4) and ink-jet printing (Agilent, among others), (5) are alternative methods that add probe content onto a standard microarray. The advantage of photolithographic approach is the ability to place many more probes onto a single microarray slide or chip, which is not feasible with ink-jet printing. Since the photolithographic method is capable of providing hundreds of thousands of probes on each chip, multiple probes for individual genes are used to increase its reliability. In contrast, the ink-jet method is much more restricted with regard to the number of probes it can print on a single microarray chip.

3.3. One-Dye vs. Two-Dye Microarrays

In one-dye microarrays, a microarray experiment is performed using transcripts from a single sample (Fig. 2a). For the purpose of performing differential gene expression analysis, all samples are labeled with a single fluorescent label (usually Cy3 or Cy5). In contrast, two-dye microarrays are performed where differential gene expression is performed directly on a single microarray chip using two different fluorescent labels. Two dyes are often used so an experimental (test sample) and a control (reference) can be hybridized to the same array leading to ratios of the two colors in various proportions. For example, sample 1 of group A can be labeled with Cy3 (emission wavelength of 570 nm) while sample 1 of group B is labeled with Cy5 (emission wavelength of 670 nm). The two samples labeled with unique fluorescent markers are

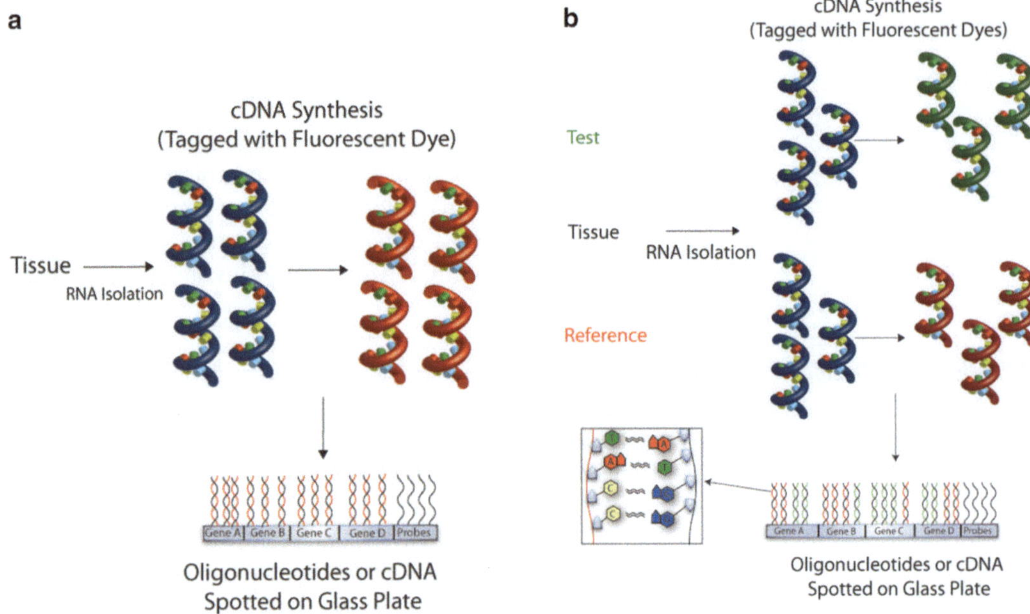

Fig. 2. (**a**, **b**) One-dye and two-dye microarray platforms. Microarrays contain thousands of probes (oligonucleotides) that can vary in length (from 25 to over 1,000 bp) and are affixed onto a solid surface. Microarray experiments can be divided into two groups based on their labeling: (**a**) one-dye or (**b**) two-dye microarray experiments. Essentially, in two-dye experiments, two samples are labeled each with a distinct dye (e.g., one sample with Cy3-dye and the other with Cy5-dye), producing a ratio unit measurement, whereas in a one-dye experiment, an absolute unit of measurement is generated.

then combined on a single microarray chip and hybridized with affixed complementary microarray probes (Fig. 2b). Since Cy3 emits green light and Cy5 emits red light, the combined emission would indicate the abundance of one over the other (e.g., orange would indicate more red fluorescence than green while yellow would indicate more green fluorescence than red). This ratio of green and red fluorescence, after accounting for possible loading error, would indicate the differential expression profile between the example group A and group B. The major assumption is that the abundance of mRNA corresponding to a certain gene is positively correlated with the expression of a certain gene. However, it has been found that one-dye microarray platforms provide more consistent results than two-dye microarray platforms (3) (see Note 4).

3.4. 2D- vs. 3D-Microarrays

The intrinsic nature of 2D-microarray surface is the limitation on the density of the probes that can be printed in a given area. As a result, the hybridization generated from the probe–transcript interaction on a 2D surface has intrinsically low signal to noise ratio (SNR), contributing to decreased sensitivity and dynamic range. Generally, a standard microarray platform has a dynamic range of about 2.5–3.5 logs (6), in contrast, real-time PCR can have a dynamic range as high as 7 logs. A novel way to address this shortcoming is achieved through 3D microarrays (7, 8) where each gene-specific probe is secured onto the walls of microchannels, therefore resulting in greater probe density within a given field (since the device used to capture the image/fluorescence detects in a 2D plane) (see Note 5). The close proximity of the probes to target transcripts (due to the architecture of the microchannels) and the ability to use microfluidics also allows for greatly reduced hybridization times when compared to a 2D surface (9). Finally, enzymatic reactions, such as chemiluminescence can be used to substitute for fluorescence. Considering that Cy5, a commonly used fluorescent dye in microarray, is susceptible to ozone (10), the ability to use chemiluminescence provides a viable alternative to generate consistent microarray data. 3D-microarrays are ideal for customized arrays or for gene expression analysis in pathways of interest as each array supports up to 500 probes. However, 3D-microarray systems usually allow simultaneous multisample processing. For example, the Ziplex® System (Axela, Toronto, ON, Canada) is a multiplex gene expression platform that combines total assay integration using their proprietary flow-through chip technology that allows a researcher to processing eight unique samples within a few hours (11).

3.5. Particle/Bead Microarrays

Another method for enhancing the transcript capturing density (thereby enhancing SNR) is illustrated by Illumina®'s BeadArray™ technology (12). In particle/bead-array technology, beads are coupled to an "address" oligomer of 29 bases that is, in turn, linked to a 50-mer oligo probe. Each bead (approximately 3 μm) is covered with more than 100,000 probes, providing a 3D surface within a given area. The small bead size also allows for greater number of features per microarray slide. In fact, Illumina®'s BeadArray™ platform (HumanHT-12 v4 Expression BeadChip, Illumina®, San Diego, CA) allows as many as 12 simultaneous sample analyses on a single slide.

3.6. Types of Gene Expression Analysis

In addition to conventional gene expression analysis, other aspects of gene expression can be analyzed by the use of microarrays. Listed below are four types of commercially available microarray chips that cater to specific aspects of gene expression analysis.

3.6.1. Splicing/Fusion Analysis

Although the human genome consists of approximately three billion base pairs of DNA, it only codes for about 25,000 genes. Each gene is often capable of producing different proteins with different functions due to alternative splicing. Another way to increase the diversity of proteins is found in gene fusion, which is known to be responsible for some cancers. A microarray can be used to detect alternative splicing variants and fusion genes by probing for exon junctions and fusion junctions, respectively, of mature transcripts.

3.6.2. Single Nucleotide Polymorphism Analysis

It is possible to be heterogeneic for the same gene due to single nucleotide polymorphisms (SNPs) (acquired by inheritance or mutation) where the alleles may differ by a single nucleotide. Even though the allelic difference may be innocuous in some cases, SNPs can contribute to disease susceptibility by affecting either protein function or abundance (13–17). A high-density SNP microarray can be designed to detect not only SNPs, but also other variations in genetic material. Unlike conventional chromosomal microarrays that only detect loss or gain of genetic material, SNP microarrays are able to detect copy number neutral loss of heterozygosity and uniparental disomy, which are found in tumors (18). Current consensus supports SNP analysis as a prerequisite for providing personalized medicine-based therapy. As such, drug efficacy is being evaluated in the context of SNPs to correlate differences in individual response to therapy.

3.6.3. Tiling/Full Coverage Analysis

A DNA microarray, in general, probes for annotated genes; in contrast, a tiling array or a high-density whole genome array allows unbiased detection of an unknown or a lowly expressed genome (19). The array consists of either partially overlapping or nonoverlapping probes that span the entire genome. Tiling arrays are particularly useful in addressing DNA–protein interaction studies. Prefabricated commercial tiling array chips exist for gene expression analysis including Chromatin ImmunoPrecipitation (ChIP)-chip, transcriptome mapping, MeDIP-chip, and Dnase Chip, as well as SNP and DNA methylation analysis.

3.6.4. DNA/RNA–Protein Interactions

The interaction of nucleic acids and proteins plays an important role in biological systems, including DNA–protein interactions (in transcriptional regulation and replication), rRNA–protein interactions (in translation), hnRNA–spliceosome interactions, as well as miRNA processing by the Dicer complex or the identification of miRNA target transcripts (20).

Even though the above arrays are commercially available, recent advances allow individual laboratories to customize arrays to their own needs. Namely, the Geniom® One (Febit, Inc, Lexington, MA), is a stand-alone system that allows a researcher to (1) print oligonucleotides (from 25 to 85 mers) on a microfluidics biochip consisting of eight channels that can hold 15,000 features (therefore, affording the ability to run eight samples simultaneously or run one sample for 120,000 unique features), (2) hybridize samples, and (3) detect and analyze the signal intensity. By automating most of the processes, human error is reduced, thus, minimizing the level of variation in the data.

3.7. Microarray Databases

Gene expression data derived from microarrays can be obtained in Web supplements to journal publications or in public repositories. Numerous microarray repository/database exists; most notably the Gene Expression Omnibus (21) by the National Center for Biotechnology Information (NCBI) and ArrayExpress (22, 23) by the European Bioinformatics Institute (EBI). In this context, it is important that this information be archived in standardized fashion (see Note 6). This effort toward standardization has been initiated by the Microarray Gene Expression Data (MGED) Society (24), which has taken the initiative to develop and enforce guidelines, formats, and tools for submission of microarray data (25). This allows researchers to share common information and make valid comparisons among experiments. MGED is an international organization of scientists involved with gene expression profiles. Their primary contributions are proposed standards for publication and data communication. MGED proposed Minimal Information About a Microarray Experiment (MIAME) as a potential publication standard (26).

3.8. Cross-Platform Studies

The diversity of platforms and microarray data raise questions of whether and how data from different platforms can be compared and combined. Early studies comparing Stanford type cDNA arrays to Affymetrix oligonucleotide arrays demonstrated poor consistency between the two platforms (27). The interplatform inconsistency resulted from factors inherent to probe design (GC-content, probe length, signal intensity, etc.). The importance of probe design was further supported by other studies showing improved consistency when the two platforms target a gene in overlapping regions of the transcript (3, 28, 29). Because of the diversity of technical and analytical sources that can affect the results of an experiment and hence affect comparison among experiments, standardization within a single platform may be insufficient. Results from cross-platform comparisons have been mixed (30–33). Nonetheless, several comparison studies involving microarrays have justified guarded optimism for the reproducibility of measurements across platforms, while also indicating the need for further large-scale comparison studies (34, 35).

Kuo et al. were the first group to present a large-scale comprehensive cross-platform comparison of DNA microarrays (3). Their results demonstrated that greater interplatform consistency was observed in highly expressing genes than in low expressing genes (3). When the same microarray experiments were performed in different laboratories, there was greater interlaboratory variability than intralaboratory variability, demonstrating users also play a role in generating different gene expression measurements (3). The results suggested that there are many platforms available that provide good quality data, especially on highly expressed genes, and that, among these platforms, there is generally good agreement.

Another large initiative was the MicroArray Quality Control (MAQC) project (36), spearheaded by the Food and Drug

Administration (FDA). The MAQC attempted to develop the following:

- Provide quality control (QC) tools to the microarray community to avoid procedural failures.
- Develop guidelines for microarray data analysis by providing the public with large reference datasets along with readily accessible reference RNA samples.
- Establish QC metrics and thresholds for objectively assessing the performance achievable by various microarray platforms.
- Evaluate the advantages and disadvantages of various data analysis methods.

The MAQC study involved six FDA Centers, major providers of microarray platforms and RNA samples, the Environmental Protection Agency, the National Institute of Science and Technology, academic laboratories, and other stakeholders. Two human reference RNA samples were selected (see Note 7), and differential gene expression levels between the two samples were measured by microarrays and other technologies [e.g., Quantitative Real-Time Polymerase Chain Reaction (QRT-PCR)]. The resulting microarray datasets were used for assessing the precision and cross-platform/laboratory consistency of microarray results, and the QRT-PCR datasets enabled evaluation of the nature and magnitude of systematic biases that existed between microarrays and QRT-PCR. The availability of the well-characterized RNA samples combined with the resulting microarray and QRT-PCR datasets, which have been made readily accessible to the scientific community, allow individual laboratories to more easily identify and correct procedural failures. As shown by the MAQC consortium, sufficient consistency is seen in intraplatform and interplatform comparisons (37).

3.9. Cutting Edge Microarray Technologies

As discussed above, a microarray provides a flexible platform for revealing many aspects of gene expression and chromosomal characteristics. However, the vast majority of microarray platforms are designed to address one specific aspect of a gene (such as its level of expression, transcript variability, allelic heterogeneity, etc.) using a high-throughput approach. Figure 3 lists a description of commonly used commercially available microarray platforms including those discussed in this section. A new strategy in microarray design involves multiplexing. For example, the NanoString® Technologies nCounter™ Analysis System (38) allows a researcher to multiplex up to 800 gene transcripts in a single reaction without amplification. Other recent technologies incorporates QRT-PCR into the microarray format (see Note 8), like the OpenArray® (Life Technologies™ Corporation, Carlsbad, CA) system and Fluidigm® (39) platforms. The OpenArray® allows a researcher to

Company	Platform	Application	Customizable	Sample Type	Input Amount	Dynamic Range (log)	Probe length	1 dye	2 dye	Website
Agilent Technologies	SurePrint G3	Comparative Genomic Hybridization	Yes	RNA	10-200ng	5-6	60mer	Yes	Yes	http://www.genomics.agilent.com
		Copy Number Variation								
		ChIP-chip/DNA Methylation								
		DNA Methylation								
		Whole Genome Expression	No							
	Human Gene Expression 4X44k	Genome Expression	Yes	RNA	10-200ng	5-6	60mer	Yes	Yes	
		Comparative Genomic Hybridization								
	Whole Human Genome	Genome Expression	Yes	RNA	10-200ng	5-6	60mer	Yes	Yes	
	Human ENCODE ChIP-on-chip	ChIP-chip	Yes	RNA	10-200ng	5-6	60mer	Yes	Yes	
Affymetrix®	Human Exon Array	Whole Genome Expression	No	RNA	100ng-2µg	4	25mer	Yes	Yes	http://www.affymetrix.com/estore/index.jsp
		Splice Variants	No							
	Human Gene Array	Whole Genome Expression	No	RNA	100ng-2µg	4	25mer	Yes	Yes	
	Human Genome Array	Tiling Array	No	DNA						
Roche Nimblegen	Eukaryotic Array	Gene Expression	Yes	RNA	10µg	log2 of 4 - 16	60mer	Yes	Yes	http://www.nimblegen.com
	Comparative Genomic Hybridization	Cytogenetics	Yes	RNA	10µg	log2 of 4 - 16	50-75mer	Yes	Yes	
		Copy Number Variation								
		Comparative Genomic Hybridization								
		Tiling								
	Human 2.1M WG-T	Whole Genome Tiling ChIP	No	DNA	6.5µg/IP	log2 of 4 - 16	50-75mer	Yes	Yes	
		Whole Genome Tiling DNA Methylation	No							
Illumina®	Human HT-12 Expression BeadChip	Whole Genome Expression	Yes	RNA	50-500ng	3-4	50mer	Yes	No	http://www.illumina.com
		cDNA-mediated Annealing, Selection, Extension, and Ligation (DASL)	Yes	RNA	10-100ng					
Miltenyi Biotech	PIQOR™	Focus Array	Yes	RNA	250ng-40µg	5-6	200-400mer	Yes	Yes	http://www.miltenyibiotec.com/en/default.aspx
Phalynx Biotech Group	Human OneArray™	Whole Genome Expresion	No	RNA	5-10µg	2-3	60mer	Yes	Yes	http://www.phalanxbiotech.com/main.html
Fluidigm®	Digital Array™ Chip	Copy Number Variation	Yes	RNA	<1ug	6				http://fluidigm.com/
	Dynamic Array™ Chip	Single Cell Gene Expression								
NanoString Technologies®	nCounter™ Gene Expression	Gene Expression	Yes	RNA	100-250ng	5-6	50mer			http://www.nanostring.com/
	nCounter™ miRNA Expression Assay Kits	Human miRNA Quantification	Yes	RNA	100-250ng	5-6	50mer			
	nCounter™ Custom Copy Number Variation	Copy Number Variation	Yes	DNA	100-250ng	5-6	50mer			
Febit	Geniom® Biochip	Gene Expression	Yes	RNA	500ng	3-4	70mer	Yes	No	http://www.febit.com/
Axela	Xceed Molecular - Ziplex® System - 3 Dimensional TipChip	Gene Expression	Yes	RNA	500ng	3	35-50mer	Yes	No	http://www.axela.com/
High Throughput Genomics	qNPA™ ArrayPlate	Gene Expression	Yes	RNA	5ng	3-4	50mer	Yes	No	http://www.htgenomics.com/
Life Technologies™	OpenArray® Assays	Gene Expression	Yes	RNA	50ng	6		Yes	No	http://www.lifetechnologies.com/home.html

Fig. 3. List of commercially available microarray platforms. Attributes of the table are: company name, platform, application, whether the platform is customizable, sample type, input amount, dynamic range, probe length, one-dye or two-dye platform, and company Web site.

perform QRT-PCR on 3,072 unique features simultaneously (33 nl reactions), thereby bypassing the validation process entirely. The Fluidigm® platform can perform up to 2,304 or 9,216 reactions simultaneously on their 48.48 (10 nl reactions) and 96.96 (5 nl reactions) dynamic arrays, respectively.

However, microarrays are likely to be substituted by sequencing technologies. In fact, second generation sequencing has already surpassed microarray hybridization in ChIP assays. In ChIP-chip assay, DNA pulled down by immunoprecipitation needs to be identified by hybridization to a known oligo probe. Since the DNA is unknown, several thousands of oligo-probes are used for hybridization (see Subheading 3). In ChIP-seq, however, the DNA is sequenced directly using second generation sequencing (40, 41). The resulting analysis then reveals the identity of the region to which the transcription factor binds, the relative changes in transcription factor binding (as evidenced by the abundance of the sequenced region), as well as the detection of mutations in a given site. Furthermore, the technological and economical advances made in second generation sequencing make ChIP-seq a much more attractive option.

In summary, microarray technologies have revolutionized genomic research in the past decade and virtually every domain of biological science has been impacted by this technology. The area has evolved significantly from home-grown spotted arrays to commercial quality controlled microarrays. Nevertheless, currently the microarray field has been gradually giving way to the next wave of sequencing technologies. It would be interesting to see the future role of microarrays play out as DNA sequencing technologies under development promise to bring huge strides in sequencing speed and cost reduction in the next decade.

4. Notes

1. There is a wide selection of microarray scanners, calibrating your scanner is a critical step for determining the dynamic range, detection limit and uniformity of microarray scanners. In addition, this step will also detect laser channel cross-talk and laser stability.
2. As a suggestion, if using TRIzol-isolated (Life Technologies™ Corporation, Carlsbad, CA) RNA for cDNA synthesis, it is beneficial to perform a secondary cleanup step. Immediately after the ethanol precipitation step in the TRIzol procedure, proceed with a cleanup kit according to the manufacturer's recommendations.

3. Pure and intact RNA and cDNA should have A260/A280 and A260/A230 ratios of at least 1.8. In addition, they should appear intact when analyzed by gel electrophoresis or using an Agilent 2100 Bioanalyzer (Agilent Technologies, Santa Clara, CA).
4. One-dye microarray experiments have shown to be more consistent than two-dye microarray experiments. The strength lies in the fact that an aberrant sample cannot affect the raw data derived from other samples, because each array chip is exposed to only one sample. The disadvantage is that, when compared to the two-dye system, the one-dye approach requires twice as many microarrays to compare samples within an experiment.
5. In 3D-microarrays, because the surfaces have much higher binding capacity, they can offer more reactive sites to bind to the target, which greatly improves the sensitivity of the microarray.
6. Most journals require that authors submitting manuscripts that describe results of their microarray experiments make the raw and normalized data and protocol descriptions available in MIAME-compliant format in either of the two main public data repositories [Gene Expression Omnibus (GEO) from NCBI or ArrayExpress from EBI].
7. The Universal Human Reference RNA and Human Brain Reference Total RNA reference samples presented in the MAQC project are both commercially available from Agilent and Life Technologies, respectively. The accessibility of these samples permits the evaluation of new microarray platforms as they emerge in terms of their reproducibility and quality of their results.
8. The advantage of high-throughput QRT-PCR strategies has been the small reaction volumes that are needed and significant reduction in reagent costs. This has been tremendously useful in cases where the starting material is limited. When the reaction volumes are in the nanoliter levels, liquid handlers are needed.

Acknowledgments

This work was conducted with support from Harvard Catalyst – The Harvard Clinical and Translational Science Center (NIH Award #UL1 RR 025758 and financial contributions from Harvard University and its affiliated academic health care centers). The content is solely the responsibility of the authors and does not necessarily represent the official views of Harvard Catalyst, Harvard University and its affiliated academic health care centers, the National Center for Research Resources, or the National Institutes of Health.

Alexander J. Trachtenberg and Jae-Hyung Robert Chang contributed equally to this work.

References

1. Gentleman RC, Carey VJ, Bates DM et al (2004) Bioconductor: open software development for computational biology and bioinformatics. Genome Biol 5:R80.
2. Kulesh DA, Clive DR, Zarlenga DS et al (1987) Identification of interferon-modulated proliferation-related cDNA sequences. Proc Natl Acad Sci U S A 84: 8453–8457.
3. Kuo WP, Liu F, Trimarchi J et al (2006) A sequence-oriented comparison of gene expression measurements across different hybridization-based technologies. Nat Biotechnol 24:832–840.
4. Fodor SP, Read JL, Pirrung MC et al (1991) Light-directed, spatially addressable parallel chemical synthesis. Science 251:767–773.
5. Lausted C, Dahl T, Warren C et al (2004) POSaM: a fast, flexible, open-source, inkjet oligonucleotide synthesizer and microarrayer. Genome Biol 5:R58.
6. Baum M, Bielau S, Rittner N et al (2003) Validation of a novel, fully integrated and flexible microarray benchtop facility for gene expression profiling. Nucleic Acids Res 31: e151.
7. Ruano JM, Benoit VV, Aitchison JS et al (2000) Flame hydrolysis deposition of glass on silicon for the integration of optical and microfluidic devices. Anal Chem 72: 1093–1097.
8. Benoit V, Steel A, Torres M et al (2001) Evaluation of three-dimensional microchannel glass biochips for multiplexed nucleic acid fluorescence hybridization assays. Anal Chem 73:2412–2420.
9. Hokaiwado N, Asamoto M, Tsujimura K et al (2004) Rapid analysis of gene expression changes caused by liver carcinogens and chemopreventive agents using a newly developed three-dimensional microarray system. Cancer Sci 95: 123–130.
10. Fare TL, Coffey EM, Dai H, et al (2003) Effects of atmospheric ozone on microarray data quality. Anal Chem 75:4672–4675.
11. Quinn MC, Wilson DJ, Young F et al (2009) The chemiluminescence based Ziplex automated workstation focus array reproduces ovarian cancer Affymetrix GeneChip expression profiles. J Transl Med 7:55.
12. Gunderson KL, Kruglyak S, Graige MS et al (2004) Decoding randomly ordered DNA arrays. Genome Res 14:870–877.
13. Bond GL, Hu W, Levine A (2005) A single nucleotide polymorphism in the MDM2 gene: from a molecular and cellular explanation to clinical effect. Cancer Res 65:5481–5484.
14. Guilford P, Hopkins J, Harraway J et al (1998) E-cadherin germline mutations in familial gastric cancer. Nature 392:402–405.
15. Imyanitov EN (2009) Gene polymorphisms, apoptotic capacity and cancer risk. Hum Genet 125:239–246.
16. Lindblad-Toh K, Tanenbaum DM, Daly MJ et al (2000) Loss-of-heterozygosity analysis of small-cell lung carcinomas using single-nucleotide polymorphism arrays. Nat Biotechnol 18:1001–1005.
17. Reddy EP (1983) Nucleotide sequence analysis of the T24 human bladder carcinoma oncogene. Science 220:1061–1063.
18. Tuna M, Knuutila S, Mills GB (2009) Uniparental disomy in cancer. Trends Mol Med 15:120–128.
19. Mockler TC, Chan S, Sundaresan A et al (2005) Applications of DNA tiling arrays for whole-genome analysis. Genomics 85:1–15.
20. Nonne N, Ameyar-Zazoua M, Souidi M et al (2010) Tandem affinity purification of miRNA target mRNAs (TAP-Tar). Nucleic Acids Res 38:e20.
21. Wheeler DL, Church DM, Lash AE et al (2001) Database resources of the National Center for Biotechnology Information. Nucleic Acids Res 29:11–16.
22. Brazma A, Parkinson H, Sarkans U et al (2003) ArrayExpress – a public repository for microarray gene expression data at the EBI. Nucleic Acids Res 31:68–71.
23. Brooksbank C, Camon E, Harris MA et al (2003) The European Bioinformatics Institute's data resources. Nucleic Acids Res 31:43–50.
24. Ball CA, Sherlock G, Parkinson H et al (2002) Standards for microarray data. Science 298:539.

25. Ikeo K, Ishi-i J, Tamura T et al (2003) CIBEX: center for information biology gene expression database. C R Biol 326:1079–1082.
26. Brazma A, Hingamp P, Quackenbush J et al (2001) Minimum information about a microarray experiment (MIAME) – toward standards for microarray data. Nat Genet 29:365–371.
27. Kuo WP, Jenssen TK, Butte AJ et al (2002) Analysis of matched mRNA measurements from two different microarray technologies. Bioinformatics 18: 405–412.
28. Mecham BH, Klus GT, Strovel J et al (2004) Sequence-matched probes produce increased cross-platform consistency and more reproducible biological results in microarray-based gene expression measurements. Nucleic Acids Res 32:e74.
29. Carter SL, Eklund AC, Mecham BH et al (2005) Redefinition of Affymetrix probe sets by sequence overlap with cDNA microarray probes reduces cross-platform inconsistencies in cancer-associated gene expression measurements. BMC Bioinformatics 6:107.
30. Bammler T, Beyer RP, Bhattacharya S et al (2005) Standardizing global gene expression analysis between laboratories and across platforms. Nat Methods 2: 351–356.
31. Larkin JE, Frank BC, Gavras H et al (2005) Independence and reproducibility across microarray platforms. Nat Methods 2:337–344.
32. Wang H, He X, Band M et al (2005) A study of inter-lab and inter-platform agreement of DNA microarray data. BMC Genomics 6:71.
33. Zhu B, Ping G, Shinohara Y et al (2005) Comparison of gene expression measurements from cDNA and 60-mer oligonucleotide microarrays. Genomics 85:657–665.
34. Barnes M, Freudenberg J, Thompson S et al (2005) Experimental comparison and cross-validation of the Affymetrix and Illumina gene expression analysis platforms. Nucleic Acids Res 33:5914–5923.
35. Sherlock G (2005) Of fish and chips. Nat Methods 2:329–330.
36. Casciano DA, Woodcock J (2006) Empowering microarrays in the regulatory setting. Nat Biotechnol 24:1103.
37. Shi L, Reid LH, Jones WD et al (2006) The MicroArray Quality Control (MAQC) project shows inter- and intraplatform reproducibility of gene expression measurements. Nat Biotechnol 24:1151–1161.
38. Geiss GK, Bumgarner RE, Birditt B et al (2008) Direct multiplexed measurement of gene expression with color-coded probe pairs. Nat Biotechnol 26:317–325.
39. Spurgeon SL, Jones RC, Ramakrishnan R (2008) High throughput gene expression measurement with real time PCR in a microfluidic dynamic array. PLoS One 3:e1662.
40. Robertson G, Hirst M, Bainbridge M et al (2007) Genome-wide profiles of STAT1 DNA association using chromatin immunoprecipitation and massively parallel sequencing. Nat Methods 4:651–657.
41. Park PJ (2009) ChIP-seq: advantages and challenges of a maturing technology. Nat Rev Genet 10:669–680.

Chapter 2

The KEGG Databases and Tools Facilitating Omics Analysis: Latest Developments Involving Human Diseases and Pharmaceuticals

Masaaki Kotera, Mika Hirakawa, Toshiaki Tokimatsu, Susumu Goto, and Minoru Kanehisa

Abstract

In this chapter, we demonstrate the usability of the KEGG (Kyoto encyclopedia of genes and genomes) databases and tools, especially focusing on the visualization of the omics data. The desktop application KegArray and many Web-based tools are tightly integrated with the KEGG knowledgebase, which helps visualize and interpret large amount of data derived from high-throughput measurement techniques including microarray, metagenome, and metabolome analyses. Recently developed resources for human disease, drug, and plant research are also mentioned.

Key words: Pathway map, KEGG orthology, BRITE hierarchy, KEGG API, KegArray

1. Introduction

"Omics" is a general term for a research field of life science analyzing massive amounts of interactions of biological information objects, including genome, transcriptome, proteome, metabolome, and many other derivatives. As omics data has been rapidly accumulating as the result of recent development of high-throughput measurement techniques, the needs for omics-data integration have been becoming more important. In general, bioinformatics techniques have been developed and utilized to computationally process a vast amount of biological data. However, only the collection and computation of these data is not sufficient to understand the complete and dynamic system of life programmed in the genome sequence. These data must be described as the knowledge on life science, i.e., network diagram of various interactions such as cellular functions,

Junbai Wang et al. (eds.), *Next Generation Microarray Bioinformatics: Methods and Protocols*, Methods in Molecular Biology, vol. 802, DOI 10.1007/978-1-61779-400-1_2,

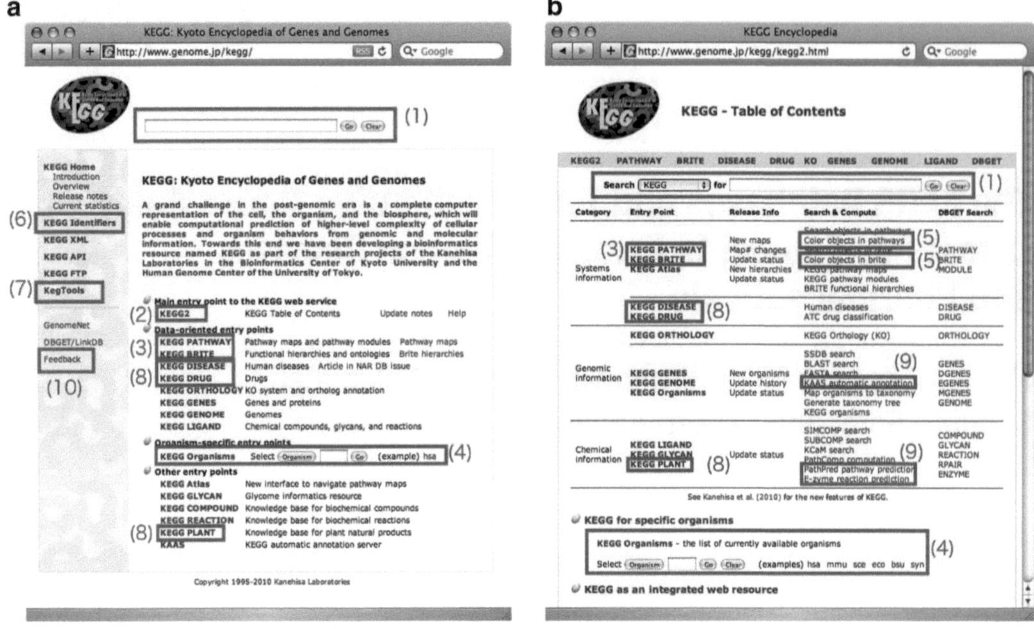

Fig. 1. Overview of the KEGG homepage and sitemap. (**a**) KEGG homepage. (**b**) KEGG2: sitemap. (1) Search boxes. (2) Link to KEGG2. (3) KEGG PATHWAY/BRITE. (4) KEGG Organisms: entry points for the genome-sequenced organisms (see Note 1). The user can limit the search only in an organism of interest (see Note 2). (5) Tools to customize PATHWAY/BRITE, with which the user can color the objects of interest (see Subheading 2.2). (6) KEGG Identifiers. The gene accession numbers from the outside databases can be converted to the corresponding KEGG IDs from here (see Note 3). The users can also obtain the multiple KEGG entries simultaneously (see Note 4). (7) KegTools: Desktop applications, KegHier, KegArray, and KegDraw can be downloaded from here (see Subheadings 2.1 and 2.3, and Note 5, respectively). (8) KEGG DISEASE/DRUG/PLANT. (9) KAAS, PathPred, and E-zyme tools to create new pathways (see Notes 5 and 6). (10) Feedback: Any questions or comments are appreciated (see Note 7).

signaling/metabolic pathways, and enzyme reactions. Thus, we have been focusing on generating the integrated knowledge database named KEGG (Kyoto encyclopedia of genes and genomes) (1) by the high-quality manual curation.

KEGG can be seen as an efficient viewer of living systems. The main page is given in ref. 2 (Fig. 1), and it can also be reached from GenomeNet (3). KEGG and GenomeNet have a search option named "*dbget*" (4), by which the user can use any term without knowing the database structure, just like to "*google*" without knowing how web pages are linked to each other in the Internet. The user can find many similar search boxes in many different pages in KEGG, which can generally be used in the same way, with the mere differences in the selection of databases being searched and the display style. The user need not know which database contains the data of interest, since the dbget searches all relevant data throughout all databases. This integrity is a big advantage with which the user cannot only look up the data of interest, but can also trace the links to collect and understand the relevant information.

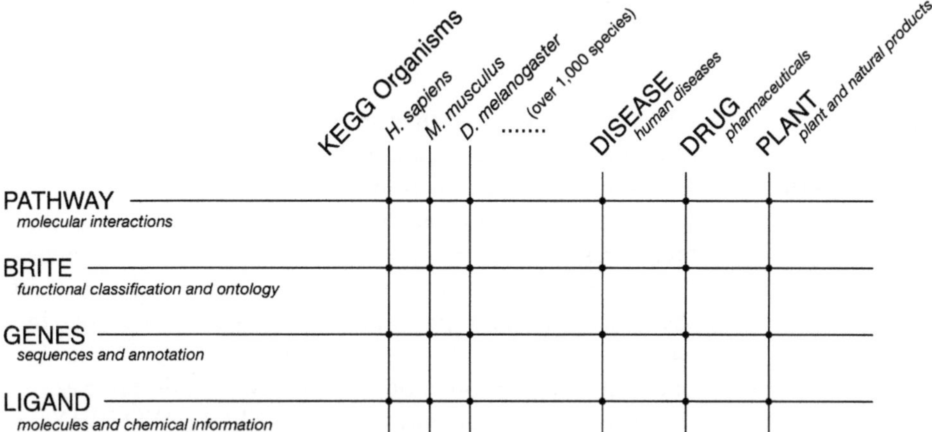

Fig. 2. Grid-shaped structure of the KEGG data. KEGG has a variety of entry points from which the user can start searching or analyzing data, depending on the various perspective. For example, PATHWAY contains molecular interaction data such as metabolic or regulatory pathways throughout all the genome-sequenced organisms, which we refer to as "reference pathways" (Fig. 3a). The user can also limit the pathway for only a specified organism (see Note 1), or can compare the pathways in different organisms (see Subheading 2.1). The DISEASE category of the PATHWAY database (or the PATHWAY category of the DISEASE database) can be regarded as the human pathways that are perturbed by diseases. The DRUG and PLANT categories of the PATHWAY database are the collections of pathway maps specialized for pharmaceuticals and plants, respectively. These relationships also apply for other databases such as BRITE, GENES, and LIGAND. This figure is illustrated simply for the explanation: the actual structure is a little more complicated. For example, chemical compounds in LIGAND are also hierarchically classified in BRITE. Similarly, GENES are grouped by KO (KEGG Orthology), which is also hierarchically classified in BRITE.

At the first sight, the KEGG data structure seems quite complicated, because there are many Web pages (which we refer to as "entry points") focusing on different objects and different purposes, even though they occasionally reach the same data. However, this becomes actually advantageous when the user learns the basics about the KEGG data structure. Figure 2 describes the grid-shaped relationships of the KEGG data. KEGG can be divided into four main databases: PATHWAY, BRITE, GENES, and LIGAND, from one perspective. GENES consists of genes and genomes (see Note 8 for details), while LIGAND contains the other objects, e.g., metabolites and reactions (5). PATHWAY describes intermolecular networks such as regulatory or metabolic pathways, and BRITE is a collection of hierarchical classifications (ontology) of biological or pharmaceutical vocabularies. In other words, GENES and LIGAND are the databases of "components," while PATHWAY and BRITE are those of "circuits" of living systems. On the other hand, the recently developed resources, e.g., DISEASE, DRUG, and PLANT, view the data in different ways. They focus on human diseases, pharmaceutical compounds, and plants, respectively, with the same usability of GENES, LIGAND, PATHWAY, and BRITE. Thus, the user can use the same data and tools with the most efficient way depending on the situation and purpose.

2. Methods

2.1. Experience the Structure of PATHWAY/BRITE

KEGG PATHWAY (6) had started as a computational description of metabolic pathways, and still keeps growing and expanding to represent the phenomenon (such as metabolism, cellular processes, and human diseases) manually compiled from published literatures. KEGG has about 400 maps where the genes from genome-sequenced organisms are assigned, and the number of the organisms and pathway maps keeps increasing. In other words, the user is able to compare the genomes in the viewpoint of about 400 phenomenon just by viewing this database.

Browsing the pathway map using KEGG PATHWAY is similar to searching a restaurant using the Internet. The user might want to view and understand the content (the collection of the genes, proteins, and small molecules) and context (their interaction) in the organism of interest. The user might input the name of the restaurant into the search box, or narrow down the search area from the map. The KEGG PATHWAY can be used just in the same way, i.e., the user can search the gene or any substances in whichever pathway, or browse many pathways in a specified organism, or compare the specified pathway in many species, just by choosing options or clicking links.

KEGG PATHWAY entries generally do not focus on a specific organism. Reference pathways are defined as the combined pathways that are present in a number of organisms and are consensus among many published papers. Only the reference pathway map is manually drawn; all other organism-specific maps are computationally generated. The KEGG pathway map is manually drawn with in-house software called KegSketch, which generates the KGML (KEGG Markup Language; see ref. 7) file. This xml files contain graphics information and also KEGG entry, relation, and reaction information.

GENES and PATHWAY can be viewed in two different ways (Fig. 2): the limited search in an organism of interest, and the comprehensive search throughout all genome-sequenced organisms. The former method is explained in Note 1. Here, we explain the latter method. Figure 3a is a screenshot of the inositol phosphate metabolism pathway, which can be seen by clicking one of the links on the PATHWAY main page. In this graphic, rectangles and circles represent gene products (mostly proteins) and other molecules (mostly metabolites), respectively. This black-and-white graphic is one of the reference pathways for which no organism has been specified.

The user can view the organism-specific pathways by using the pull-down menu. Figure 3b is taken as an example PATHWAY page of a specified organism. The colored rectangles in this page

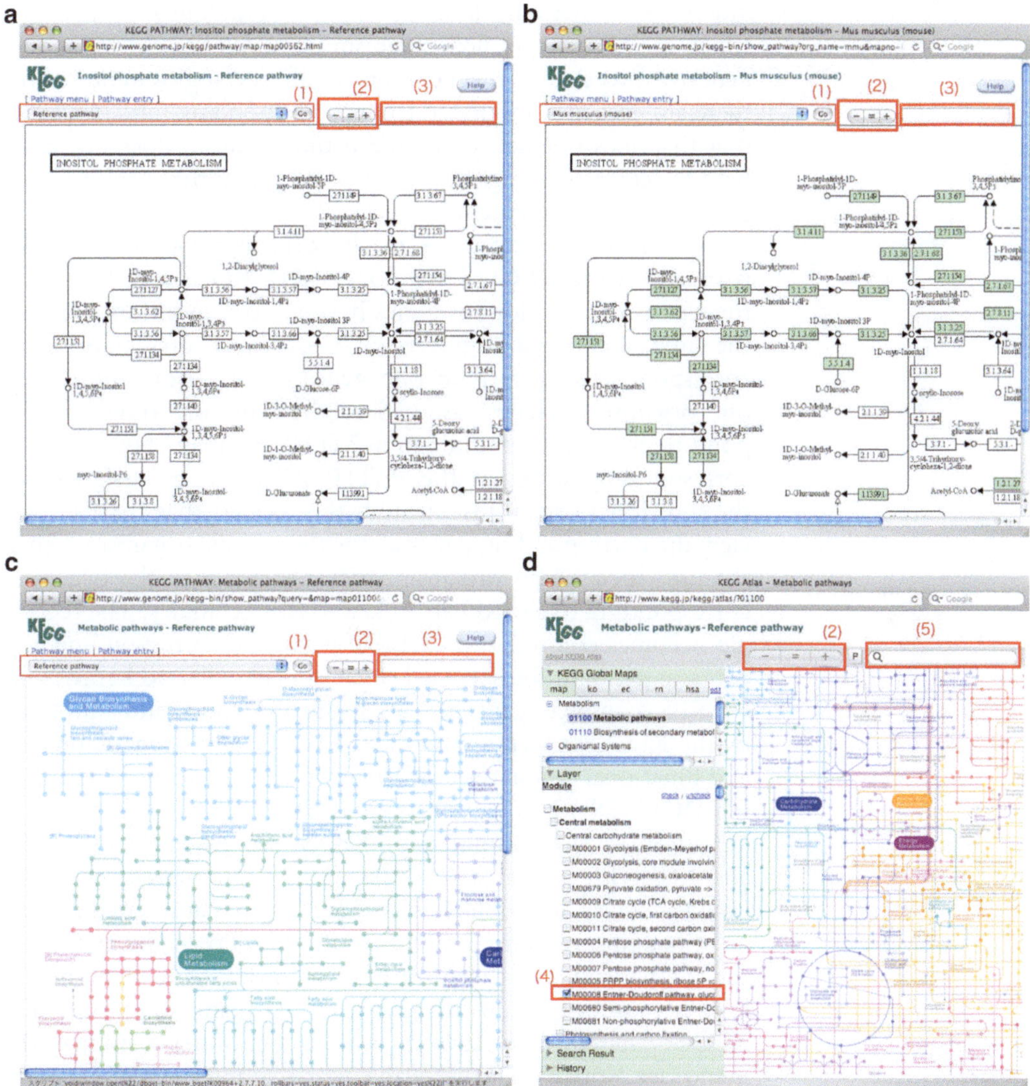

Fig. 3. KEGG PATHWAY and Atlas. (**a**) KEGG PATHWAY map of inositol phosphate metabolism as a reference pathway. Chemical compounds are represented as *circles*, and gene products (such as enzyme proteins) are represented as *rectangles*. (**b**) The same map with the genes information deduced from mice genome. (**c**) An example global map. Chemical compounds are represented as *dots*, and enzyme reactions are represented as *lines*. Different categories of pathways are drawn in different colors in a map. (**d**) KEGG Atlas. (1) The pull-down menu to choose an organism. If the user selects "reference pathway" in the menu, the rectangles provide the links to other objects that are not specific to an organism, such as enzymes, reactions, and KO (KEGG Orthology). The user can customize the selection of organism in the menu (see Note 2). (2) The graphics can be zoomed in or out by clicking these buttons. (3) Input any term in this search box, and the corresponding objects are highlighted, if any. (4) KEGG Modules, manually defined tighter functional units for pathways and protein complexes, can be selected to emphasis the part of the global map of interest. (5) Search box accepting any term to navigate the Atlas.

indicate that there are links to the corresponding GENE pages, which means the specified organism possesses the corresponding genes or proteins in the genome. White rectangles indicate that there are no genes annotated to the corresponding function. Note that this does not necessarily mean the organism does not really have the corresponding genes. It is possible that the corresponding genes have not been identified yet.

Coloring the rectangles in the organism-specific pathways is based on the KEGG Orthology (KO). KO is a collection of the classes of orthologous genes having a common function and the same evolutional origin. An orthology (KO entry) in principle corresponds to more than one genes derived from more than one organisms. Genes assigned to the same orthology correspond to the same rectangle in a PATHWAY map (Fig. 3a). The corresponding genes in the PATHWAY maps are assigned for the individual organisms through the KO, so that the user can view the corresponding pathway for the specific organism. When the user specifies an organism, then the genes in the organism corresponding to the KO are linked to the rectangles. The rectangle becomes colored and clickable when the corresponding KO contains genes in the specified organism (Fig. 3b). KO entries for GENES (complete genomes) are manually defined and annotated by the KEGG expert curators based on the phylogenetic profiles and functional annotations of the genes. On the other hand, KO for DGENES (draft genomes) and EGENES (EST sequences) are automatically annotated by KAAS (see Note 6). DGENES and EGENES have relatively less number of colored rectangles (and less links) due to the less number of genes annotated to KO.

Changing organisms by using the pull-down menu enables the comparison of pathways among organisms. The menu is very long because it contains the entire set of organisms registered in KEGG. Therefore, we provide a useful option to customize the menu (see Note 2). The user can emphasize any genes or chemical compounds using any color to customize the pathway map for presentation (see Subheading 2). KEGG PATHWAY is also useful for understanding the relationships of the genes identified in experiments such as microarray analysis. The user can quickly obtain the graphics representing the functions to which the genes up- (down-) regulated in microarray experiments are related (see Subheading 3).

KEGG PATHWAY recently incorporated new types of pathway maps, named "Global Maps" (Fig. 3c), which are also reachable from the PATHWAY top page. The user can map any set of genes to grasp the overview by using the Global Maps. We expect this will become more valuable for the interpretation of metagenome and pangenome studies. We also developed a new graphical interface, KEGG Atlas (8), to map smaller functional units (such as pathway maps and pathway modules) in the Global Maps with zooming and navigation capabilities (Fig. 3d).

KEGG BRITE (9) represents the hierarchy of vocabularies used in papers, references, and academic communities. It contains the widely accepted classifications derived from other databases or references, and hierarchical classifications that we originally compiled (see Subheading 4 and Fig. 8c for a DISEASE example), as well as the hierarchy of the substances defined in KEGG (such as KO). The BRITE functional hierarchies contain tab-delimited fields, which can be handled by the desktop application KegHier (downloadable from the KEGG homepage; see Fig. 1).

2.2. Customize the PATHWAY/BRITE as You Like

The user can color KEGG PATHWAY/BRITE as necessary. As explained above, when the user specifies an organism, the gene products are colored in pathway maps (Fig. 3b). There is also an option to specify multiple organisms at a time (see Note 9). In addition, when the user inputs the term of interest into the search box, the corresponding objects are colored (as explained in Fig. 3). Here, we provide more flexible options to color PATHWAY (10) or BRITE (11). Figure 4a is reachable from the KEGG sitemap (see Fig. 1b). The user can easily find any objects of interest (genes, metabolites, etc.) in the KEGG PATHWAY or BRITE by coloring them (Fig. 4c, d). The objects have to be specified by the KEGG IDs. Therefore, if the objects of interest are represented by the identifiers of other databases, they have to be converted into the KEGG IDs (see Note 3).

Another flexible option is available through the KEGG API (12). KEGG API is a Web service to use the KEGG system from the user's program via SOAP/WSDL. The service enables the user to develop software that accesses and manipulates a massive amount of online KEGG contents that are constantly refreshed. KEGG API provides many useful functions, including those for coloring pathways that colors the given objects on the pathway map with the specified colors and returns the URL of the colored image.

For the users who would like to deal with the pathways that are not still present in KEGG PATHWAY, we provide a number of options. See Note 5 for details.

2.3. Use the KegArray Application

KegArray is a Java application that provides an environment to analyze either transcriptome/proteome and metabolome data. Closely integrated with the KEGG database, KegArray enables the user to easily map those data to KEGG resources including PATHWAY, BRITE, and genome maps. It can be downloaded from the KegTools page (13) linked from the KEGG homepage (Fig. 1a).

KegArray can read the transcriptome data format of the KEGG EXPRESSION database (14) or tab-deliminated text similar to the

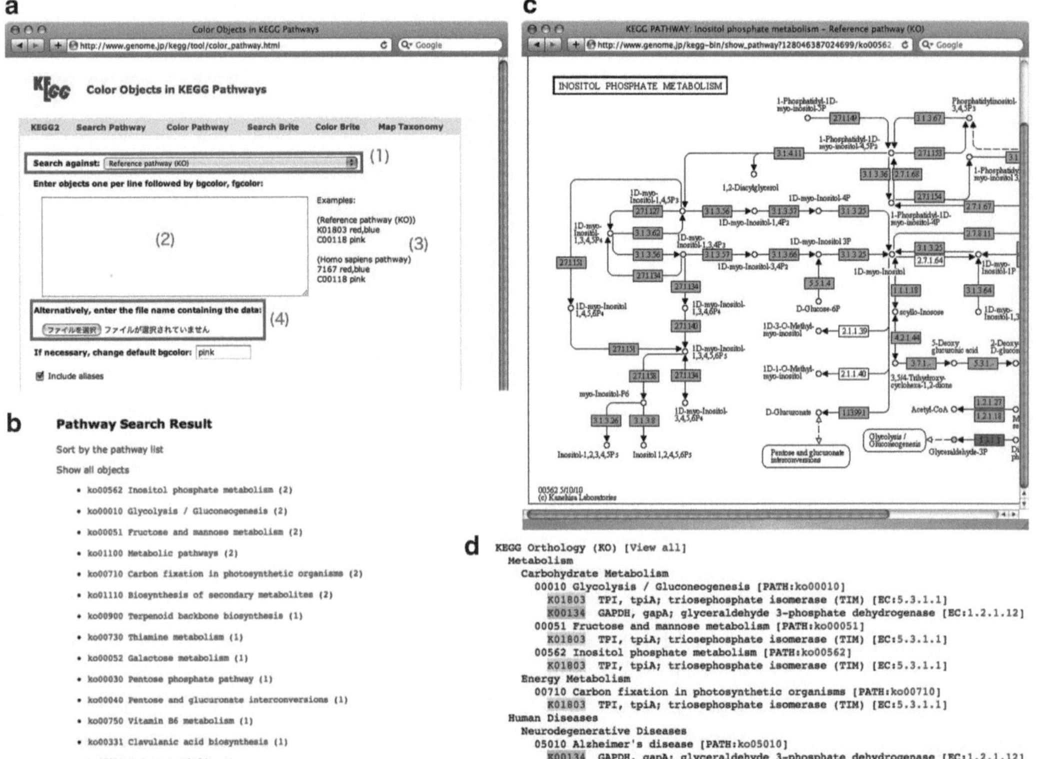

Fig. 4. Color objects in PATHWAY/BRITE. (**a**) The page for coloring the KEGG pathways. (1) An organism or a reference pathway has to be specified in this menu. (2) Input the list of the genes by KEGG IDs and colors for them. (3) Examples of the inputs are shown here. (4) The input data can be also uploaded from here. (**b**) After clicking the "Exec" button, the list of the PATHWAY maps containing the input objects is displayed. (**c**) One of the pathways derived from the resulting list. The graphics of the maps are automatically generated as gif files, which will be removed from the KEGG server within few hours. If the user wants to preserve the graphics, they should be downloaded to the local computer. (**d**) An example result of coloring the BRITE functional hierarchy. The user can grasp the genes of interest at a sight, with using different colors for different groups as the user wants.

EXPRESSION format. Each entry of EXPRESSION consists of brief descriptions about experiment, reference information, and a set of intensity values or ratios of two-channels derived from a DNA microarray. Examples for intensity values and for expression ratios between two-channels are given in Fig. 5a, b, respectively. KegArray also deals with the metabolome data, although only ratio values can be available as shown in Fig. 5c. To convert data in Microsoft Excel format for KegArray, the user needs to order the columns as in the KegArray format in advance and save them as a tab-delimited text.

Once KegArray is launched, the user can see the KegArray control panel (Fig. 6), where there are two tabs to select "Gene/Compound" or "Clustering" on the top. In the "Gene/Compound" pane, the user can load a data file of transcriptome and/or metabolome experiments from the local computer or the KEGG EXPRESSION database, by clicking the "Local" or "GenomeNet"

a

#organism: syn						
#ORF	x	y	Control-sig	Control-bkg	Target-sig	Target-bkg
slr1485	1	2	1037.13	502.62	1593.30	695.25
slr1119	1	3	1261.63	494.72	2685.37	742.87
sll0708	1	4	922.97	561.38	1598.37	727.28
sll1120	1	6	2152.80	560.96	2591.23	771.07
sll1734	1	7	1918.47	574.57	5968.97	823.66
:	:	:	:	:	:	:
:	:	:	:	:	:	:

b

#organism: syn			
#ORF	X	Y	Ratio
slr1485	1	2	0.610282
slr1119	1	3	2.360655
sll0708	1	4	0.842321
sll1120	1	5	0.769038
:	:	:	:
:	:	:	:

c

# COMPOUND	ratio
C00668	1.2
C00221	0.5
C01172	2.2
C00118	1.0
:	:
:	:

Fig. 5. Example input files for KegArray. All lines beginning with the "#" character (other than the "#organism:" or "#source:" line) are regarded as comments and skipped by KegArray. The organism information is necessary to identify the ORFs. The organism should be provided by the three-letter (or four-letter) KEGG Organism code (see Note 1). The lines in tab-delimited format below the #ORF section contain gene expression profile data. (**a**) Table representing intensity values: First column represents the KEGG GENES ID, the unique identifier of the ORF in the organism. The second and third columns are for specifying the location (*X*- and *Y*-axis coordinates, respectively) of the ORF on the DNA microarray. The fourth and fifth columns are the signal intensity and the background intensity of the control channel, respectively. The sixth and seventh columns are the signal intensity and the background intensity of the target channel, respectively. (**b**) Table representing ratio values: The first column is for the KEGG GENES ID. The second and the third columns are *X*- and *Y*-axis coordinate information of the ORF on the microarray, respectively. The fourth column describes the ratio value between control channel and target channel. (**c**) Table representing metabolome data: The first column represents KEGG COMPOUND ID, and the second column represents the relative amount of the target compound compared with the control.

buttons, respectively. The user can obtain the list of up- or down-regulated genes (or compounds) by choosing the option from the menu. The number of listed genes can be modified by changing the value in the box at upper-right of the pop-up table. The up- or downregulated genes (or compounds) can be mapped onto PATHWAY, Genome map, and BRITE for the user to understand the result (as the examples shown in Fig. 7).

In the "Clustering" pane, the user can load several data files of transcriptome experiments and set an intensity threshold. Once the user selects more than one data files, the "Clustering" button becomes active. Clicking this button performs hierarchical clustering of the gene expression profiles constructed from the files listed. A tree-view window is shown when the calculation is completed. The user can change the number of clusters (1–6) by

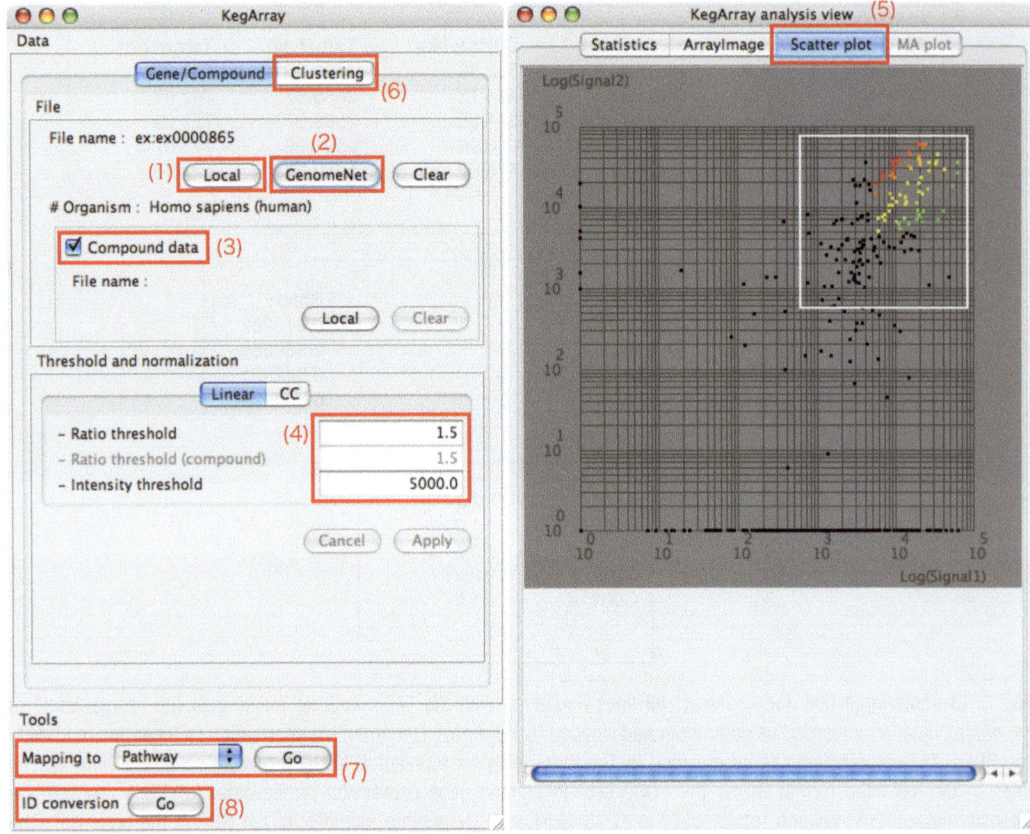

Fig. 6. Screenshots of the KegArray control panels. (1) The "Local" button opens a pop-up window to select a data file on your local disk. The data file should comply with the format described in Fig. 5. (2) The "GenomeNet" button opens a pop-up window to retrieve the data stored in the GenomeNet EXPRESSION database. Available entry IDs are listed in the window, and once you select one, its description will be displayed. (3) The "Compound data" box should be checked (default) for loading metabolome data. (4) There are three input boxes to specify the parameters for the confidence lines discriminating the regulated genes/compounds from unregulated ones. (5) The scatter plot of the data is shown in this pane. The colors of spots represent levels of increase or decrease of the target gene expressions against the control. The coloring scheme can be changed in the preference menu. (6) The "Clustering" pane. (7) Mapping to PATHWAY, Genome Map, and BRITE. (8) ID conversion tool (see Note 3).

specifying the number in the input box at the top of the tree-view window. Different clusters are shown in different colors. Clicking the "Set results" button saves the color-coding for further analysis using the Tools section.

2.4. Overview the DISEASE/DRUG Resources

Before closing this chapter, we briefly explain recently released three resources for specific requirements: DISEASE, DRUG, and PLANT. DISEASE database contains information of human molecular system perturbed by gene mutation, infection of pathogens, etc. DRUG database contains information of pharmaceutical compounds, identified with the chemical structures and classified hierarchically based on various perspectives: the Anatomical Therapeutic Chemical (ATC) Classification System, US pharmacopeia

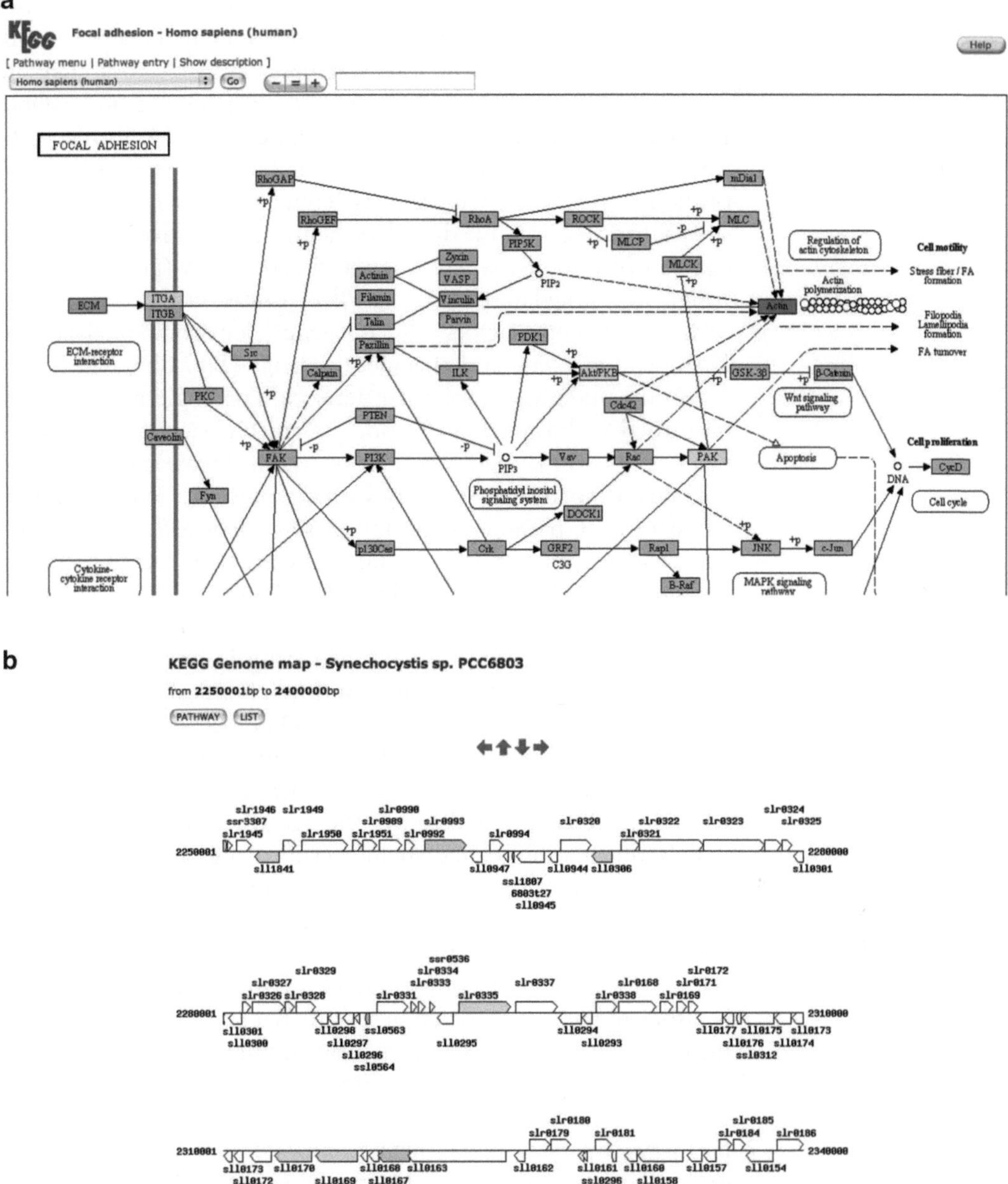

Fig. 7. Mapping microarray data onto PATHWAY/GENOME/BRITE. KegArray has options to visualize the up- or down-regulated genes on various KEGG objects, i.e., (**a**) PATHWAY, (**b**) GENOME, and BRITE. The input data does not have to be from microarray experiments; KegArray can be used as a visualization tool of gene functions as long as the data complies the format described in Fig. 5.

(USP) classification, Therapeutic category of drugs in Japan, etc. Plant species produce those with medical, nutritional, and environmental values, which is one of the motivations for us to produce the PLANT resources and the EDRUG database.

KEGG DISEASE (15) is a new collection of disease entries capturing knowledge on genetic and environmental perturbations. There are a number of disease databases available, but they are mostly descriptive databases for humans to read and understand. Disease information in KEGG is in more computable forms, pathway maps, and gene/molecule lists. The Human Diseases category of the KEGG PATHWAY database contains multifactorial diseases such as cancers, immune disorders, neurodegenerative diseases, and circulatory diseases, where known disease genes (genetic perturbants) are marked in red (Fig. 8a). Each disease entry contains a list of known genetic factors (disease genes), environmental factors, diagnostic markers, and therapeutic drugs (Fig. 8b), which may reflect the underlying molecular network. For single-gene diseases, perturbed pathway maps are not drawn, but causative genes are mapped to normal pathway maps through disease entries. It also contains some infectious diseases where molecular interaction networks of both pathogens and humans are depicted. Diseases with known genetic factors and infectious diseases with known pathogen genomes are being organized in KEGG DISEASE and classified in the BRITE hierarchy (Fig. 8c).

KEGG DRUG (16) is a unified drug information resource that contains chemical structures and/or chemical components of all prescription and over-the-counter (OTC) drugs in Japan, most prescription drugs in the USA, and many prescription drugs in Europe. All the marketed drugs in Japan are fully represented in KEGG DRUG and linked to the package insert information (labels information). These include crude drugs and TCM (Traditional Chinese Medicine) drugs, which are popular in Japan and some of which are specified in the Japanese Pharmacopeia. Each KEGG DRUG entry distinguishes the chemical structure of chemicals or the chemical component of mixtures and crude drugs. It is associated with generic names, trade names, efficacy, and target information, as well as information about the history of drug development. KEGG DRUG contains information about three types of molecular networks. The first is the drug degradation pathways by drug-metabolizing enzymes. The second is the molecular interaction network involving target and other molecules. The drug–target relationship is not simply a molecule–molecule relationship. The target is given in the context of KEGG pathways, enabling the analysis of drugs as perturbants to molecular systems. The last molecular network is the one representing drug development history (17). Many marketed drugs have been developed from lead compounds or existing drugs by introducing chemical structure transformations retaining the core chemical structures. KEGG DRUG structure maps graphically illustrated knowledge on such drug development in a manner similar to the KEGG pathway maps.

a

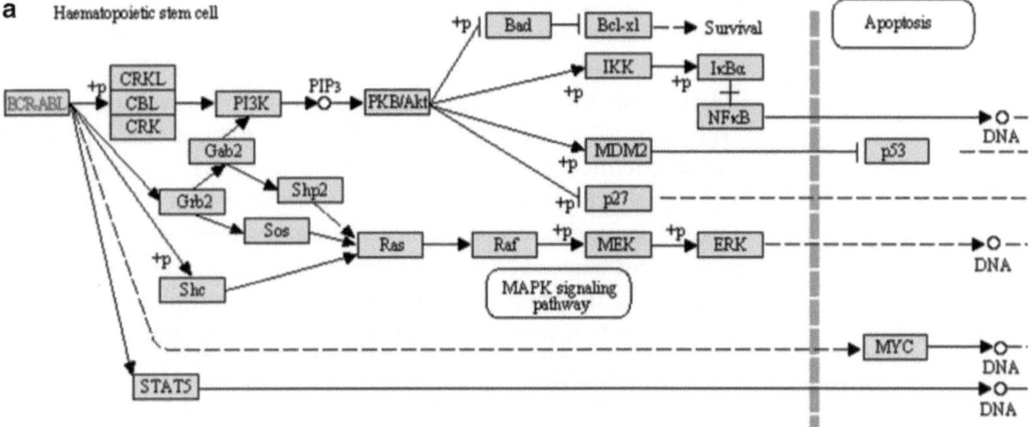

b

Entry	H00004 Disease
Name	Chronic myeloid leukemia (CML)
Description	Chronic myelogenous leukemia (CML) oric stem cell of the bone marrow and is cha numbers of granulocytes in the blood. M lineages are involved in the process of differentiation. On the cellular level, chromosome abnormality, the t(9; 22) re the Philadelphia (Ph) chromosome. The B molecular rearrangement between the c-A and the BCR (breakpoint cluster region) fusion gene codes for an 8-kb mRNA and higher and aberrant tyrosine kinase act counterpart. While fusion of c-ABL and cause of the chronic phase of CML, proc other molecular changes. Common seconda in TP53, RB, and p16/INK4A, or overexpr Additional chromosome translocations ar t(3;21)(q26;q22), which generates AML1-
Category	Cancer
Pathway	hsa05220 Chronic myeloid leukemia
Gene	BCR-ABL (translocation) [HSA:613 25] [K EVI1 (overexpression) [HSA:2122] [KO:KC AML1 (translocation) [HSA:861] [KO:K08 p16/INK4A (mutation) [HSA:1029] [KO:K06 p53 (mutation) [HSA:7157] [KO:K04451] RB1 (mutation) [HSA:5925] [KO:K06618]
Carcinogen	1,3-Butadiene [CPD:C16450] Rubber industry
Marker	BCR-ABL (translocation) [HSA:613 25] WT1 [HSA:7490]
Drug	Imatinib mesylate (Gleevec) [DR:D01441] Hydroxyurea [DR:D00341] Interferon-alpha [DR:D00745 D02745 D03

c

- ▾ **Diseases with Known Genetic Factors**
  - ▾ **Cancers**
    - ▸ Cancers of the nervous system
    - ▸ Cancers of the digestive system
    - ▾ Cancers of haematopoietic and lymphoid tissues
      - H00003 Acute myeloid leukemia (AML) [PATH:hs
      - H00001 Acute lymphoblastic leukemia (ALL) (P
      - H00002 Acute lymphoblastic leukemia (ALL) (P
      - H00004 Chronic myeloid leukemia (CML) [PATH:
      - H00005 Chronic lymphocytic leukemia (CLL)
      - H00007 Hodgkin lymphoma
      - H00006 Hairy-cell leukemia
      - H00008 Burkitt lymphoma
      - H00009 Adult T-cell leukemia
      - H00010 Multiple myeloma
      - H00011 Lymphoplasmacytic lymphoma
      - H00012 Polycythemia vera
    - ▸ Cancers of the breast and female genital organs
    - ▸ Cancers of soft tissues and bone
    - ▸ Skin cancers
    - ▸ Cancers of the urinary system and male genital
    - ▸ Cancers of endocrine organs
    - ▸ Head and neck cancers
    - ▸ Cancers of the lung and pleura
  - ▸ **Nervous system diseases**
  - ▸ **Immune system diseases**
  - ▸ **Circulatory system diseases**
  - ▸ **Urinary system diseases**
  - ▸ **Metabolic diseases**
  - ▸ **Genetic disorders**
- ▾ **Infectious Diseases with Known Pathogen Genomes**
  - ▸ **Bacterial infections**
  - ▸ **Viral infections**
  - ▸ **Mycosis**
  - ▸ **Parasitic infections**

Fig. 8. KEGG DISEASE. KEGG DISEASE describes human diseases in computable forms. This figure illustrates chronic myeloid leukemia in the following three representations. (**a**) Human diseases are described as perturbed states of human molecular network. If some genes are known to be related with the disease, they are highlighted in colors. The user can look up the genes by clicking the corresponding *rectangles*. (**b**) Even if the mechanism is not known, the list of the known information, such as mutated genes, is still valuable. The user can obtain further information by clicking the links. (**c**) Diseases are organized and classified in the BRITE hierarchy, where the disease in question is marked in *red*. The user can view the detail of the disease by clicking the accession number (e.g., H00004), and look up diseases in other categories by clicking the *triangles*.

KEGG PLANT is a new resource for plant research, especially for understanding relationships between genomic and chemical information of natural products from plants. This is part of the EDRUG database (18), a collection of natural products such as crude drugs and essential oils. Plants are known to produce diverse chemical compounds including those with medicinal and nutritional properties. The available complete genomes for plants are very limited in comparison to other organism groups such as animals and bacteria. Thus, massive EST datasets have been established for a number of plant species to generate the EGENES database (19) where EST contigs are treated as genes and automatically annotated with KAAS (see Note 6). We have been expanding the repertoire of KEGG pathway maps for plant secondary metabolism, as well as developing the Global Maps and several category maps. The category maps are used to classify plant secondary metabolites as part of the BRITE hierarchy.

In this chapter, we introduced main KEGG resources and their usability. Emphasis was put on the usage for omics studies; however, the KEGG resources are applicable for a variety of studies on life sciences. These useful characteristics of KEGG enable the user to find new idea or to determine future direction for omics analysis. For further reading, we recommend two publications of Wheelock et al. (20, 21) explaining other KEGG contents that are not mentioned in this chapter.

3. Notes

1. *KEGG Organisms and GENOME.* KEGG Organism page (23) contains a list of organisms with complete genomes (Fig. 9a). A KEGG Organism code of a complete genome consists of three alphabets, while the code of a draft genome and EST sequences consists of four alphabets beginning with "d" and "e," respectively. KEGG Organism codes are used for specifying organisms, and also used as the headers of the pathway map IDs (e.g., hsa00010). We recently started incorporating metagenome and pangenome sequences as well, in order to meet the future needs of environmental and health problems. KEGG Organism page (Fig. 9a) contains the links to the metagenome and pangenome data. In addition to the three- or four-letter organism codes, we introduced T numbers for specifying genomes including metagenomes.

   When the user is interested in only one organism, it is efficient to jump to the corresponding GENOME page of interest. Clicking the "mmu," for instance, in the KEGG

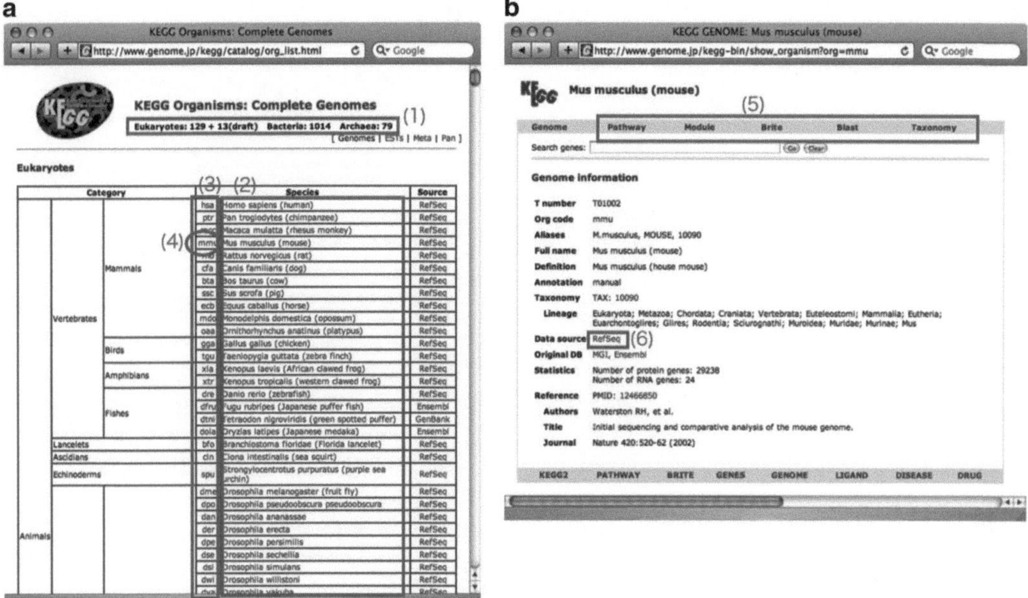

Fig. 9. KEGG Organisms and GENOME. (**a**) KEGG Organism page. (1) Statistics of the genome sequences registered in KEGG. (2) The scientific names and common names of organisms, providing the links to the corresponding search pages for GENES. (3) KEGG Organism codes, providing the links to the corresponding GENOME pages. (4) Clicking this link leads the user to the GENOME page of mouse genome. (**b**) An example GENOME page. (5) Links to the organism-specific pathways, modules, BRITE hierarchies, BLAST searches, and taxonomy information. (6) The sequence data is downloadable from the link at the "Data source".

Organism page (Fig. 9a) takes the user to the GENOME page specific for mouse *Mus musculus* (Fig. 9b). KEGG provides this type of pages for all registered organisms. The user can also reach to this page from the KEGG GENOME page (24).

2. *Find Organisms More Easily.* KEGG has already included more than 1,000 organisms, which makes it hard for the user to find the organisms of interest. Therefore, KEGG provides some options by which the user limits only the organisms of interest (Fig. 10). Once the user selects this option, it keeps working as long as the cookie retains.

3. *Accession ID Conversion to the KEGG IDs.* KEGG entries have unique identifiers (KEGG IDs), which can be used for coloring the PATHWAY maps and the BRITE hierarchy (see Subheading 2). KEGG ID consists of the abbreviated name of the subdatabase and the identifier of the entry connected with a colon (:), e.g., cpd:C00103, where "cpd" means the KEGG COMPOUND database, and "C00103" means the ID number of alpha-D-glucose 1-phosphate. Another example is hsa:4357, where "hsa" means the KEGG Organism code (see Note 1) of human (or, in other words, the human-specific GENOME database), and "4357" means the GENES ID.

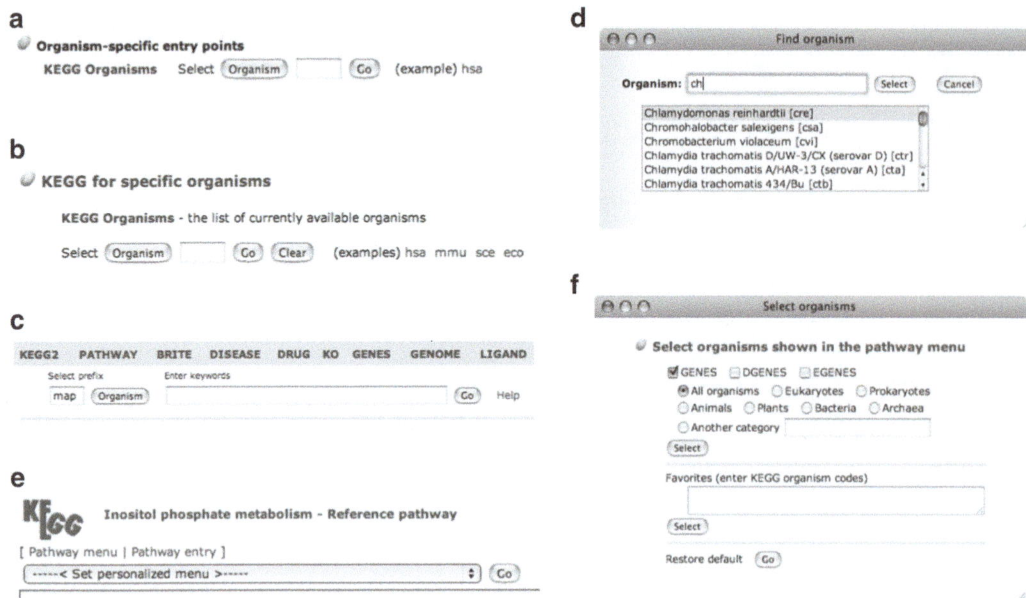

Fig. 10. Finding or limiting organisms. Organism search options are located in various pages such as (**a**) the KEGG homepage (Fig. 1), (**b**) the KEGG sitemap (Fig. 2), and (**c**) the KEGG PATHWAY page. If the user knows the KEGG Organism code for the organism of interest, input the code in the box to reach the GENOME page (Fig. 9b). In the case, the user does not remember the code, click the "Organism" button to pop up the "Find organism" window. (**d**) This window can be used as a dictionary, and also a reverse dictionary, of the scientific name of organisms and the corresponding KEGG Organism codes. The user need not complete the spell of organism names; the search engine complements the name, as shown in this figure. This window works even after other Web pages are closed, so this can still be used for looking up the organisms. (**e**) Every PATHWAY page (Fig. 3a) has a pull-down menu to select an organism from more than 1,000 organisms with complete genomes. For the user feeling difficulty in finding an organism of interest, there are options to sort organisms in alphabetical order and to generate the personalized menu. Select "< Set personalized menu >" and click "Go," and the "Select organism" window pops up. (**f**) The user can generate the personalized menu by specifying organisms of interest. These settings are preserved in the user's browser and are used next time.

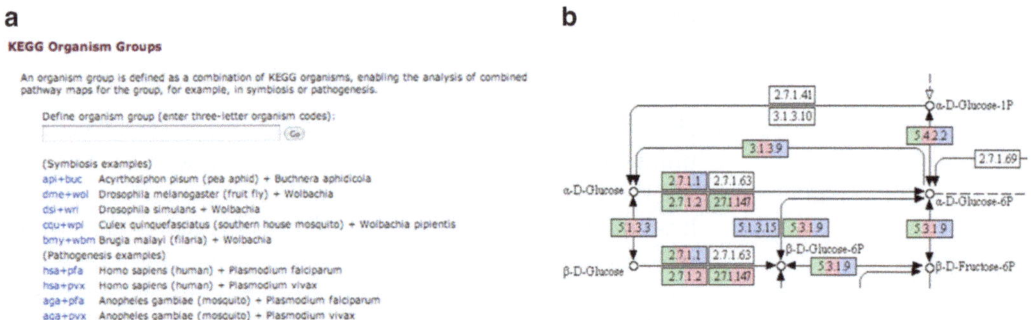

Fig. 11. KEGG Organisms groups. (**a**) The option to specify two or more organisms in the middle of the KEGG GENOME page. (**b**) Using the option provides multicolor pathway maps representing the gene products from the specified organisms.

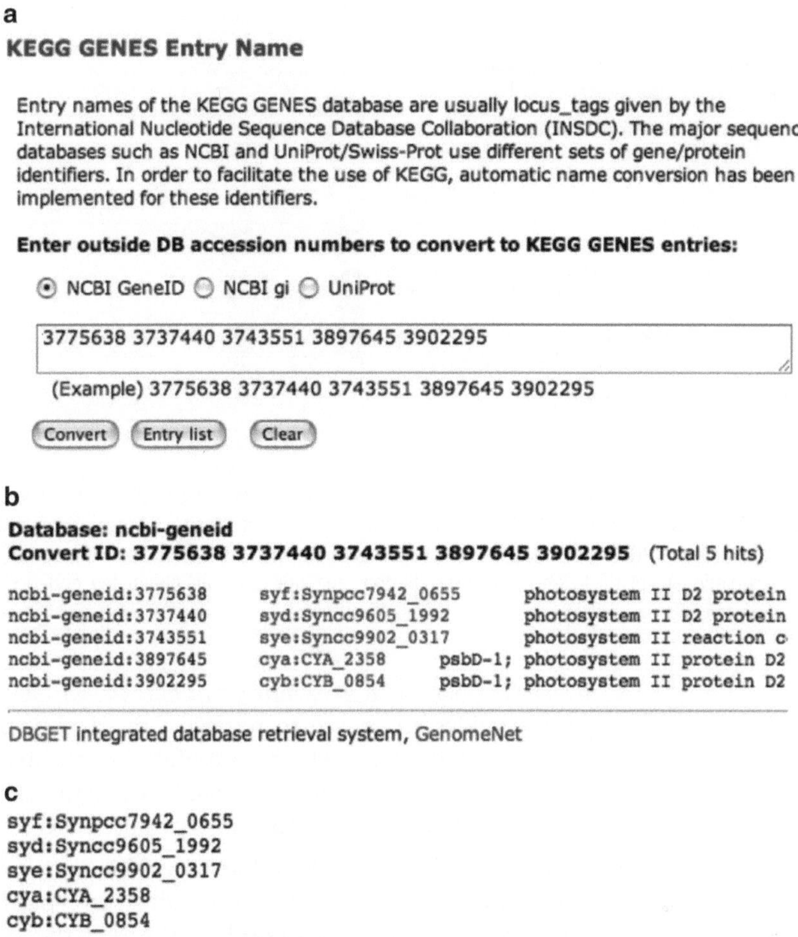

a

KEGG GENES Entry Name

Entry names of the KEGG GENES database are usually locus_tags given by the International Nucleotide Sequence Database Collaboration (INSDC). The major sequence databases such as NCBI and UniProt/Swiss-Prot use different sets of gene/protein identifiers. In order to facilitate the use of KEGG, automatic name conversion has been implemented for these identifiers.

Enter outside DB accession numbers to convert to KEGG GENES entries:

◉ NCBI GeneID ○ NCBI gi ○ UniProt

3775638 3737440 3743551 3897645 3902295

(Example) 3775638 3737440 3743551 3897645 3902295

Convert | Entry list | Clear

b

Database: ncbi-geneid
Convert ID: 3775638 3737440 3743551 3897645 3902295 (Total 5 hits)

```
ncbi-geneid:3775638    syf:Synpcc7942_0655          photosystem II D2 protein
ncbi-geneid:3737440    syd:Syncc9605_1992           photosystem II D2 protein
ncbi-geneid:3743551    sye:Syncc9902_0317           photosystem II reaction c
ncbi-geneid:3897645    cya:CYA_2358        psbD-1; photosystem II protein D2
ncbi-geneid:3902295    cyb:CYB_0854        psbD-1; photosystem II protein D2
```

DBGET integrated database retrieval system, GenomeNet

c

```
syf:Synpcc7942_0655
syd:Syncc9605_1992
sye:Syncc9902_0317
cya:CYA_2358
cyb:CYB_0854
```

Fig. 12. The accession ID conversion tool. The user can see the KEGG Identifiers page [28] by clicking one of the links of the KEGG homepage (Fig. 1a). (**a**) In the middle of the page, the accession numbers from outside databases can be converted to the corresponding KEGG entries. (**b**) Click the "Convert" button to obtain this page, showing external-DB IDs, the corresponding KEGG IDs, and brief annotations. (**c**) Click the "Entry list" button, and obtain the list that can be directly used as an input of coloring the KEGG objects (see Subheading 2.2 and Fig. 4a).

The abbreviated name of the subdatabases in KEGG can be looked up at ref. 27, and the format of the KEGG IDs can be seen at the KEGG Identifier page (28).

The user needs KEGG GENES and COMPOUND IDs to color the PATHWAY maps. If the user only has the list of NCBI gene IDs or UniProt IDs, they can be converted to the corresponding KEGG IDs using the option in the KEGG Identifiers page (Fig. 12). Entry list style (Fig. 12c) is recommended because it can be simply pasted in the input box of the color objects page (Fig. 4a). KegArray (Subheading 3) also has an option to convert the external database IDs to the KEGG GENES IDs, which are necessary for mapping the array data to the KEGG resources such as pathway maps.

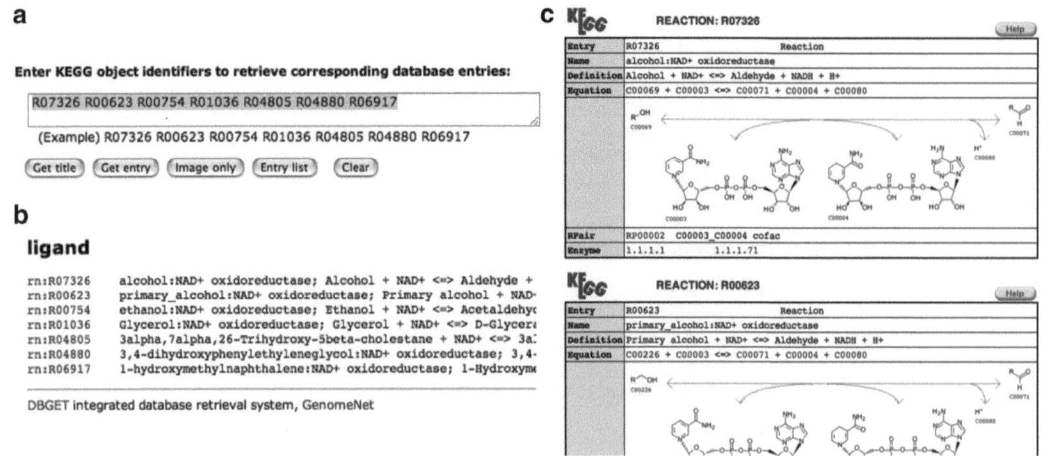

Fig. 13. Retrieving multiple KEGG entries simultaneously. We provide a convenient way to simultaneously view a number of objects indicated by KEGG identifiers. (**a**) In the middle of the KEGG Identifiers page, there is an input form. (**b**) Input some KEGG IDs and click the "Get title" button, and the user can obtain the list of IDs and the corresponding titles (descriptions or annotations). (**c**) Click the "Get entry" button, and the user can obtain the corresponding entries simultaneously in a page.

4. *Retrieving Several KEGG Entries at a Time.* KEGG Identifiers page provides an option to retrieve a number of the KEGG entries at a time (Fig. 13). This is useful when the user is using a Web browser. When retrieving more KEGG entries is preferred, go to the KEGG FTP site (36) or try to use KEGG API (12).

5. *Create New Pathway Maps That Are Not Present in KEGG.* Even though KEGG keeps incorporating novel pathways published recently, there is a good chance that the user finds a pathway that is not still present in KEGG. If this is the case, sending us a request is highly appreciated (see Note 7). In some cases, however, the user might need to create new pathway maps that are not present in KEGG. Such cases are divided into two types. In the first type of cases, the steps of the pathway are already described in a KEGG PATHWAY map, although they are not attached to the corresponding genes derived from an organism of interest. In the second type of cases, some (or all) of the steps are not described in the KEGG PATHWAY maps because they are still unpublished or unknown. KEGG provides KAAS to address the first type of cases, as explained in Note 6.

   To address the second type of cases, PathPred (29) and E-zyme (30, 31) are available (see Fig. 1b). When the user obtains a chemical structure of a metabolite for which the biosynthesis/biodegradation pathway is unknown, PathPred automatically suggests possible pathways. The suggested pathway includes the steps with the plausible EC numbers (enzyme classification IDs established by IUBMB), which are predicted by E-zyme. E-zyme is also available to suggest possible EC numbers for a given (partial) enzyme reaction equation. PathPred and E-zyme require chemical structures as input. If the chemical compounds are registered in KEGG, then the user can use the corresponding KEGG IDs. In the case the user wants to input the chemical compound that is not present in KEGG, or the user does not know the corresponding KEGG ID, we recommend to use KegDraw, a desktop application designed for drawing and searching chemical structures. This application has options to incorporate the chemical structures predefined in KEGG, as well as to edit the structures. It is notable that this application is also capable of drawing glycan structures (32). The edited structures of compounds and glycans are also used as queries of the similarity search programs SIMCOMP/SUBCOMP (33, 34) and KCaM (35), respectively.

6. *KAAS Automatic Annotation.* KAAS (KEGG Automatic Annotation Server) (25) has been used for annotating DGENES, EGENES, and MGENES in KEGG. The public version of KAAS is available to annotate any groups of gene sequences, when the user wants to display the genes in the organism that is not still a member of the KEGG Organisms, or when the user has a set of sequences for which the corresponding IDs are not known. This service is of particular value when the user has a draft genome, EST, or the sequence sets obtained from microarray analysis. Note that KAAS uses BLAST search; therefore, the user should examine the quality and the length of the input sequences just as when using BLAST. Multiple FASTA format is used as an input. KAAS accepts both nucleic and amino acid sequences; however, the two types of sequences should not be mixed in one file.

   The user can jump to the KAAS page (26) by clicking one of the links in the KEGG sitemap (Fig. 1b). It is recommended that the user specify a set of organisms that are evolutionally close to the input organism, because the KAAS searches the similar sequences in KO. It may take a while depending on the data size or the status of the server, therefore an e-mail will be sent later to inform the URL to access the result page, containing the corresponding KO list. The automatically colored PATHWAY pages are obtained according to the result. It is recommended that the user download the result since they will be removed from KEGG server in a few days. The results can also be seen in the BRITE form, where the annotated functions such as enzymes, transcription factors, and receptors are listed hierarchically to help understand the overview of the gene set.

7. *Feedback*. We appreciate any suggestions, questions and comments on the KEGG data and tools. We intend that KEGG keeps incorporating more and more genomes, pathways, the BRITE hierarchies, etc. Suggesting something that should be added to KEGG is also greatly appreciated. Please send a message to the feedback form (37).
8. *KEGG GENES*. KEGG GENES (22) is a database of the genes derived from all organisms with the sequenced genomes publicly available. GENES contains nucleic and amino acid sequences, identifiers in KEGG and other databases and the functional KEGG annotation. For eukaryotes, there are DGENES and EGENES databases containing draft genomes and EST sequences, respectively. We also started to collect and annotate metagenome information that is stored as MGENES. Gene and genome sequences have been retrieved from Refseq in NCBI, and other public databases of the genome-sequencing organizations.
9. *KEGG Organism Groups*. We also recently defined KEGG Organism Groups, combinations of organisms, enabling the analysis of the combined pathways generated as the results of symbiosis or pathogenesis. The combined pathways can be obtained using the search box located in the middle of the KEGG GENOME page (see ref. 24; Fig. 11a). For example, when the user inputs "hsa + pfa," meaning human (*Homo sapiens*) plus a pathogen (*Plasmodium falciparum 3D7*), this option provides the two-colored pathways. These two colors represent the gene products from the two organisms.

   In fact, this option is not limited only for symbiosis and pathogenesis, but this accepts any combinations of genomes. For instance, the query "hsa + mmu + dme," which means human (*H. sapiens*) + mouse (*M. musculus*) + fruit fly (*Drosophila melanogaster*), provides the three-colored map (Fig. 11b) that is useful to compare the three pathways in a map.

Acknowledgments

The computational resources were provided by the Bioinformatics Center, Institute for Chemical Research, Kyoto University. The KEGG project is supported by the Institute for Bioinformatics Research and Development of the Japan Science and Technology Agency, and a grant-in-aid for scientific research on the priority area "Comprehensive Genomics" from the Ministry of Education, Culture, Sports, Science and Technology of Japan.

References

1. Kanehisa M, Goto S, Furumichi M et al (2010) KEGG for representation and analysis of molecular networks involving diseases and drugs. Nucleic Acids Res 38:D355-360.
2. KEGG Home Page. http://www.kegg.jp/.
3. GenomeNet. http://www.genome.jp/.
4. Fujibuchi W, Sato K, Ogata H et al (1998) KEGG and DBGET/LinkDB: Integration of biological relationships in divergent molecular biology data. In: Knowledge Sharing Across Biological and Medical Knowledge Based Systems. Technical Report WS-98-04, pp 35–40, AAAI Press.
5. Goto S, Okuno Y, Hattori M et al (2002) LIGAND: database of chemical compounds and reactions in biological pathways. Nucleic Acids Res 30:402–404.
6. KEGG PATHWAY. http://www.kegg.jp/kegg/pathway.html.
7. KEGG Markup Language. http://www.genome.jp/kegg/xml/.
8. Okuda S, Yamada T, Hamajima M et al (2008) KEGG Atlas mapping for global analysis of metabolic pathways. Nucleic Acids Res 36: W423-426.
9. KEGG BRITE. http://www.genome.jp/kegg/brite.html.
10. PATHWAY color Page. http://www.genome.jp/kegg/tool/color_pathway.html.
11. BRITE color Page. http://www.genome.jp/kegg/tool/color_brite.html.
12. KEGG API. http://www.genome.jp/kegg/soap/.
13. KegTools Page. http://www.genome.jp/kegg/download/kegtools.html.
14. KEGG EXPRESSION database. http://www.genome.jp/kegg/expression/.
15. KEGG DISEASE. http://www.genome.jp/kegg/disease/.
16. KEGG DRUG. http://www.genome.jp/kegg/drug/.
17. Shigemizu D, Araki M, Okuda S et al (2009) Extraction and analysis of chemical modification patterns in drug development. J Chem Inf Model 49:1122–1129.
18. EDRUG database. http://www.genome.jp/kegg/drug/edrug.html.
19. Masoudi-Nejad A, Goto S, Jauregui R et al (2007) EGENES: transcriptome-based plant database of genes with metabolic pathway information and expressed sequence tag indices in KEGG. Plant Physiol 144:857–866.
20. Wheelock CE, Wheelock AM, Kawashima S et al (2009) Systems biology approaches and pathway tools for investigating cardiovascular disease. Mol Biosyst 5:588–602.
21. Wheelock CE, Goto S, Yetukuri L et al (2009) Bioinformatics strategies for the analysis of lipids. Methods Mol Biol 580:339–368.
22. KEGG GENES. http://www.genome.jp/kegg/genes.html.
23. KEGG Organism Page. http://www.genome.jp/kegg/catalog/org_list.html.
24. KEGG GENOME Page. http://www.genome.jp/kegg/genome.html.
25. Moriya Y, Itoh M, Okuda S et al (2007) KAAS: an automatic genome annotation and pathway reconstruction server. Nucleic Acids Res 35:W182-185.
26. KAAS Page. http://www.genome.jp/tools/kaas/.
27. DBGET Page. http://www.genome.jp/dbget/.
28. KEGG Identifier Page. http://www.genome.jp/kegg/kegg3.html.
29. Moriya Y, Shigemizu D, Hattori M et al (2010) PathPred: an enzyme-catalyzed metabolic pathway prediction server. Nucleic Acids Res 38:W138-143.
30. Kotera M, Okuno Y, Hattori M et al (2004) Computational assignment of the EC numbers for genomic-scale analysis of enzymatic reactions. J Am Chem Soc 126:16487–16498.
31. Yamanishi Y, Hattori M, Kotera M et al (2009) E-zyme: predicting potential EC numbers from the chemical transformation pattern of substrate-product pairs. Bioinformatics 25: i179-186.
32. Hashimoto K, Kanehisa M (2008) KEGG GLYCAN for integrated analysis of pathways, genes, and structures. In: Taniguchi N, Suzuki A, Ito Y, Narimatsu H, Kawasaki T, Hase S (eds.) Experimental Glycoscience. pp 441–444, Springer.
33. Hattori M, Okuno Y, Goto S et al (2003) Development of a chemical structure comparison method for integrated analysis of chemical and genomic information in the metabolic pathways. J Am Chem Soc 125: 11853–11865.
34. Hattori M, Tanaka N, Kanehisa M et al (2010) SIMCOMP/SUBCOMP: chemical structure search servers for network analyses. Nucleic Acids Res 38:W652-656.
35. Aoki KF, Yamaguchi A, Ueda N et al (2004) KCaM (KEGG Carbohydrate Matcher): a software tool for analyzing the structures of carbohydrate sugar chains. Nucleic Acids Res 32:W267-272.
36. KEGG FTP Site. http://www.genome.jp/kegg/download/.
37. KEGG Feedback. http://www.genome.jp/feedback/.

Chapter 3

Strategies to Explore Functional Genomics Data Sets in NCBI's GEO Database

Stephen E. Wilhite and Tanya Barrett

Abstract

The Gene Expression Omnibus (GEO) database is a major repository that stores high-throughput functional genomics data sets that are generated using both microarray-based and sequence-based technologies. Data sets are submitted to GEO primarily by researchers who are publishing their results in journals that require original data to be made freely available for review and analysis. In addition to serving as a public archive for these data, GEO has a suite of tools that allow users to identify, analyze, and visualize data relevant to their specific interests. These tools include sample comparison applications, gene expression profile charts, data set clusters, genome browser tracks, and a powerful search engine that enables users to construct complex queries.

Key words: Database, Microarray, Next-generation sequence, Gene expression, Epigenomics, Functional genomics, Data mining

1. Introduction

The Gene Expression Omnibus (GEO) database (1) was launched in 2000 by the National Center for Biotechnology Information (NCBI) to support the storage, use, and dissemination of high-throughput gene expression data (2). High-throughput methodologies have evolved considerably since GEO's inception to include both array- and sequence-based methodologies that generate a wide variety of functional genomics data types. Due to GEO's flexible design and ability to store diverse data structures, GEO's current holdings are much more diverse than implied by its name. Table 1 illustrates the diversity and relative quantities of both array- and sequence-based functional genomics studies that are currently represented in GEO.

Most data in GEO represent original research that is submitted by scientists who are publishing their work in a journal that requires

Junbai Wang et al. (eds.), *Next Generation Microarray Bioinformatics: Methods and Protocols*, Methods in Molecular Biology, vol. 802, DOI 10.1007/978-1-61779-400-1_3,

Table 1
Listing of GEO study types and the number of Series records with those types, correct at the time of writing

Application	Technology	Number of series
Expression profiling	By array	17,988
Noncoding RNA profiling	By array	348
Genome binding/occupancy profiling	By array	73
Genome variation profiling	By array	314
Methylation profiling	By array	46
Protein profiling	By protein array	31
SNP genotyping	By SNP array	151
Genome variation profiling	By SNP array	272
Expression profiling	By genome tiling array	305
Noncoding RNA profiling	By genome tiling array	82
Genome binding/occupancy profiling	By genome tiling array	849
Genome variation profiling	By genome tiling array	410
Methylation profiling	By genome tiling array	118
Expression profiling	By high-throughput sequencing	134
Noncoding RNA profiling	By high-throughput sequencing	234
Genome binding/occupancy profiling	By high-throughput sequencing	250
Methylation profiling	By high-throughput sequencing	31
Expression profiling	By SAGE	206
Expression profiling	By RT-PCR	25
Expression profiling	By MPSS	21

The types describe both the general application (e.g., expression profiling) as well as the technology (e.g., high-throughput sequencing). Users can retrieve studies of a particular type using the "DataSet Type" field in the *GEO DataSets* query interface

its contributors to deposit data in a public repository as a condition of publication. Consequently, GEO now has supporting data for over 10,000 published manuscripts. In total, GEO is currently comprised of data from almost half a million public samples representing over 1,300 different organisms submitted by over 8,000 laboratories, and the submission rate exceeds 10,000 new sample deposits per month. GEO has been under constant development to keep up with the growing diversity of data and to provide useful tools to help researchers effectively query the database in order to identify data that are relevant to a specific area of interest (3). This chapter addresses the practical aspects of effectively utilizing

GEO search mechanisms to find and retrieve data of interest, and explores the use of tools developed for visualizing and interpreting specific data types.

2. Methods

2.1. "GEO Accession" Query Box

This is a simple retrieval mechanism that works with Series (GSExxx), Sample (GSMxxx), Platform (GPLxxx), and DataSet (GDSxxx) accession numbers (see Note 1) to retrieve the queried entry. This feature is used primarily for straightforward retrievals of data that has been quoted in a publication when one has possession of an accession number and wishes to retrieve the corresponding GEO entry. To retrieve an entry using an accession number: (a) go to the GEO home page (1), (b) enter the accession number to be retrieved in the "GEO accession" query box, (c) Click "GO." The "GEO accession" query box is also available at the top of most GEO pages.

2.2. *Searching* Entrez GEO DataSets *and* Entrez GEO Profiles

NCBI has a powerful search and retrieval system called Entrez that can be used to search the content of its network of integrated databases (4). This system can be used to query individual databases or all databases from a single interface (5). GEO data are available in two separate Entrez databases referred to as *GEO DataSets* and *GEO Profiles.*

2.2.1. Entrez GEO DataSets

The *Entrez GEO DataSets* search interface is directly accessible at ref. 6. This "study-level" database is where users can search for studies relevant to their interests. The database stores all original submitter-supplied records, as well as curated gene expression DataSets. As explained in Subheading 3, while *GEO DataSets* can be searched using many different attributes including organism, DataSet type, supplementary file types and authors, it is also possible to retrieve useful data simply by entering relevant keywords. For example, to find studies that examine lung cancer, just type "lung cancer" into the search box. Retrievals include a summary of each study matching the search criteria and a listing of the Samples they include.

2.2.2. Entrez GEO Profiles

The *Entrez GEO Profiles* search interface is directly accessible at ref. 7. This "gene-level" database is where users can search for specific genes of interest, either across all DataSet records or within specific DataSets. The database stores individual gene expression profiles from curated DataSets (see Note 1; *GEO Profiles* are generated only for DataSet entries, so only a subset of GEO data is represented as profiles). As explained in Subheading 3, while *GEO Profiles* can be searched using many different attributes including gene names, GenBank accession numbers, Gene Ontology (GO) terms, or genes flagged as being differentially expressed, it is also possible to retrieve useful data simply by entering relevant keywords. For example, to find profiles for gene Nqo1, just type "Nqo1" into the search box. Retrievals include gene names and individual thumbnail images that depict the expression values of a particular gene across each Sample in a DataSet (Fig. 1). Experimental context is provided in the bars at the foot of the charts making it possible to see at a glance whether a gene is expressed differentially across experimental conditions. Clicking on the thumbnail image enlarges the chart to reveal the full profile details, expression values, and the DataSet subsets that reflect experimental design.

2.3. Advanced Entrez queries

As mentioned in the previous section, Entrez searches may be effectively performed by simply entering appropriate keywords and phrases into the search box. However, given the large volumes of data stored in these databases, it is often useful to perform more refined queries in order to filter down to the most relevant data. GEO data are indexed under many different fields. This enables sophisticated queries to be performed by restricting searches to specific fields and combining terms with Boolean operators (AND, OR, NOT) using the following syntax:

term[field] *OPERATOR* term[field]

A query tutorial page (8) was recently released to explain to users how to build complex, fielded queries in the *GEO DataSets* and *GEO Profiles* databases. The tutorial includes an exhaustive listing of the field qualifiers that are available for each database, as well as clickable examples to demonstrate their use (see Note 2). Furthermore, new tools are available on "Advanced Search" and "Limits" pages, which are linked from the Entrez home pages, to assist users to quickly construct multipart, fielded queries.

1. *Search Builder*: This section includes a complete listing of all the fields that can be searched, and the values indexed under each field. To use, the following basic steps are performed: (a) select a search field from the drop-down menu, (b) type a search term – OR – select search term from list after clicking "Show Index," (c) choose desired Boolean operator (AND, OR, NOT) and click "Add to Search Box," (d) repeat steps a–c for additional search terms until query has been completed, and (e) execute search by clicking "Search" (alternatively, click "Preview" to see the result count of your query in the Search History section).
2. *Limits*: This section presents a specific box for several of the most popular and useful search fields. The user simply enters keywords, or selects search terms from the drop-down menus, hits "Add to Search Box" and the query is automatically constructed.

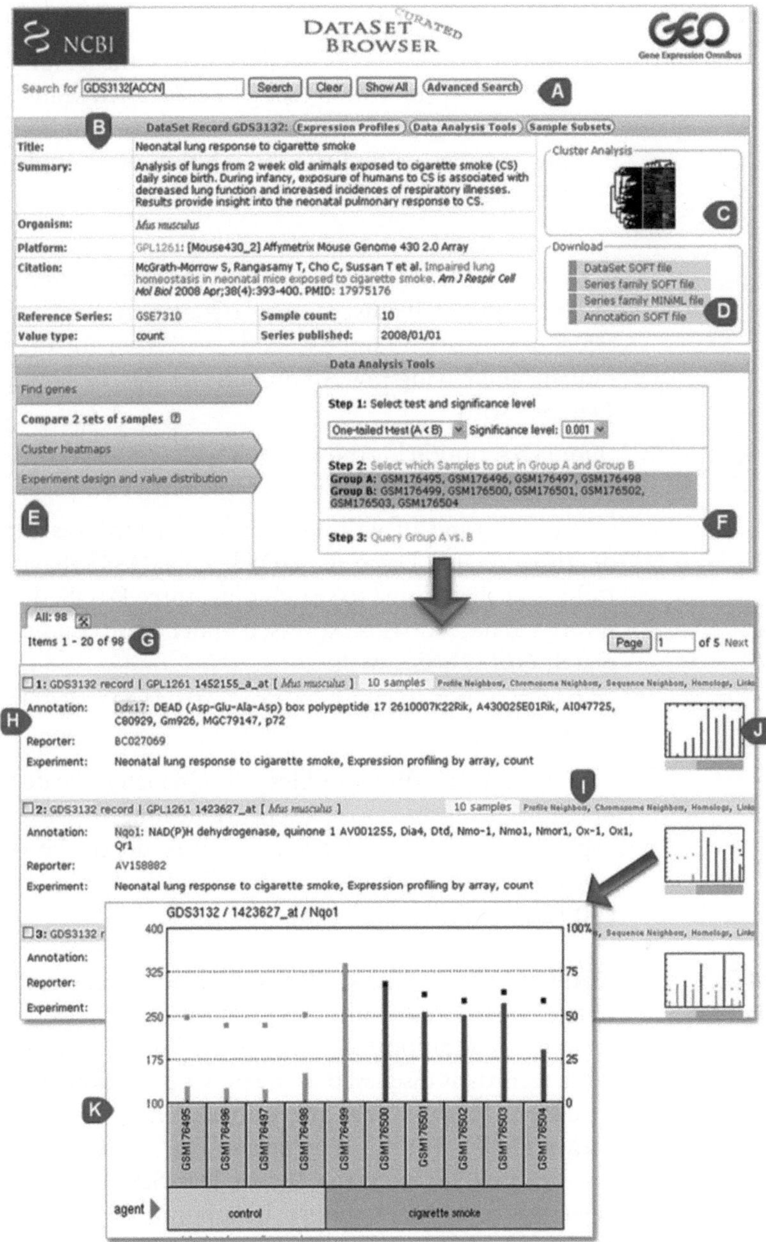

Fig. 1. Screenshot of a *GEO DataSet* record, data analysis tools, and corresponding *GEO Profiles*. (A) DataSet Browser search box. (B) Area containing descriptive information about that DataSet, including the title, summary, organism, and citation (*see* ref. 27 for this example). (C) Thumbnail image of cluster heatmap. Click the image to be directed to the full interactive cluster from where regions may be selected and exported. (D) Download section containing various file format options; mouseover each option for the description of content. (E) Data Analysis Tools options. Select from "Find genes," "Compare 2 sets of Samples," "Cluster heatmaps," and "Experiment design and value distribution." (F) "Compare two sets of Samples" analysis. In this example, the user has opted to perform a one-tailed *t*-test in order to find genes more highly expressed in mouse lung Samples exposed to cigarette smoke, compared to controls. (G) Results of the previous *t*-test; 98 genes were retrieved in this case. (H) Gene annotation area. (I) "Neighbors" links that connect the targeted profile to genes related by expression pattern (Profile neighbors), sequence similarity (Sequence neighbors), or physical proximity (Chromosome neighbors). (J) Thumbnail image of gene expression profile. (K) Full profile image that in this example depicts how gene Nqo1 is more highly expressed in smoke-exposed Samples compared to controls. Each bar in the chart represents the expression level of Nqo1 in a Sample. The bars at the foot of the chart represent the experimental variables, in this case "control" or "cigarette smoke".

3. *Search History*: This section stores the results of previous searches for up to 8 h (see Note 3). Each search is assigned a number, e.g., "#2." Users can use these numbers to construct new queries or find the intersection of multiple queries, e.g. (#2 NOT #3) AND human.

Users typically perform multiple searches of both *GEO DataSets* and *GEO Profiles* to arrive at the data they are interested in. For example, if a user wants to locate studies that examine the effect of smoke on lung tissue, derived from any organism except human, and having raw Affymetrix .cel files, he could search *GEO DataSets* with:

(lung[Description] AND smok*[Description]) NOT human [Organism] AND cel[Supplementary Files]

At the time of writing, this search retrieves three independently generated DataSets: GDS3622, GDS3548, and GDS3132. If the user then wants to search these three DataSets to see how his favorite gene, Nqo1, is expressed under these conditions, he could search *GEO Profiles* with:

(GDS3622 OR GDS3548 OR GDS3132) AND Nqo1[Gene Symbol]

This returns three profiles, all of which indicate that Nqo1 is upregulated upon smoke exposure in lung. If the user wants to explore any of these DataSets in more depth, he could use the advanced data mining tools described in Subheadings 4 and 5 and Fig. 1.

2.4. Advanced Data Mining Features for GEO DataSets

As discussed in Note 1, DataSet records are assembled by GEO staff using the data and information derived from select Series records. In addition to querying the *GEO DataSets* interface for these records as discussed in the previous section, it is also possible to directly browse and query these entries using the "DataSet Browser" (9) (Fig. 1). The Search bar at the top of the browser can be used to filter the list of DataSets by entering relevant keywords (e.g., heart, mouse, lymphoma, GPL81, etc.). Selecting a row in the browser displays the corresponding DataSet record in the panel below.

DataSet records have integrated "Data Analysis Tools" (Fig. 1) that facilitate examination and interrogation of the data in order to identify potentially interesting genes. These tools include:

Find genes: Allows users to retrieve specific expression profiles in that DataSet using gene names or symbols, or to retrieve expression profiles that have been flagged as potentially showing differential expression across experimental variables.

Compare two sets of Samples: Allows users to retrieve expression profiles based on specified statistical parameters. Users select which Samples to include in their comparison, the type of statistical comparison to be performed, and the significance level or cut-off to apply.

Cluster heatmaps: Allow users to visualize several types of precomputed cluster heatmaps of data and to select regions of interest for further study. GEO cluster heatmap images are interactive; cluster regions of interest may be selected, enlarged, charted as line plots, viewed in *GEO Profiles*, and the original data downloaded.

Experimental design and value distribution: Provides users with a graphic representation of the study's experimental design showing experimental subsets, and a box and whiskers plot displaying the distribution of expression values of each Sample within the DataSet.

2.5. Advanced Data Mining Features for GEO Profiles

The *GEO Profiles* results page (Fig. 1) includes features that enable users to identify additional gene expression profiles based on similarity to a given profile of interest, and to link to related information in other NCBI Entrez databases.

Profile neighbors: Retrieves profiles with similar patterns of expression within the same DataSet. This feature assists in the identification of genes that may show coordinated regulation.

Chromosome neighbors: Retrieves profiles for up to 20 of the closest-found chromosome neighbors within the same DataSet. This feature assists in the identification of available data for genes within the same chromosomal region.

Sequence neighbors: Retrieves profiles based on BLAST nucleotide sequence similarity across all DataSets. This feature assists in the identification of profiles representing sequence homologs and orthologs.

Homologs: Retrieves profiles that belong to the same HomoloGene group across all DataSets. HomoloGene is a NCBI resource for automated detection of homologs among the annotated genes of several completely sequenced eukaryotic genomes.

2.6. Programmatic Access to GEO DataSets and GEO Profiles

The *GEO DataSets* and *GEO Profiles* databases can be accessed programmatically using a suite of programs collectively referred to as the Entrez Programming Utilities (E-utilities). GEO has a help page (10) describing some common examples and uses but more advanced users, for example, those wishing to perform sophisticated retrievals using Perl scripts, should consult the E-utilities help page (11) for further guidance.

2.7. GEO BLAST Query

This feature, linked from the GEO home page, allows users to retrieve gene expression profiles based on BLAST (12) nucleotide sequence similarity. Entered nucleotide sequences or accession identifiers are queried against nucleotide sequences corresponding to the GenBank identifiers represented on microarray Platforms of DataSet entries. The initial output of a GEO BLAST query is similar to conventional BLAST output showing significant alignments between query and subject sequences. On the BLAST output page, users can click the "E" icon to view *GEO Profiles* corresponding to a particular subject sequence of interest. This query method can be used to find GEO data representing sequence homologs and orthologs, or for gaining insight into potential roles of uncharacterized nucleotide sequences.

2.8. Specialized Resources for Next-Generation Sequence Data

Increasingly, the microarray community is switching to next-generation sequence technologies to perform functional genomics analyses. Table 1 lists the major categories of sequence study types handled by GEO. GEO hosts the processed and analyzed sequence data, together with descriptive information about the Samples, protocols, and study; raw data files are brokered to NCBI's Sequence Read Archive (SRA) database. Next-generation sequence studies can be located in *GEO DataSets* using the same search strategies as described for array-based studies. However, sequence data present new challenges in terms of data analysis and visualization. As a first step, hundreds of GEO Samples have been selected for integration into NCBI's new Epigenomics resource (13). This resource maps the sequence reads to genomic coordinates to generate data "tracks" that can be viewed using genome browsers. Multiple tracks can be viewed side-by-side, allowing data for specific genes to be visualized and compared across different Samples (Fig. 2). The GEO records selected for this advanced processing can be identified using the following cross-database search in *GEO DataSets*: "gds epigenomics"[Filter].

In addition, GEO has a new centralized page (14) dedicated to the organization and presentation of next-generation sequence data derived from the NIH RoadMap Epigenomics Project. Features available on this page include the ability to link to the original GEO records, filter for records based on keywords, download data, and view selected Samples as tracks on either the NCBI Sequence Viewer or the UCSC Genome Browser (15).

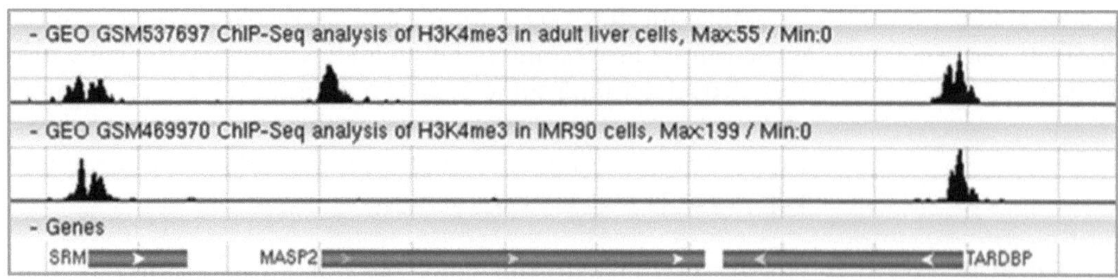

Fig. 2. Chromatin immunoprecipitation sequence (ChIP-seq) tracks displayed in NCBI's Sequence Viewer. Histone H3 lysine 4 trimethylation (H3K4me3) peaks are typically observed at the 5′ end of transcriptionally active genes. In this example, there is a clear peak next to MASP2 in the adult liver cells (top track, GEO Sample GSM537697) but not in the IMR90 cells (lower track, GEO Sample GSM469970).

2.9. Data Download

Data are made available for bulk download in several formats from the GEO FTP site (16) (see Note 4). There are currently five DATA/ subdirectories:

SeriesMatrix/: This directory contains tab-delimited value-matrices generated from the VALUE column of the Sample tables of each Series entry. Files also include Series and Sample metadata and are ideal for opening in spreadsheet applications such as Microsoft Excel. Most users find SeriesMatrix files the most convenient format for handling data that have not been assembled into a DataSet.

SOFT/: This directory contains files in "Simple Omnibus Format in Text" (SOFT). SOFT files are generated for DataSet entries, as well as for Series and Platform entries (subdirectories are included for each entry type). The Series and Platform files are actually "family files" that include the metadata and complete data tables of all related entries in the family. In contrast, the DataSet SOFT files include the metadata of the DataSet entry only, plus a matrix table containing the extracted gene annotations and Sample values used in *GEO Profiles*.

MINiML/: This directory includes files in MINiML (MIAME Notation in Markup Language) format. MINiML is essentially an XML rendering of SOFT format, and the files provided here are the XML-equivalents of the Series and Platform family files provided in the SOFT/ directory.

Supplementary/: This directory contains supplementary files organized according to the entry type (Platforms, Samples, Series). Platform supplementary files are typically related to the array design (e.g., .gal, .bpmap, or .cdf), Sample supplementary files are typically native files representing raw (e.g., .cel, .gpr, or . txt) (see Note 5) or processed data (e.g., .chp, .bed, .bar, .wig, or . gff), and Series files would typically include results of upper-lever analyses such as ANOVA tables or significant genes lists. In addition, there is a compressed archive for each Series entry (GSExxxx_RAW.tar) that is composed of the supplementary files gathered from all related Samples and Platforms. The "RAW" part of the name is a misnomer since these files often include more than just raw data, but they enable users to download all supplementary files associated with a given Series entry in one step.

Annotation/: This directory includes gene annotations for Platforms that participate in DataSet entries and, consequently, *GEO Profiles*. The annotations are derived by extracting stable sequence tracking identifiers directly from GEO Platform tables (e.g., GenBank accession numbers, clone identifiers, etc.) and using them to retrieve up-to-date gene annotations from the Entrez Gene and UniGene databases. This helps to ensure that the gene annotations associated with *GEO Profiles* are as up-to-date as possible.

3. Conclusions

Functional genomics assays employing microarrays and next-generation sequencing have become standard tools in biological research. Deposition of such data sets in public repositories is mandated by many journals for the purpose of allowing the research community to access and critically evaluate the data discussed in manuscripts. This requirement has resulted in astonishing growth in the number of studies and data types that are now available in the GEO database.

This chapter provides an overview of strategies for navigating the data in GEO and locating information relevant to the users' particular interests. Approaches include simple and complex text-based searches, tools that identify genes with specific patterns of expression, as well as various easily interpretable graphical renderings of select data. GEO is a well-used resource, typically receiving over 40,000 Web hits and 10,000 bulk downloads per day. A review of the literature reveals that the community is applying GEO data to their own studies in diverse ways; see ref. 17 for a listing of over 1,000 papers that cite usage of GEO data. It is clear that researchers use these data to address questions far beyond those for which the original studies were designed to address. Examples include using GEO data to test new algorithms (18), functionally characterize genes (19), create new added-value targeted databases (20), perform massive meta-analyses across thousands of independently generated assays (21), and identify diagnostic protein biomarkers for disease (22).

GEO will continue to support these endeavors by improving the utility of the data in several ways, including enhancing data annotation standards, expanding integration with related resources, and by developing new analysis tools that can be used by as many users as possible.

4. Notes

1. Entry types, accession codes, and their relationships to each other are described in detail at ref. 23. There are three primary entry types, referred to as Platform (GPLxxx), Sample (GSMxxx), and Series (GSExxx) entries. Platform entries are used to list the elements being detected by the experiment, e.g., oligonucleotide sequences, gene symbols, or representative GenBank accession numbers. Sample entries are used to describe the biomaterials under investigation and the treatments to which they were subjected, and to provide access to

the associated hybridization protocols and measurements. Series entries are used to group experimentally related Samples and provide summary and design details. A fourth entry type, referred to as DataSets (GDSxxx), is assembled by the GEO curation staff from the three primary entries. DataSet entries contain essentially the same data and information as in the three primary entries, but the format has been arranged such that the submitter-supplied normalized data can be visualized and interrogated using downstream analysis tools. Only array-based expression data are currently considered for DataSet creation, and not all expression data qualify (for instance, due to having experimental designs or data processing methods that are incompatible with GEO tools). Furthermore, many expression studies have not yet been reviewed by the curation staff for DataSet creation. The net result is that only about 20% of the expression data in GEO are currently represented as DataSets and analyzable using GEO's analysis tools.

2. It is critical to recognize that some Entrez fields can only be searched using a fixed list of controlled terms while others are free text fields that can be searched with any keyword or quoted phrase. The query tutorial page distinguishes between "fixed list" and "free text" fields, but acquiring the list of searchable terms for fixed list fields requires using the "Show Index" feature available on the "Advanced Search" pages. For instance, to see a list of fixed terms for the "Entry Type" field:

   (a) Go to the *GEO DataSets* advanced search page (24).

   (b) Select the "Entry Type" field from the drop-down list in the Search Builder section.

   (c) Click "Show Index."

   The results are shown in Fig. 3. This result indicates that the *GEO DataSets* "Entry Type" field can be queried only for "gds," "gpl," and "gse" terms. The numbers in parentheses are the total number of each entry type. For example, all DataSet

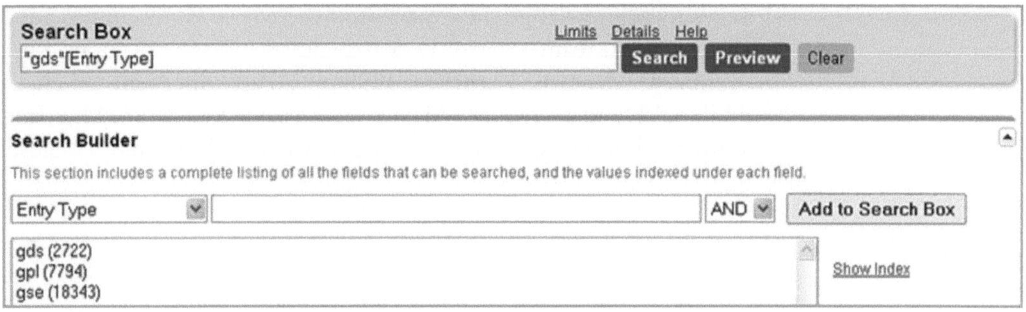

Fig. 3. Screenshot of Search Builder results, demonstrating fixed list terms for the "Entry type" field.

entries can be retrieved by searching *GEO DataSets* with "gds [Entry Type]." "Show Index" can be used to see a listing of the indexed terms for any field listed in the drop-down list, but is mostly useful for identifying searchable terms for fixed list fields.

3. To save Entrez searches indefinitely, create a My NCBI account (25). When logged in, after performing your query you should see a "Save Search" option next to the search box. In addition, you will be presented with the option to receive e-mail alerts when new data matching your search criteria have been added to the database.
4. FTP directory content and file formats are described in detail in the README file (26). In many cases, direct links to the FTP site are provided on records. For instance, Series and Platform entries contain a direct link to their corresponding SOFT and MINiML family files, and SeriesMatrix files. Supplementary files are directly accessible using the links provided at the foot of Series, Sample, and Platform entries, and DataSet entries contain links to the DataSet SOFT file, Series family SOFT and MINiML files, and the annotation SOFT file. SOFT and MINiML formats can also be exported using the toolbar located at the top of Series, Sample, and Platform records. Furthermore, document summaries can be exported from the *GEO DataSets* and *GEO Profiles* result pages by setting the tool bar at the head of the page to "Send to: File."
5. Studies that have supplementary files of specific types may be identified by constructing a query using the [Supplementary Files] field in *GEO DataSets*. This is useful for users who want to identify, download, and reanalyze, for example, all .cel files for a specific Affymetrix platform.

Acknowledgments

This chapter is an official contribution of the National Institutes of Health; not subject to copyright in the USA. The authors unreservedly acknowledge the expertise of the whole GEO curation and development team – Pierre Ledoux, Carlos Evangelista, Irene Kim, Kimberly Marshall, Katherine Phillippy, Patti Sherman, Michelle Holko, Dennis Troup, Maxim Tomashevsky, Rolf Muertter, Oluwabukunmi Ayanbule, Andrey Yefanov, and Alexandra Soboleva.

Funding

This research was supported by the Intramural Research Program of the NIH, National Library of Medicine.

References

1. http://www.ncbi.nlm.nih.gov/geo/
2. Edgar R, Domrachev M, Lash AE (2002) Gene Expression Omnibus: NCBI gene expression and hybridization array data repository. Nucleic Acids Res 30:207–210
3. Barrett T, Troup DB, Wilhite SE et al (2009) NCBI GEO: archive for high-throughput functional genomic data. Nucleic Acids Res 37:D885–890
4. Sayers EW, Barrett T, Benson DA et al (2009) Database resources of the National Center for Biotechnology Information. Nucleic Acids Res 37:D5–15
5. http://www.ncbi.nlm.nih.gov/gquery/
6. http://www.ncbi.nlm.nih.gov/gds/
7. http://www.ncbi.nlm.nih.gov/geoprofiles/
8. http://www.ncbi.nlm.nih.gov/geo/info/qqtutorial. html
9. http://www.ncbi.nlm.nih.gov/sites/GDSbrowser/
10. http://www.ncbi.nlm.nih.gov/geo/info/geo_paccess.html
11. http://www.ncbi.nlm.nih.gov/books/NBK25501/
12. Altschul SF, Gish W, Miller W et al (1990) Basic local alignment search tool. J Mol Biol 215:403–410
13. Fingerman IM, McDaniel L, Zhang X et al (2011) NCBI Epigenomics: A new public resource for exploring epigenomic datasets. Nucleic Acids Res 39:D908–12
14. http://www.ncbi.nlm.nih.gov/geo/ roadmap/epigenomics/
15. Rhead B, Karolchik D, Kuhn RM et al (2010) The UCSC Genome Browser database: update 2010. Nucleic Acids Res 38:D613–619.
16. ftp://ftp.ncbi.nih.gov/pub/geo/DATA/
17. http://www.ncbi.nlm.nih.gov/geo/info/ucitations.html
18. Bhattacharya A, De RK (2008) Divisive Correlation Clustering Algorithm (DCCA) for grouping of genes: detecting varying patterns in expression profiles. Bioinformatics 24:1359–1366
19. Pierre M, DeHertogh B, Gaigneaux A et al (2010) Meta-analysis of archived DNA microarrays identifies genes regulated by hypoxia and involved in a metastatic phenotype in cancer cells. BMC Cancer 10:176
20. Ogata Y, Suzuki H, Sakurai N et al (2010) CoP: a database for characterizing co-expressed gene modules with biological information in plants. Bioinformatics 26:1267–1268
21. Liu S (2010) Increasing alternative promoter repertories is positively associated with differential expression and disease susceptibility. PLoS One 5:e9482
22. Chen R, Sigdel TK, Li L et al (2010) Differentially Expressed RNA from Public Microarray Data Identifies Serum Protein Biomarkers for Cross-Organ Transplant Rejection and Other Conditions. PLoS Comput Biol 6: e1000940
23. http://www.ncbi.nlm.nih.gov/geo/info/overview.html
24. http://www.ncbi.nlm.nih.gov/gds/advanced/
25. http://www.nlm.nih.gov/pubs/techbull/jf05/jf05_myncbi.html#register
26. ftp://ftp.ncbi.nih.gov/pub/geo/README.TXT
27. McGrath-Morrow S, Rangasamy T, Cho C et al (2008) Impaired lung homeostasis in neonatal mice exposed to cigarette smoke. Am J Respir Cell Mol Biol 38:393–400

Part II

Microarray Data Analysis (Top-Down Approach)

Chapter 4

Analyzing Cancer Samples with SNP Arrays

Peter Van Loo, Gro Nilsen, Silje H. Nordgard, Hans Kristian Moen Vollan, Anne-Lise Børresen-Dale, Vessela N. Kristensen, and Ole Christian Lingjærde

Abstract

Single nucleotide polymorphism (SNP) arrays are powerful tools to delineate genomic aberrations in cancer genomes. However, the analysis of these SNP array data of cancer samples is complicated by three phenomena: (a) aneuploidy: due to massive aberrations, the total DNA content of a cancer cell can differ significantly from its normal two copies; (b) nonaberrant cell admixture: samples from solid tumors do not exclusively contain aberrant tumor cells, but always contain some portion of nonaberrant cells; (c) intratumor heterogeneity: different cells in the tumor sample may have different aberrations. We describe here how these phenomena impact the SNP array profile, and how these can be accounted for in the analysis. In an extended practical example, we apply our recently developed and further improved ASCAT (allele-specific copy number analysis of tumors) suite of tools to analyze SNP array data using data from a series of breast carcinomas as an example. We first describe the structure of the data, how it can be plotted and interpreted, and how it can be segmented. The core ASCAT algorithm next determines the fraction of nonaberrant cells and the tumor ploidy (the average number of DNA copies), and calculates an ASCAT profile. We describe how these ASCAT profiles visualize both copy number aberrations as well as copy-number-neutral events. Finally, we touch upon regions showing intratumor heterogeneity, and how they can be detected in ASCAT profiles. All source code and data described here can be found at our ASCAT Web site (http://www.ifi.uio.no/forskning/grupper/bioinf/Projects/ASCAT/).

Key words: Cancer, Tumor, SNP arrays, ASCAT, Allelic bias, Aneuploidy, Intratumor heterogeneity

1. Introduction

Single nucleotide polymorphism (SNP)-based DNA microarrays represent a powerful technology, allowing simultaneous measurement of the allele-specific copy number at many different single nucleotide polymorphic loci in the genome. A SNP is a single base

Junbai Wang et al. (eds.), *Next Generation Microarray Bioinformatics: Methods and Protocols*,
Methods in Molecular Biology, vol. 802, DOI 10.1007/978-1-61779-400-1_4,

locus in the genome that occurs in the population in two different variants, for example, some individuals can have a cytosine base (C) at that locus, while other individuals have a guanine base (G). Calling one of the allelic variants as A and the other as B, the fact that our DNA contains one paternal and one maternal copy means we may obtain genotypes AA (homozygous A), AB (heterozygous), or BB (homozygous B) for any given SNP locus. By measuring thousands or even millions of such SNP loci, a considerable part of the genome that is variable in the population can effectively be arrayed. At present, SNP array platforms are available from Affymetrix (1) and Illumina (2). Current Affymetrix SNP array technology is based on hybridization to oligonucleotides, arrayed in a regular and predefined pattern on glass slides, while Illumina technology is based on in situ single nucleotide extension reactions on bead arrays. However, despite these substantial technological differences, the resulting data show that similar properties and techniques developed on one technology are in general applicable to the other technology, after an appropriate data transformation.

Cancer genomes often show numerous DNA sequence changes, ranging in size from single nucleotide mutations to gains, amplifications, insertions or deletions of large chromosomal fragments, and even whole-genome duplications (3, 4). For this reason, genotypes in cancer are no longer limited to AA, AB, or BB, but can also be, e.g., A, BBB, AAB, or ABBB. The SNP array data contain in principle all the necessary information to deduce these more complex genotypes, but three phenomena can complicate the analysis in practice:

Aneuploidy: Owing to a multitude of aberrations, the total amount of DNA in a tumor cell can differ significantly from the normal state of two copies of each chromosome. This is called aneuploidy (compared to the normal state of diploidy). Aneuploidy makes it difficult to determine the normal reference state, as the average signal strength does not necessarily correspond to two copies, as in noncancer genomes. Hence, aneuploidy should be explicitly accounted for in the data analysis.

Nonaberrant cell admixture: A cancer biopsy always contains some nonaberrant cells. These nonaberrant cells can be nontumoral cells in the tumor microenvironment (e.g., fibroblasts, endothelial cells, infiltrating immune cells) (5), normal cells in nontumoral regions of the biopsy, or possibly a subpopulation of tumor cells with no visible aberrations. The measured signal will therefore reflect a combination of aberrant and nonaberrant cells and will be more similar to the signal of a normal sample than would have been the case for a homogeneous sample of tumor cells. The amount of nonaberrant cell admixture may differ significantly between cancer samples (from less than 10% to more than 80%), necessitating separate calculation of the fraction of nonaberrant cells for each assayed sample.

Intratumor heterogeneity: Different cells in a cancer biopsy may harbor different aberrations. In a recent study (6), multiple separable populations of breast cancer cells were found in more than half of the breast carcinomas, but the major cancer cell populations within any given tumor were limited to one, two, or three different subclones. These typically shared many aberrations, indicating that they had a common ancestor. As a result of this intratumor heterogeneity, for some loci, unambiguous genotypes cannot be obtained, even when accounting for nonaberrant cell admixture and aneuploidy.

Numerous data analysis tools for SNP array data exist, including many tools specifically aimed at analyzing cancer samples. Examples of automated SNP array data analysis methods that account for nonaberrant cell admixture in tumor samples are genoCNA (7) and BAFsegmentation (8). Two tools that take tumor aneuploidy into account are OverUnder (9) and PICNIC (10). Methods that automatically account for both tumor aneuploidy and nonaberrant cell admixture are GAP (genome alteration print) (11) and ASCAT (allele-specific copy number analysis of tumors) (12). These methods match the data from one sample to discrete allele-specific copy number states, thus determining tumor ploidy and aberrant tumor cell fraction, as well as copy numbers and genotypes across the genome. GAP uses pattern recognition on copy number and allelic imbalance profiles, while ASCAT directly models allele-specific copy number as a function of the SNP data, the tumor ploidy, and the aberrant cell fraction, and subsequently selects the solution that is closest to nonnegative integer copies at all assayed loci in the genome. Finally, regions subject to intratumor heterogeneity can be predicted from the output of both methods as outlier regions after the optimal genome-wide fit has been obtained.

Here, we focus on the analysis of SNP array data of cancer samples using ASCAT. We first introduce the structure of SNP array data, and explain how nonaberrant cell admixture and tumor aneuploidy influence the signal. Next, a breast cancer example dataset is analyzed using ASCAT. The data is subsequently visualized, filtered for germline heterozygous loci, and segmented. Finally, the actual ASCAT algorithm is applied and the output is discussed.

2. Materials

All source code and data described here can be found at our ASCAT Web site (13) (see Note 1). *R* is required for application of the ASCAT algorithm. ASCAT version 2.0 is used.

3. Methods

3.1. SNP Array Data of Cancer Samples

SNP array data consist of two data tracks (Fig. 1a): the total signal intensity and the allelic contrast. The total signal intensity is represented by Log R and shows the total copy number on a

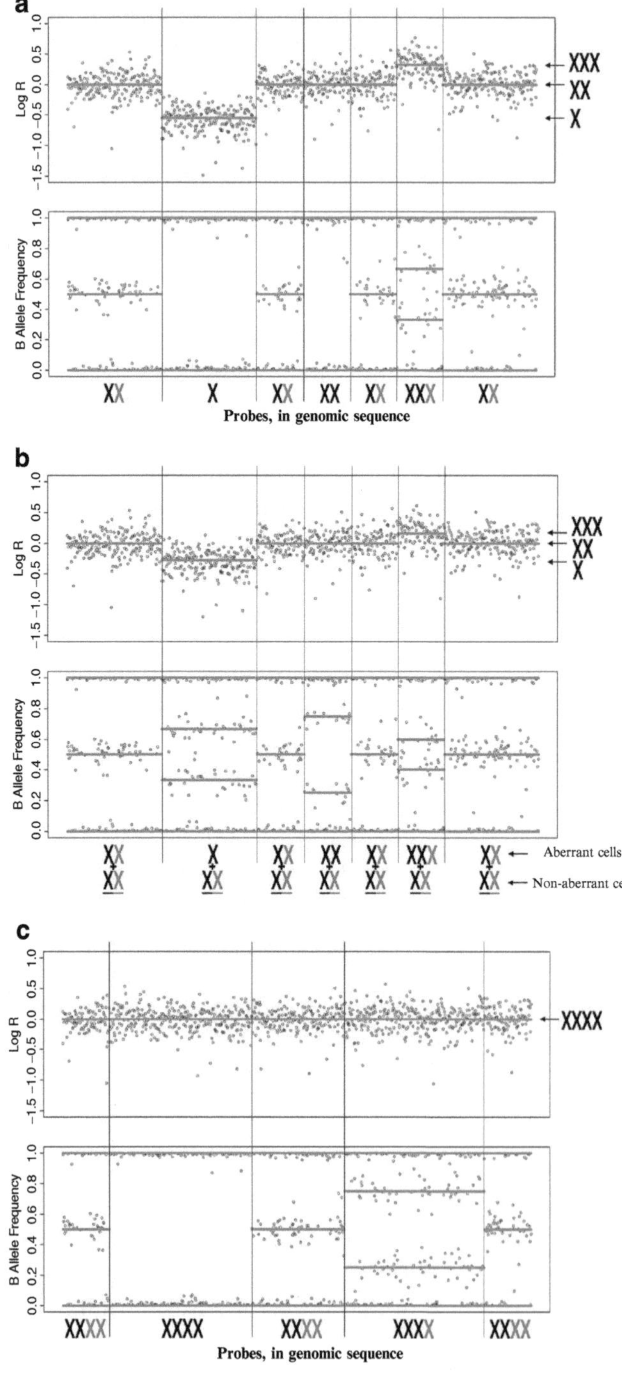

Fig. 1. The structure of SNP array data. (**a**) Log R (*top*) and BAF data (*bottom*). The Log R data track shows the copy number, with the lines close to 0 corresponding to normal

logarithmic scale. The allelic contrast is represented by the B allele frequency (BAF) and shows the relative presence of each of the two alternative nucleotides at each SNP locus profiled (see Note 2). In a diploid sample, a locus with two identical copies will appear with a Log *R* value close to 0, and a BAF value either close to 0 (genotype AA) or close to 1 (genotype BB). A heterozygous locus (genotype AB) will appear as a BAF close to 0.5. From these SNP array data, different genomic aberrations (gains, losses, copy-number-neutral events) can be delineated, as exemplified in Fig. 1a.

Most cancers show evidence of nonaberrant cell admixture (Fig. 1b). This is most evident in the BAF track, where it can be most clearly illustrated in regions with deletions. In case of a

Fig. 1. (Continued) (copy number 2), the decrease to −0.55 corresponding to a deletion (copy number 1) and the increase to 0.4 to a duplication (copy number 3). Both the raw data and the data after application of a segmentation algorithm are shown. The BAF data track shows three bands for normal regions (genotypes AA, AB, and BB with BAF of 0, 0.5, and 1, respectively). In these regions, 1 copy from each parent is inherited (shown at the *bottom*). In the deleted region, only A and B genotypes occur (BAF of 0 and 1, respectively), and in the duplicated region, the four bands correspond to AAA (BAF = 0), AAB (BAF = 0.33), ABB (BAF = 0.67), and BBB (BAF = 1) genotypes. Finally, the middle region shows copy-number-neutral loss-of-heterozygosity (LOH): only AA and BB genotypes are found and hence both copies of this region originate from the same parent (also called uniparental disomy). (**b**) Toy example of SNP array data of a cancer sample showing 50% nonaberrant cell admixture (compare to (**a**), which shows the same example without nonaberrant cell admixture). Notice the lower range of the Log *R* track and the particular differences in the BAF track. In the region deleted in the tumor cells, two extra bands are observed, corresponding to mixture of A genotypes in the tumor cells, admixed with nonaberrant cells with an AB genotype (BAF = 0.33) and B genotypes in the tumor cells, admixed with nonaberrant cells with AB genotype (BAF = 0.67). Similarly, the region showing copy-number-neutral LOH also shows two extra bands (AA mixed with AB at BAF = 0.25 and BB mixed with AB at BAF = 0.75). Finally, in the duplicated region, the bands are shifted compared to the homogeneous case shown in (**a**). (**c**) Toy example of SNP array data of an aneuploid sample. Based on the Log *R* track, the entire stretch of DNA shown has an identical copy number. However, the BAF track shows clear differences in allelic contrast. Three regions show an allelic balance (two homozygous bands at BAF = 0 and BAF = 1, and one heterozygous band at BAF = 0.5), one region shows complete LOH (only the two homozygous bands at BAF = 0 and BAF = 1 are present), and one region shows partial LOH (two "homozygous" bands at BAF = 0 and BAF = 1, and two partially heterozygous bands at BAF = 0.25 and BAF = 0.75). These data cannot be explained under a hypothesis of copy numbers 1, 2, or 3 and hence, this entire region is most likely copy number 4. The regions showing allelic balance have two copies from each parent, the region showing complete LOH has four identical copies, and the region showing partial LOH has three copies from one parent and one copy from the other parent. The two partially heterozygous bands correspond to AAAB (BAF = 0.25) and ABBB (BAF = 0.75) genotypes.

hemizygous deletion (one of the copies is lost) in a homogeneous (and diploid) sample, only two bands are expected in the BAF track: one at 0, corresponding to A genotypes, and one at 1, corresponding to B genotypes. In tumor samples, two extra bands are observed (Fig. 1b), corresponding to an AB genotype in the host, where A (top line) or B (bottom line) has been lost in the tumor. This results in a mixture of tumor cells with B genotypes and admixed nonaberrant cells with AB genotypes (top line) and a mixture of tumor cells with A genotypes and admixed nonaberrant cells with AB genotypes (bottom line). The closer both lines are, the higher the relative signal of nonaberrant cells. In the Log *R* track, nonaberrant cell admixture is visible as an "inflation" of the signals: while in a homogeneous sample, Log *R* drops considerably in case of a hemizygous deletion (to −0.55 in case of Illumina SNP arrays (2)), this drop is lower when nonaberrant cell admixture is observed (Fig. 1b and Table 1). Also for other aberrations, an influence of nonaberrant cell admixture can be seen. For example, for duplications, Log *R* is lower and BAF for "ABB" and "AAB" genotypes is closer together than for homogeneous samples. In addition, many cancers show aneuploidy, resulting in a shift of the Log *R* track compared to diploid samples, while the BAF track is not affected (Fig. 1c, Table 1). In the next sections, we will apply our ASCAT suite of tools (12) (version 2.0, see Note 3) to an example series of breast carcinomas. The added value of using a tool like ASCAT for the analysis of cancer SNP array data is illustrated in Fig. 2. ASCAT calculates the tumor ploidy and the aberrant cell fraction, and subsequently outputs an ASCAT profile, containing the allele-specific copy numbers across the genome, calculated specifically for the aberrant tumor cells and correcting for both aneuploidy and nonaberrant cell infiltration (Fig. 2).

3.2. Data Loading and Visualization

The example SNP array data consists of four files, containing Log *R* and BAF data derived from tumor samples and matched germline samples. Each is a tab-separated file, containing one data column for each sample, a header containing sample names and three columns describing the SNP loci [containing an identifier (in this case, the RS identifier of the SNP) and the genomic location (chromosome and base pair position on the chromosome)]. BAF data has by definition a range between 0 and 1, while Log *R* can in theory range between $-\infty$ and $+\infty$ (although the large majority of the values will be between −1 and 1). Both data tracks may contain NA values (see also Note 4).

First, the ASCAT libraries must be loaded (in *R*):

```
source(ascat.R)
```

Table 1
Influence of infiltration of nonaberrant cells and of aneuploidy of the aberrant tumor cells on Log *R* and BAF data from Illumina SNP arrays

			Genotype tumor (BAF)		
		Log *R*	host: AA	host: AB	host: BB
No infiltration of nonaberrant cells, aberrant cells diploid	Normal, 2 copies	0	AA (0)	AB (0.5)	BB (1)
	Deletion, 1 copy	−0.55	A (0)	A (0) B (1)	B (1)
	Duplication, 3 copies	0.4	AAA (0)	AAB (0.33) ABB (0.67)	BBB (1)
Infiltration of nonaberrant cells	Normal, 2 copies	0	AA (0)	AB (0.5)	BB (1)
	Deletion, 1 copy	>−0.55	A (0)	A ($0 < x < 0.5$) B ($0.5 < x < 1$)	B (1)
	Duplication, 3 copies	<0.4	AAA (0)	AAB ($0.33 < x < 0.5$) ABB ($0.5 < x < 0.67$)	BBB (1)
Aberrant cells aneuploid (>2 copies per cell)	Normal, 2 copies	<0	AA (0)	AB (0.5)	BB (1)
	Deletion, 1 copy	<−0.55	A (0)	A (0) B (1)	B (1)
	Duplication, 3 copies	<0.4	AAA (0)	AAB (0.33) ABB (0.67)	BBB (1)
Infiltration of nonaberrant cells and aberrant cells aneuploid (>2 copies per cell)	Normal, 2 copies	<0	AA (0)	AB (0.5)	BB (1)
	Deletion, 1 copy	<0	A (0)	A ($0 < x < 0.5$) B ($0.5 < x < 1$)	B (1)
	Duplication, 3 copies	<0.4	AAA (0)	AAB ($0.33 < x < 0.5$) ABB ($0.5 < x < 0.67$)	BBB (1)

Typical values of Log *R* and BAF, as well as genotypes, are shown under different scenarios, each time for regions with normal copy number (two copies), deleted regions (one copy), and duplicated regions (three copies)

Next, the data described above can be loaded into ASCAT:

```
ascat.bc = ascat.loadData(Tumor_LogR.
   txt,Tumor_BAF.txt,
Germline_LogR.txt,Germline_BAF.txt)
```

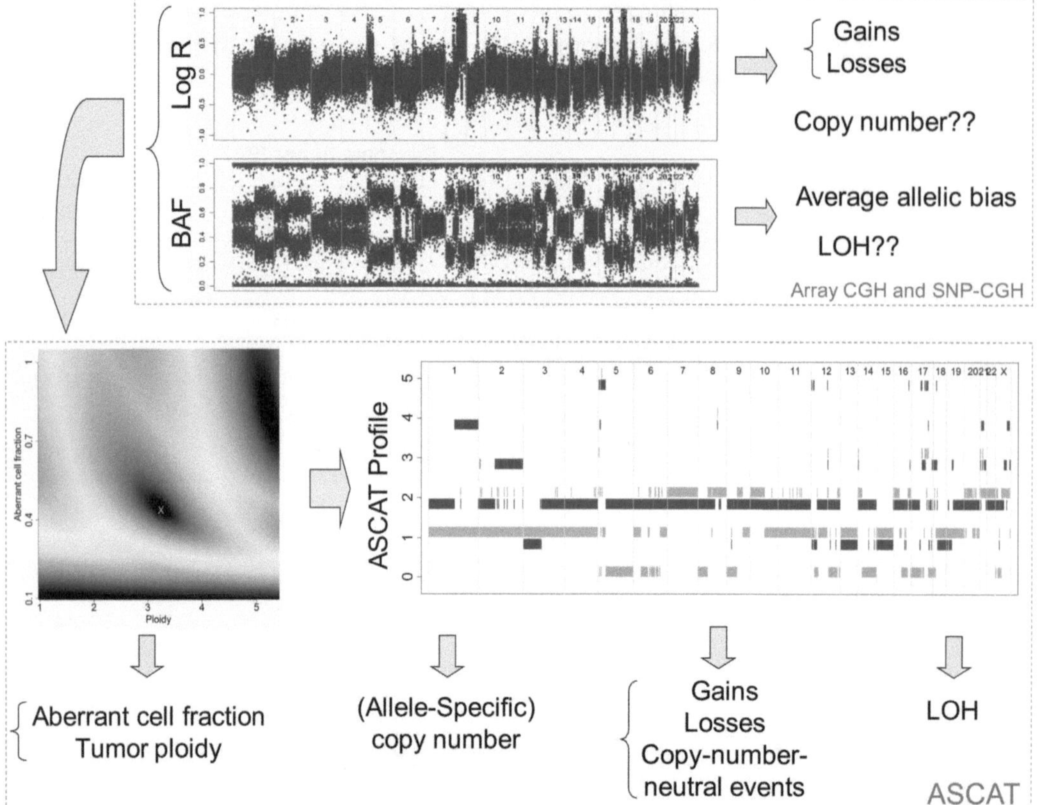

Fig. 2. The principle of data analysis using ASCAT. The result of an array-CGH experiment is a genome-wide measure of total copy number. This allows derivation of gains and losses, but in cancer samples, copy-number estimates are difficult, due to nonaberrant cell infiltration and tumor aneuploidy. SNP-CGH in addition delivers a measure of allelic contrast (BAF). From BAF, allelic bias can be derived, but, e.g., LOH is difficult to determine (due to the nonaberrant cell admixture). ASCAT calculates genome-wide allele-specific copy number profiles for tumor samples, taking into account tumor ploidy and nonaberrant cell admixture. The algorithm first determines the ploidy of the tumor cells and the fraction of aberrant cells ("sunrise plot," *bottom left*). This procedure evaluates the goodness-of-fit for a grid of possible values for both parameters. The optimal solution of tumor ploidy and percentage of aberrant tumor cells is shown by the cross. Next, an "ASCAT profile" is calculated, containing the allele-specific copy-number of all assayed loci (copy-number on the *Y*-axis vs. the genomic location on the *X*-axis; for illustrative purposes only, both lines are slightly shifted such that they do not overlap; only probes heterozygous in the germline are shown). These ASCAT profiles allow accurate derivation of gains (which can be further subdivided into, e.g., duplications, triplications, and amplifications), losses (of one or more copies), copy-number-neutral events, and LOH.

This will create a data structure containing the Log *R* and BAF data for both tumor and germline, as well as some supporting information, such as the position of each probe on the array and a list of the samples.

Next, the data can be plotted:

```
ascat.plotRawData(ascat.bc)
```

These plots are informative to evaluate the quality of the data and to double check if germline samples have not been contaminated with tumor tissue (Fig. 3).

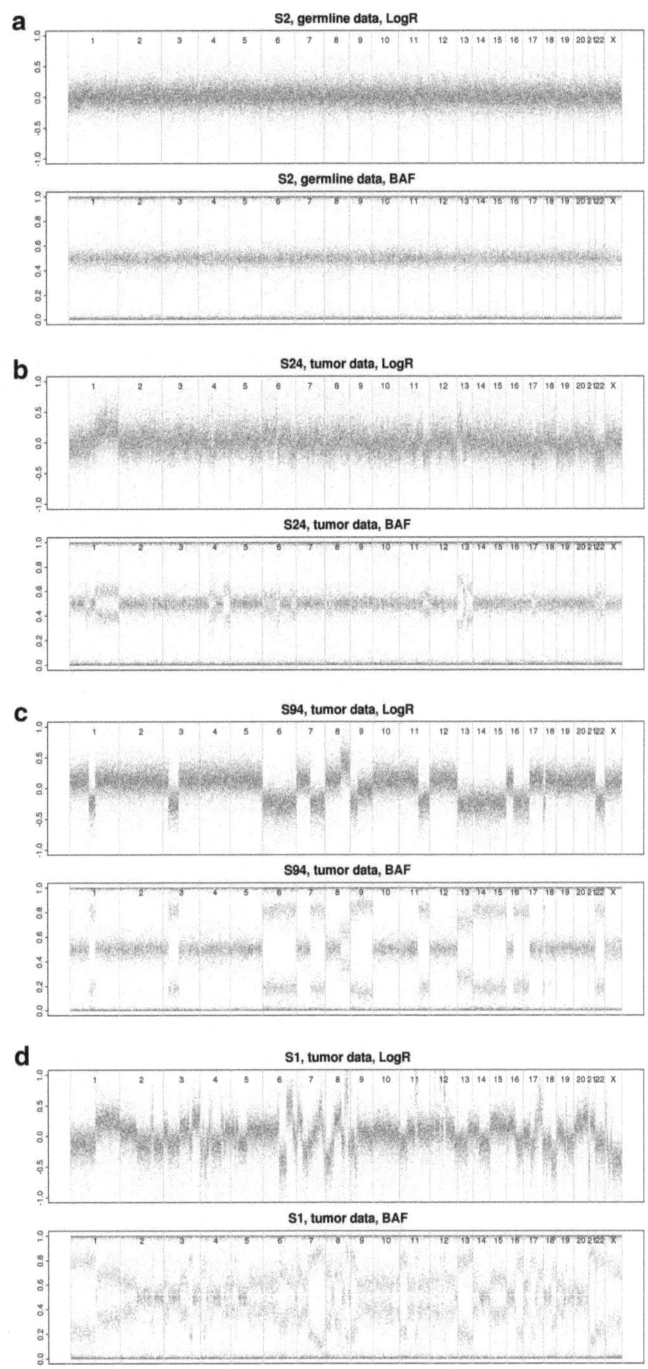

Fig. 3. Example plots of germline and tumor SNP array data. (**a**) A germline sample clearly showing a flat Log *R* profile and three bands in BAF, corresponding to AA (BAF = 0), AB (BAF = 0.5), and BB (BAF = 1) genotypes. (**b–d**) Three tumor samples showing a low aberrant cell fraction (limited range of Log *R* and BAF) (**b**), a higher aberrant cell fraction (**c**), and a higher aberrant cell fraction with extensive aberrations (**d**).

3.3. Segmentation of SNP Array Data

The loaded SNP array data can subsequently be segmented by the allele-specific piecewise constant fitting (ASPCF) algorithm (see Note 5):

```
ascat.bc = ascat.aspcf(ascat.bc)
```

In a first step, this uses the germline data to determine which SNP array probes are germline homozygous (germline genotypes AA or BB) (see Note 6). For these germline homozygous probes, the BAF data track from the tumor is uninformative for copy number determination, as, e.g., germline genotype AA cannot result in tumor genotypes containing B alleles (e.g., A, AA, AAA are possible, but, e.g., genotype AAB is not), and hence, BAF will always be close to 0. Similarly, germline genotypes BB will result in BAF close to 1 for the tumor data. In a second step, the data is segmented by the ASPCF segmentation algorithm (note that this requires aspcf.R), and the results are added to the ASCAT data structure (see Note 7).

The segmented data can subsequently be plotted, using:

```
ascat.plotSegmentedData(ascat.bc)
```

From these plots (Fig. 4), the quality of the data can be further evaluated (e.g., on samples with a serious wave artifact (14) in Log *R*, ASCAT may subsequently fail (12)).

3.4. Running the ASCAT Algorithm

The ASCAT algorithm is next applied to the segmented data:

```
ascat.output = ascat.runAscat(ascat.bc)
```

This output is saved in a data structure, and three figures are made for each tumor. The output data structure contains the aberrant cell fraction and the ploidy, and the copy numbers across the whole genome for both alleles, for each sample. In addition, a list of samples on which ASCAT analysis failed are included [this is often caused by problems with the input data, which can be traced back using the figures generated in the previous sections (Figs. 3 and 4)]. The figures include a "sunrise plot," an ASCAT profile, and a raw copy number profile, for each sample. The sunrise plot is used to determine the optimal aberrant cell fraction and ploidy of the tumor sample and contains a landscape of aberrant cell fraction and ploidy values on which the optimal solution is annotated (Fig. 5). The ASCAT profile contains the estimated allele-specific copy numbers across the genome and can be considered the key output of ASCAT analysis. From these plots, all gains and losses are visualized, as well as copy-number-neutral aberrations and loss-of-heterozygosity (LOH) (Fig. 6). An aberration reliability score for each aberration is also shown in this plot (Fig. 6). The raw copy number profile contains the total copy number, as well as the copy number of the minor allele (the allele with the lowest copy number), without rounding to nonnegative integers (Fig. 7). This plot can be used to evaluate the solution reported

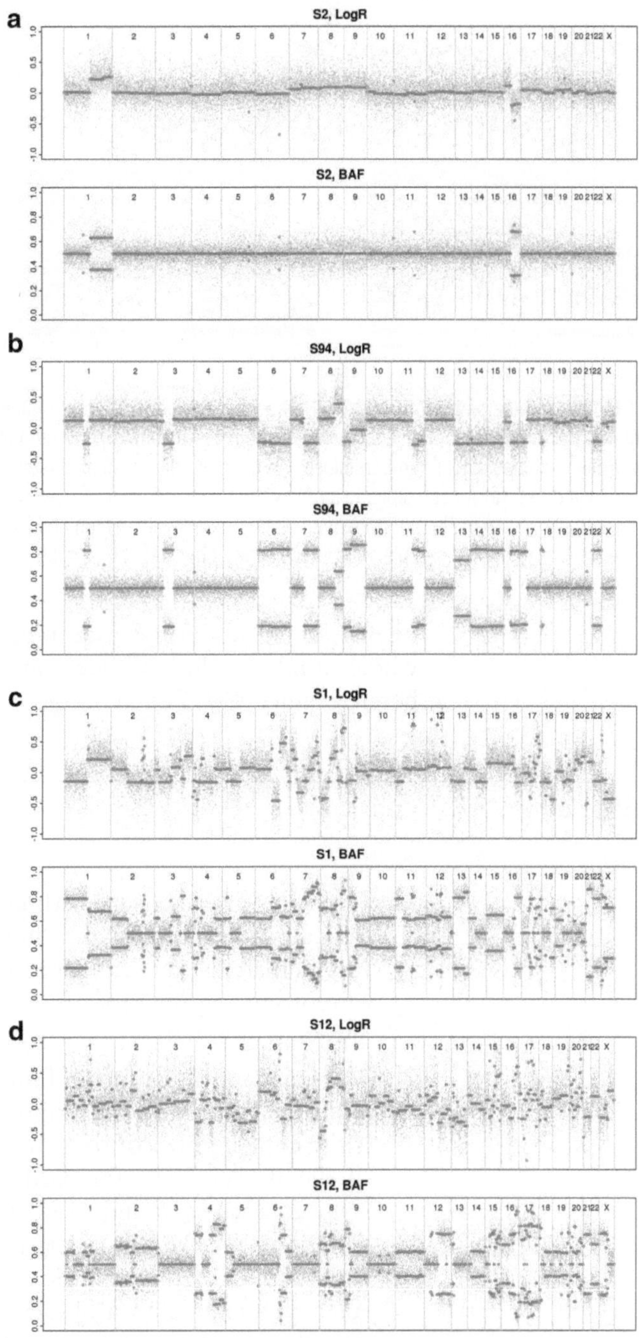

Fig. 4. Example plots of tumor SNP array data, after segmentation. The raw data is plotted, as well as the data after application of the ASPCF segmentation algorithm. (**a**) A sample with few aberrations. (**b**) A sample with more aberrations. (**c**) A highly complex sample. (**d**) A sample showing a clear wave artifact in the Log *R* data track. This is most clearly visible in segments with constant BAF but fluctuating Log *R* (which is not eliminated by the segmentation). In case of such problems, ASCAT may be unable to obtain a solution.

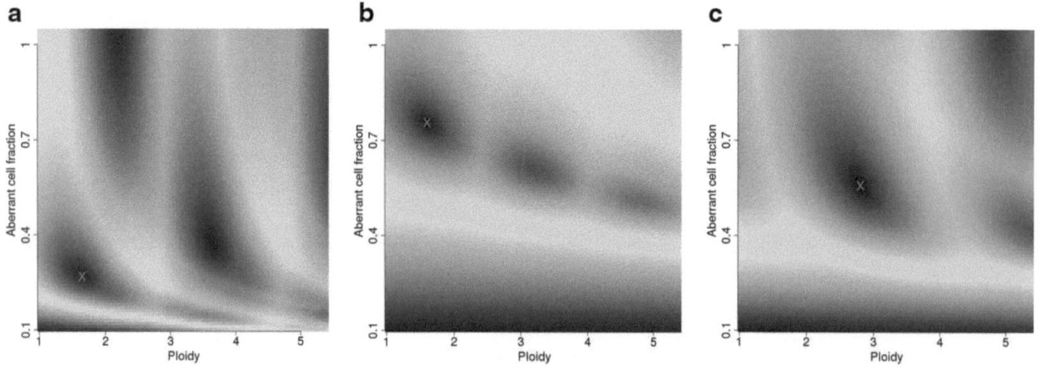

Fig. 5. Example sunrise plots from ASCAT. These plots evaluate different options for the tumor ploidy (*X*-axis) and the aberrant cell fraction (*Y*-axis). For each value plotted, the resulting copy-number profile is evaluated. When the copy-number profile matches whole numbers closely, a good match is obtained (see ref. 12 for details). The optimal match is annotated by a cross. (**a**) A near-diploid sample with a very low aberrant cell fraction (high nonaberrant cell admixture). This is the sample shown in Fig. 3b. (**b**) A near-diploid sample with a high aberrant cell fraction. This is the sample shown in Fig. 4b. (**c**) A near-triploid sample with intermediate aberrant cell fraction. This is the sample shown in Fig. 4c.

by ASCAT. In addition, by scanning for regions that do not fit the whole-number solution, one can gain insight into intratumor heterogeneity (Fig. 7c).

4. Notes

1. Owing to privacy issues with genome-wide genotyping data, data access is often limited. For the data used as an example here, a material transfer agreement is in place. For the purpose of reproducing the procedures outlined here, data access will always be granted.
2. The standard output from Illumina SNP array data is Log *R* and BAF, the latter corresponding to $n_B/(n_A + n_B)$, where n_A is the copy number of the A allele and n_B is the copy number of the B allele. The standard output from Affymetrix SNP array data is Log *R* and an Allelic Difference score that corresponds to $\log_2(n_A/n_B)$. Apart from a rescaling/transformation of this measure of allelic contrast, the choice of BAF over Allelic Difference is arbitrary. However, methods exist to directly calculate Log *R* and BAF from Affymetrix CEL files, such as PennCNV (15) and the aroma.affymetrix R package (16) as well as some commercial packages.
3. ASCAT 2.0 has evolved considerably since its inception (12). The ASPCF segmentation algorithm has been ported from MATLAB to R and now has a faster implementation that scales linearly with array density, making it highly suitable also for high-density platforms. ASCAT 2.0 is applicable to SNP array data from both Illumina and Affymetrix (see also Note 2).

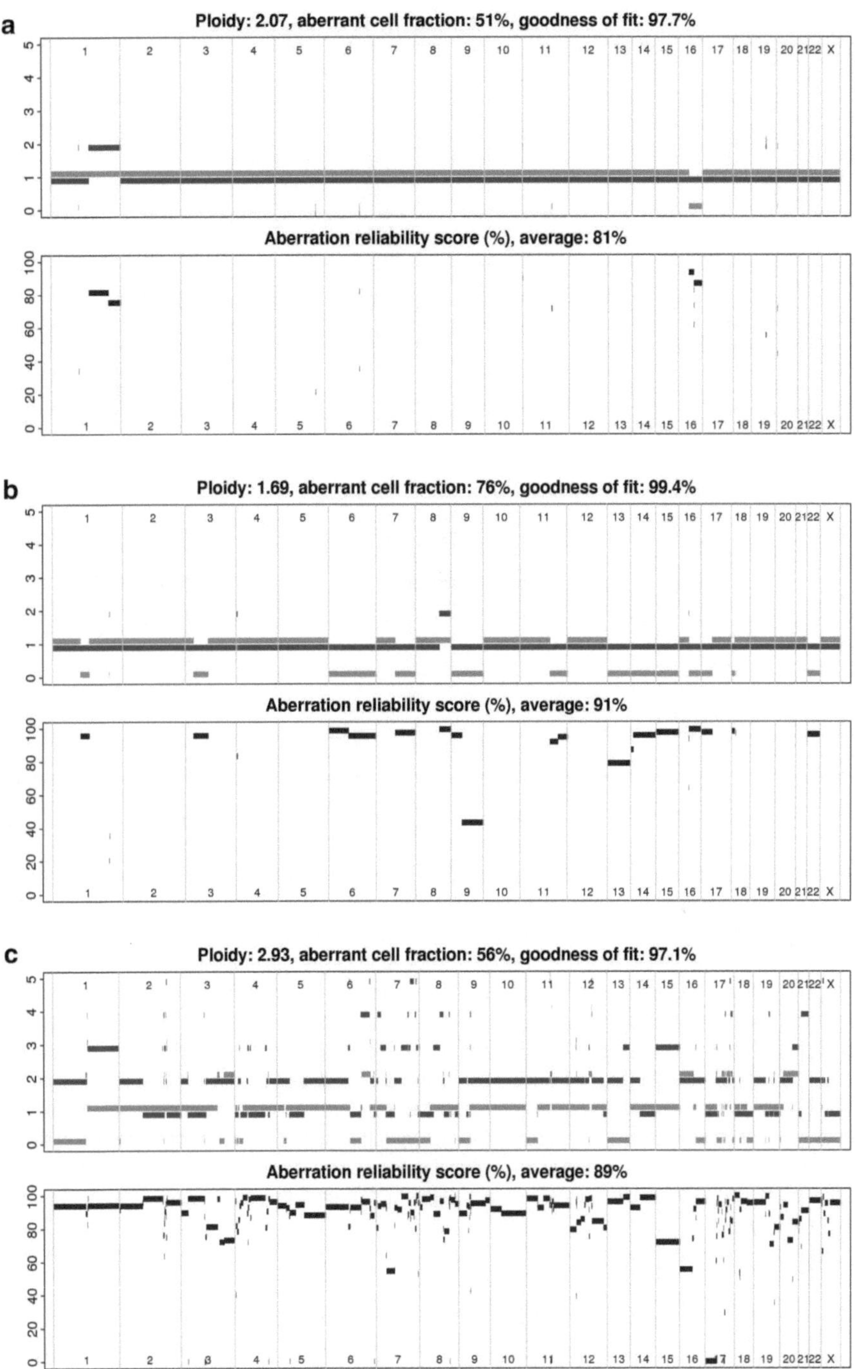

Fig. 6. Example ASCAT profiles and corresponding aberration reliability score plots. The ASCAT profiles (*top*) show the allele-specific copy number across the genome. The copy number of both alleles is shown. All estimated copy numbers are nonnegative whole numbers. Both lines are slightly shifted such that they do not overlap. The aberration reliability score plots (*bottom*) show the confidence one can have in each detected aberration, compared to the hypothesis of no aberration (see ref. 12 for details). (**a**) A sample with few aberrations (shown in Fig. 4a). A duplication of the 1q chromosome arm, as well as a hemizygous deletion (one copy lost) of 16q is immediately apparent. (**b**) A more complex sample (shown in Figs. 4b and 5b). Multiple hemizygous deletions are present, as well as a duplication at 8q. (**c**) A highly complex sample (shown in Figs. 4c and 5c). Few regions in the genome are unaffected by genomic aberrations in this sample.

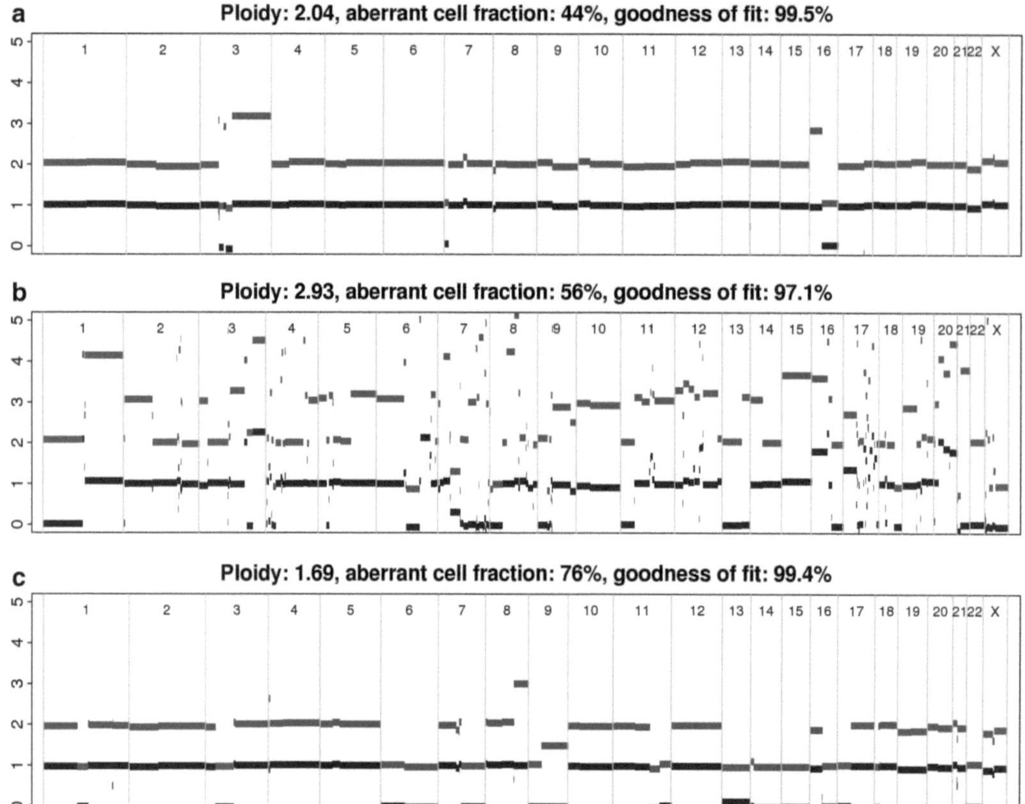

Fig. 7. Examples of raw copy number profile plots from ASCAT. These plots can be used to evaluate the solution reported by ASCAT and to gain insight into intratumor heterogeneity. The copy number of the minor allele is shown, as well as the total copy number, as directly derived from the data, without rounding to nonnegative integers. When a good solution is obtained, most or all regions should be close to whole numbers. In cases where a good global fit is obtained, yet some particular regions show copy numbers that are far from integers, these regions are likely subject to intratumor heterogeneity. (**a**) A sample showing few aberrations. All calculated copy numbers are close to integers, confirming a close fit. (**b**) A highly complex sample (shown in Figs. 4c, 5c, and 6c). Copy numbers clearly cluster close to integers. (**c**) A sample of intermediate complexity (shown in Figs. 4b, 5b, and 6b). All segments cluster close to whole numbers, except one on chromosome 9. One copy of the entire chromosome 9 has been lost in all tumor cells. In addition, the copy number of the q arm of chromosome 9 is close to 1.5, suggesting that there are two major subclones in the tumor: about 50% of the aberrant tumor cells show a gain of (the remaining copy of) 9q, while the other 50% of tumor cells do not have this gain. Due to the bad fit to whole numbers of this segment, the 9q arm also shows a clear drop in the aberration reliability score (Fig. 6b).

4. Some SNP array platforms (e.g., Affymetrix SNP 6.0) contain copy-number-only probes. These are probes in non-SNP locations. ASCAT can take these copy-number-only probes into account and calculate the total copy number at these loci. As no allelic contrast information is available, these copy-number-only probes should have NA values in their BAF data.
5. A segmentation algorithm of choice can be inserted in this step. The ASPCF segmentation algorithm, as part of the

ASCAT package, segments Log *R* and BAF simultaneously (automatically accounting for the structure and symmetry of BAF). ASPCF segment borders in Log *R* and BAF are automatically aligned (and optimized using data from both tracks). However, Log *R* and BAF can also be segmented separately using another segmentation algorithm (e.g., CBS (17)), without causing problems in later steps of the data analysis.

6. Removal of germline homozygous probes is most easily performed when matched germline samples (i.e., from the same individual) are also profiled by SNP arrays. If this material is not available, these homozygous probes can still be eliminated, e.g., by applying a threshold or by more specialized procedures. We aim to include an automated function to infer germline genotypes from tumor data in the next release of ASCAT.

7. The ASPCF segmentation algorithm is the computationally intensive step of the pipeline. However, this step can be executed in parallel, by using, e.g.,

```
ascat.bc = ascat.aspcf(ascat.bc, 1:5)
```

to segment the first five samples of a dataset. For every sample, two files are created containing the segmented Log *R* and BAF data. When these files exist upon execution of the `ascat.aspcf()` function, the results are read from disk rather than recalculating. Hence, by first splitting the segmentation over multiple processors, copying the resulting segmentation files to one directory and finally executing

```
ascat.bc = ascat.aspcf(ascat.bc)
```

this segmentation can be easily parallelized.

References

1. McCarroll SA, Kuruvilla FG, Korn JM et al (2008) Integrated detection and population-genetic analysis of SNPs and copy number variation. Nat Genet 40:1166–1174.
2. Peiffer DA, Le JM, Steemers FJ et al (2006) High-resolution genomic profiling of chromosomal aberrations using Infinium whole-genome genotyping. Genome Res 16:1136–1148.
3. Stratton MR, Campbell PJ, Futreal PA (2009) The cancer genome. Nature 458:719–724.
4. Balmain A, Gray J, Ponder B (2003) The genetics and genomics of cancer. Nat Genet 33 Suppl:238–244.
5. Witz IP, Levy-Nissenbaum O (2006) The tumor microenvironment in the post-PAGET era. Cancer Lett 242:1–10.
6. Navin N, Krasnitz A, Rodgers L et al (2010) Inferring tumor progression from genomic heterogeneity. Genome Res 20:68–80.
7. Sun W, Wright FA, Tang Z et al (2009) Integrated study of copy number states and genotype calls using high-density SNP arrays. Nucleic Acids Res 37:5365–5377.
8. Staaf J, Lindgren D, Vallon-Christersson J et al (2008) Segmentation-based detection of allelic imbalance and loss-of-heterozygosity in cancer cells using whole genome SNP arrays. Genome Biol 9:R136.
9. Attiyeh EF, Diskin SJ, Attiyeh MA et al (2009) Genomic copy number determination in cancer cells from single nucleotide polymorphism microarrays based on quantitative genotyping corrected for aneuploidy. Genome Res 19:276–283.

10. Greenman CD, Bignell G, Butler A et al (2010) PICNIC: an algorithm to predict absolute allelic copy number variation with microarray cancer data. Biostatistics 11:164–175.
11. Popova T, Manie E, Stoppa-Lyonnet D et al (2009) Genome Alteration Print (GAP): a tool to visualize and mine complex cancer genomic profiles obtained by SNP arrays. Genome Biol 10:R128.
12. Van Loo P, Nordgard SH, Lingjærde OC et al (2010) Allele-specific copy number analysis of tumors. Proc Natl Acad Sci U S A 107:16910–16915.
13. http://www.ifi.uio.no/bioinf/Projects/ASCAT
14. Marioni JC, Thorne NP, Valsesia A et al (2007) Breaking the waves: improved detection of copy number variation from microarray-based comparative genomic hybridization. Genome Biol 8:R228.
15. Wang K, Li M, Hadley D et al (2007) PennCNV: an integrated hidden Markov model designed for high-resolution copy number variation detection in whole-genome SNP genotyping data. Genome Res 17:1665–1674.
16. Bengtsson H, Irizarry R, Carvalho B et al (2008) Estimation and assessment of raw copy numbers at the single locus level. Bioinformatics 24:759–767.
17. Venkatraman ES, Olshen AB (2007) A faster circular binary segmentation algorithm for the analysis of array CGH data. Bioinformatics 23:657–663.

Chapter 5

Classification Approaches for Microarray Gene Expression Data Analysis

Leo Wang-Kit Cheung

Abstract

Classification approaches have been developed, adopted, and applied to distinguish disease classes at the molecular level using microarray data. Recently, a novel class of hierarchical probabilistic models based on a kernel-imbedding technique has become one of the best classification tools for microarray data analysis. These models were first developed as kernel-imbedded Gaussian processes (KIGPs) for binary class classification problems using microarray gene expression data, then they were further improved for multiclass classification problems under a unifying Bayesian framework. Specifically, an adaptive algorithm with a cascading structure was designed to find appropriate featuring kernels, to discover potentially significant genes, and to make optimal disease (e.g., tumor/cancer) class predictions with associated Bayesian posterior probabilities. Simulation studies and applications to publish real data showed that KIGPs performed very close to the Bayesian bound and consistently outperformed or performed among the best of a lot of state-of-the-art methods. The most unique advantage of the KIGP approach is its ability to explore both the linear and the nonlinear underlying relationships between the target features of a given disease classification problem and the involved explanatory gene expression data. This line of research has shed light on the broader usability of the KIGP approach for the analysis of other high-throughput omics data and omics data collected in time series fashion, especially when linear model based methods fail to work.

Key words: Microarray gene expression, Kernel-imbedding, Gaussian processes, Markov chains, Monte Carlo methods, Nonlinear systems

1. Introduction

Many methods for microarray gene expression data analysis have demonstrated their usefulness for a variety of class discovery/ prediction problems in biomedical applications. Two groups of machine learning methods have been studied: the wrapper methods and the filter methods. Given a set of variables, a wrapper method is designed to associate with the performance of the provided machine learner on each tested subset. A filter method,

Junbai Wang et al. (eds.), *Next Generation Microarray Bioinformatics: Methods and Protocols*, Methods in Molecular Biology, vol. 802, DOI 10.1007/978-1-61779-400-1_5, © Springer Science+Business Media, LLC 2012

on the other hand, attempts to find predictive subsets of the variables by making use of simple statistics computed from the empirical distribution. A filter method is relatively easier to implement than a wrapper method. For example, a weighted voting scheme was first introduced to discriminate each gene targeting a class prediction problem (1). Later, a different univariate ranking (UR) criterion as gene selection strategy was used to implement a few established machine learners to microarray analysis (2). Furthermore, different approaches had been studied to apply multiple hypotheses testing for microarray analysis while controlling a suitably defined error rate (3). Efron developed an empirical Bayes procedure for a multiple hypothesis testing using a local version of the false discovery rate (4). Comparatively, a wrapper method generally requires more computation, but it often can deliver better performance than a filter algorithm. For example, under a generalized linear regression model and by introducing a technique called supervised principal component analysis (SPCA), Bair, Paul, and Tibshirani presented a way to identify significant important subset of predictors (5). Based on a simple nearest centroid classifier and via a prototype shrinking strategy, Tibshirani, Hastie, Narasimhan, and Chu proposed the nearest shrunken centroids (also known as PAM) algorithm to analyze microarray data (6). Guyon, Weston, Barnhill, and Vapnik introduced the recursive feature elimination (RFE) algorithm to utilize a linear support vector machine (SVM) to select significant genes for a cancer classification problem (7). By choosing RFE and UR as the gene selection strategies, Zhu and Hastie applied the penalized logistic regression (PLR) to microarray data analysis (8).

Another approach to microarray data analysis is to build a hierarchical Bayesian model. For example, a hierarchical Bayesian mixture Gaussian model was proposed for gene expression data analysis (9). Most previous work focused on linear (or generalized linear) functions by introducing an automatic relevance determination (ARD) parameter for each gene through the covariance matrix of a Gaussian process (GP) and adopting the ordinal regression model, Chu's group proposed a Bayesian microarray analysis method (GP_ARD) for the gene selection problem (10). Based on a linear probit regression setting, a Bayesian hierarchical model for the gene selection problem was developed with a Gibbs sampler to solve it (11). An extension to a multiclass classification problem based on a multinomial probit regression model was suggested (12). Built on a linear logistic regression setting, a Bayesian approach was also applied to the gene selection problem as a way of microarray profiling (13). All these methods have been shown with various levels of effectiveness in finding significant genes in a wide range of real experiments. However, these linear models all share three limitations: first, a linear model is not necessarily always a good approximation for the underlying biological model; second, it has been argued that linear methods might be more sensitive to outlier samples (14); third, the computations of these linear model based algorithms usually involve calculating inverse of a matrix that may be singular when the number of the selected significant genes is relatively large. As an early attempt to overcome these limitations, a nonlinear term was added to the regular linear probit regression model to extend it to a generalized linear model (15). Recently, Zhao and Cheung developed the kernel-imbedded Gaussian process (KIGP) approach as a more comprehensive framework to unify both linear modeling and nonlinear modeling that can provide a performance close to the Bayesian bound (16, 17). Originated from the dual representation of linear learning methods for a binary classification problem, kernel-induced learning is one of the approaches that show promising potential to achieve this goal. It has been showed that if the regularization parameter is appropriately chosen and the dimension of the feature space is high enough, the solution of a kernel-induced SVM approaches to the Bayesian bound when the training sample size is sufficiently large (18). Bayesian probability theory can help construct a unifying framework for modeling data and can facilitate tuning of the involved parameters and hyperparameters. Bayesian inference can also provide an estimate of uncertainty in prediction, which is very beneficial for real-world decision making. MacKay developed a Bayesian learning paradigm called the "evidence framework" for neural networks (19). A Bayesian framework for SVMs and least squares SVMs (or LSSVMs) was built (20, 21). Under the Gaussian noise assumption, the mean of the posterior prediction made by a Gaussian process (GP) coincides with the optimal decision function made by an LSSVM (22, 23). GPs have been shown very effective to capture both linear and nonlinear relationships in many applications (21–24), and have a straightforward probabilistic model form. These facts formed the foundation of the KIGP approach for microarray gene expression data analysis. Via a probit regression setting, Zhao and Cheung built KIGPs to analyze microarray data of binary as well as multiclass disease classification problems (16, 17) (Notes 1 and 2). A description of the KIGP approach is provided in Subheading 2 with an illustrative example and the results of applying KIGPs are shown, further comments and remarks are given in Subheading 3.

2. Methods

2.1. The Theory

The kernel-induced SVM was developed and intensively studied. It has been successfully applied to many classification problems and is widely accepted as one of the state-of-the-art learning methods (25, 26). Essentially, the idea is to have a mapping function $\Psi(\bullet)$ that maps samples from the observation space to a proper feature space so that the target of interest for the learning/classification problem can be better represented, hence improving

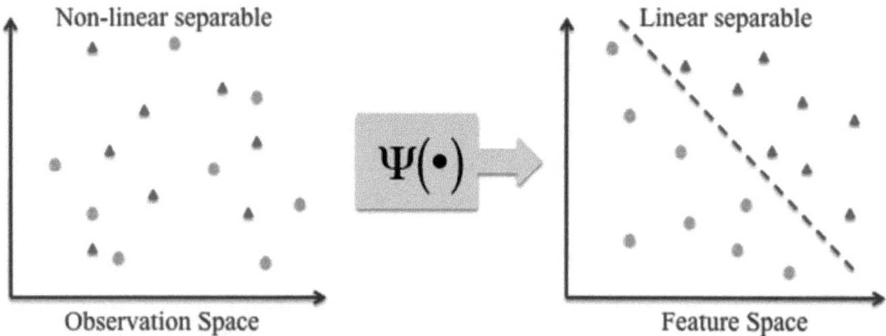

Fig. 1. Illustration of the feature mapping and feature space concept. *Green circles* and *red triangles* represent the samples of two different disease classes respectively, and Ψ(•) is the feature mapping function. These samples are nonlinear separable in the observation space but they are linear separable in the feature space.

the analyzing performance. Figure 1 illustrates the samples of two target classes in the observation space are not linearly separable. However, after the feature mapping, the two classes of samples can be well separated by a straight line in the feature space. If the mapping function is the identity function, the feature space would be exactly the observation space itself. Most linear model based methods are actually trained in the observation space, but some of the state-of-the-art methods can be categorized as learning in a feature space. For example, the feature space of a SPCA method (5) is the span of the first few principle components of the data; whereas the random forest method (27) assumes the significant gene data follow a forest structure. As for a kernel-induced learning method, it realizes training in the feature space through its implicit kernel structure.

Carefully designed imbedding techniques can offer very profound ways to tackle complex problems. One powerful imbedding technique called finite Markov chain imbedding (FMCI) has successfully widened the research in biological sequence analysis by relaxing a lot of statistical assumptions for the calculation of various distributions of runs and patterns (28–31). Extending the imbedding principle from sequence data analysis to gene expression data analysis, the KIGP approach has been developed (16, 17). Suppose we have n training samples with the class labels $\mathbf{y} = [y_1, y_2, ..., y_n]'$ in a classification problem, where $y_i \in \{1, 2, ..., M\}$ for $i = 1, 2, ..., n$, we label the base class as the class M. For each sample, there are p genes being investigated. We define the gene expression matrix $\mathbf{X}$ as

$$\mathbf{X} = \begin{bmatrix} \text{Gene } 1 & \text{Gene } 2 & \ldots & \text{Gene } p \\ X_{11} & X_{12} & \ldots & X_{1p} \\ \vdots & \vdots & \ddots & \vdots \\ X_{n1} & X_{n2} & \ldots & X_{np} \end{bmatrix}. \quad (1)$$

A gene selection vector γ is defined by

$$\gamma = [\gamma_1, \gamma_2, \ldots, \gamma_p]' \quad \text{where } \gamma_j = \begin{cases} 1 & \text{if the } j\text{th gene is selected,} \\ 0 & \text{otherwise,} \end{cases} \quad j = 1, 2, \ldots, p. \tag{2}$$

And $\mathbf{X}_\gamma$ is defined as the gene expression matrix corresponding to the selected genes in accordance to the gene selection vector γ.

$$\mathbf{X}_\gamma = \begin{bmatrix} X_{\gamma,11}, X_{\gamma,12}, \ldots, X_{\gamma,1q} \\ \vdots \\ X_{\gamma,n1}, X_{\gamma,n2}, \ldots, X_{\gamma,nq} \end{bmatrix} = \begin{bmatrix} \mathbf{x}_{\gamma 1} \\ \vdots \\ \mathbf{x}_{\gamma n} \end{bmatrix}, \tag{3}$$

where the jth column of $\mathbf{X}_\gamma$ is the ith column of the matrix $\mathbf{X}$ while the index of the jth nonzero element in the vector γ is i. In (3), there are q genes being selected out from a total of p genes; and generally $q << p$ in a typical gene selection problem.

Based on a probit regression model setting, we introduce latent vectors $\mathbf{z}_m$ and $\mathbf{t}_m$ for $m = 1, 2, \ldots, M - 1$, which are defined as

$$\begin{aligned} &[\mathbf{z}_m]_i = g_m(\mathbf{x}_{\gamma i}) + b_m + [\mathbf{e}_m]_i = [\mathbf{t}_m]_i + b_m + [\mathbf{e}_m]_i \\ &\quad \text{for } i = 1, 2, \ldots, n, \; m = 1, 2, \ldots, M - 1, \\ &\text{such that } y_i = \begin{cases} M \text{ if } [\mathbf{z}_k]_i \leq 0, \\ k \text{ if } [\mathbf{z}_k]_i > 0, \end{cases} \text{ where } k = \underset{m \in \{1, \ldots, M-1\}}{\arg\max} \{[\mathbf{z}_m]_i\}. \end{aligned} \tag{4}$$

In model (4), $\mathbf{x}_{\gamma i}$ denotes the ith row of the matrix $\mathbf{X}_\gamma$; $[\mathbf{e}_m]_i$ symbolizes the noise term that is assumed to be identically and independently distributed (IID) Gaussian with zero mean and σ^2 variance; b_m represents the intercept term; and $g_m(.)$ is chosen from a class of real-valued functions, the output of which is assumed to be a homogeneous Gaussian process. For convenience, we define $\mathbf{z} = [z_1, z_2, \ldots, z_{M-1}]$, $\mathbf{t} = [t_1, t_2, \ldots, t_{M-1}]$, $\mathbf{e} = [e_1, e_2, \ldots, e_{M-1}]$, $\mathbf{b} = [b_1, b_2, \ldots, b_{M-1}]'$. Model (4) thus becomes $\mathbf{z}_m = \mathbf{t}_m + \mathbf{e}_m + \mathbf{1}_\mathbf{n} b_m$, $m = 1, 2, \ldots, M - 1$, where $\mathbf{1}_\mathbf{n}$ is the $n \times 1$ vector of 1. Assuming the output of the discriminative mapping function $g_m(.)$ in the general model (4) is a Gaussian process, we have the following formulae for Bayesian inference:

$$\begin{aligned} &\tilde{t}_m | \tilde{\mathbf{x}}_\gamma, \mathbf{X}_\gamma, \mathbf{z}_m, b_m, \sigma^2 \sim N(f(\tilde{\mathbf{x}}_\gamma, \mathbf{X}_\gamma, \mathbf{z}_m, b_m, \sigma^2), V(\tilde{\mathbf{x}}_\gamma, \mathbf{X}_\gamma, \mathbf{z}_m, b_m, \sigma^2)) \\ &\text{where } f(\tilde{\mathbf{x}}_\gamma, \mathbf{X}_\gamma, \mathbf{z}_m, b_m, \sigma^2) = (\mathbf{z}_m - b_m \mathbf{1}_\mathbf{n})'(\mathbf{K}_{\gamma m} + \sigma^2 \mathbf{I}_n)^{-1} \mathbf{k}_m, \\ &V(\tilde{\mathbf{x}}_\gamma, \mathbf{X}_\gamma, \mathbf{z}_m, b_m, \sigma^2) = K_m(\tilde{\mathbf{x}}_\gamma, \tilde{\mathbf{x}}_\gamma) - \mathbf{k}_m'(\mathbf{K}_{\gamma m} + \sigma^2 \mathbf{I}_n)^{-1} \mathbf{k}_m, \\ &[\mathbf{K}_{\gamma m}]_{ij} = K_m(\mathbf{x}_{\gamma i}, \mathbf{x}_{\gamma j}), \; [\mathbf{k}_m]_i = K_m(\tilde{\mathbf{x}}_\gamma, \mathbf{x}_{\gamma i}), \; i, j = 1, 2, .., n, \\ &\quad m = 1, 2, \ldots, M - 1, \end{aligned} \tag{5}$$

$\tilde{t}_m = g_m(\tilde{\mathbf{x}}_\gamma)$ and $\tilde{\mathbf{x}}_\gamma = [\tilde{X}_1, \tilde{X}_2, ..., \tilde{X}_q]$ are the new testing gene expression data associated with the gene selection vector γ. Model (4) and its Bayesian inference form (5) are the key elements of the KIGP. The function $K_m(\mathbf{x}_{\gamma i}, \mathbf{x}_{\gamma j})$in (5) is a function defined in the observation space, which conceptually represents the inner product between the sample vector $\mathbf{x}_{\gamma i}$ and $\mathbf{x}_{\gamma j}$ in the relative feature space. The kernel matrix $\mathbf{K}_{\gamma m}$ has entries $[\mathbf{K}_{\gamma m}]_{ij} = \langle \Psi_m(\mathbf{x}_{\gamma i}), \Psi_m(\mathbf{x}_{\gamma j}) \rangle$ (assuming $\Psi_m(\bullet)$ is the mapping function from the observation space to the feature space for the classifier m). Common kernel functions are:

$$\text{Linear kernel}: K(\mathbf{x}_{\gamma i}, \mathbf{x}_{\gamma j}) = \langle \mathbf{x}_{\gamma i}, \mathbf{x}_{\gamma j} \rangle, \tag{6a}$$

$$\text{Polynomial kernel}: K(\mathbf{x}_{\gamma i}, \mathbf{x}_{\gamma j}) = \left(\langle \mathbf{x}_{\gamma i}, \mathbf{x}_{\gamma j} \rangle + 1 \right)^d \quad \text{where } d = 1, 2, \ldots \text{ is degree parameter}, \tag{6b}$$

$$\text{Exponential kernel}: K(\mathbf{x}_{\gamma i}, \mathbf{x}_{\gamma j}) = \exp\left(- \frac{\left\| \mathbf{x}_{\gamma i} - \mathbf{x}_{\gamma j} \right\|}{r} \right) \tag{6c}$$

where $r>0$ is the width parameter,

$$\text{Gaussian kernel}: K(\mathbf{x}_{\gamma i}, \mathbf{x}_{\gamma j}) = \exp\left(- \frac{\left\| \mathbf{x}_{\gamma i} - \mathbf{x}_{\gamma j} \right\|^2}{2r^2} \right) \tag{6d}$$

where $r>0$ is the width parameter,

$$\text{Manhattan kernel}: K(\mathbf{x}_{\gamma i}, \mathbf{x}_{\gamma j}) = \exp\left(- \frac{\left\| \mathbf{x}_{\gamma i} - \mathbf{x}_{\gamma j} \right\|_M}{r} \right) \tag{6e}$$

where $r>0$ is the width parameter.

Note: $\langle \bullet, \bullet \rangle$ is the inner product between two vectors, $\|\bullet\|$ is the L-1 norm, $\|\bullet\|^2$ is the L-2 norm, and $\|\bullet\|_M$ is the Manhattan norm of a vector. We refer a linear kernel as a LK, a polynomial kernel with degree d as PK(d), an exponential kernel with width r as EK(r), a Gaussian kernel with width r as a GK(r), and a Manhattan kernel with width r as MK(r).

The general KIGP framework is summarized in Fig. 2. With a gene selection procedure, a group of candidate significant genes is selected within an iterative updating process. For each nonbase class, $m = 1, 2, ..., M - 1$ through a feature mapping function $\Psi_m(.)$, the selected gene data are mapped to a feature space. Then the optimal classification procedure is processed in the joint feature space to determine the class of the input sample. Computationally, under the theory of having a kernel-induced feature space, we do not really do the explicit feature mapping. Instead, we equivalently train the data through a KIGP using a kernel function. The candidate training methods include SVM, LSSVM, GP, PLR, and kernel Fisher discrimination (KFD). In the KIGP approach, we focus on using the Gaussian process model.

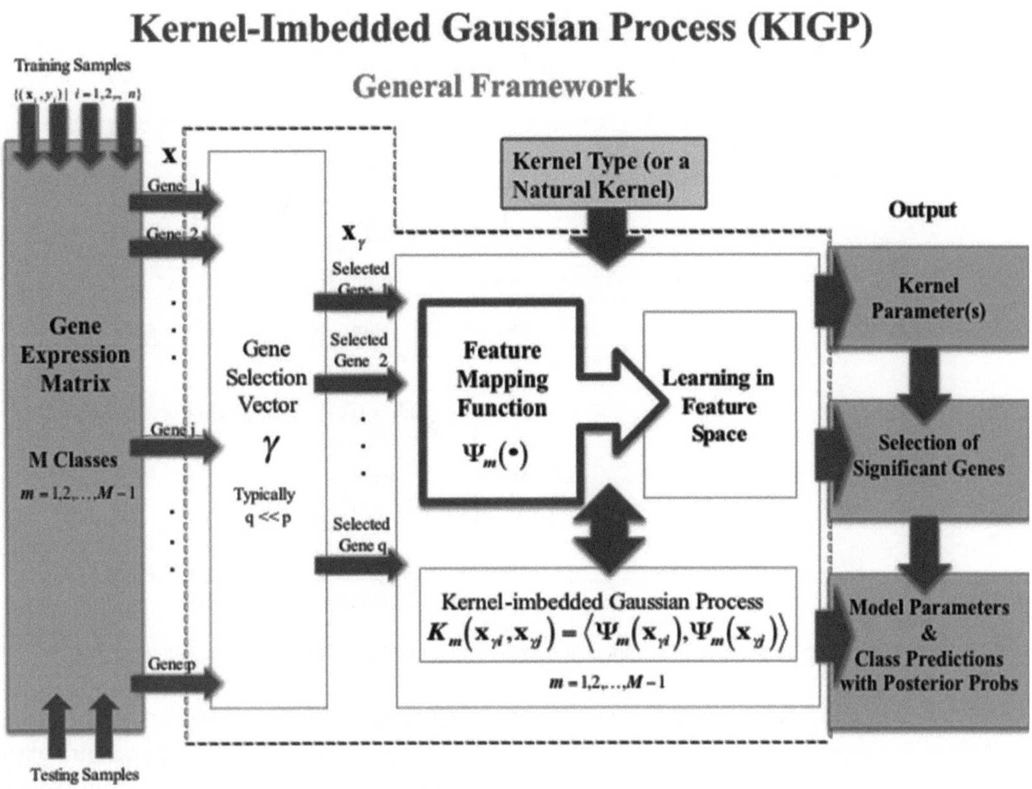

Fig. 2. Schematic workflow of the KIGP Approach. The box bounded by *dotted lines* represents the KIGP iterative learning/updating Gibbs sampling algorithm.

With a Bayesian structure, a KIGP Gibbs sampling learning algorithm is built as in Fig. 3. Complete details of the algorithm and the selection of prior distributions are described by Zhao and Cheung (16, 17). We assume the applied kernel function type is fixed and denote the kernel parameter(s) as $\theta = [\theta_1, \theta_2, ..., \theta_{M-1}]$, in which θ_m denotes kernel parameter(s) for classifier m. After the Gibbs sampling converged, the KIGP approach provides the optimal kernel type, the associated optimal kernel parameter estimate (s), model parameter estimates, selection of significant genes, and class predictions for the testing samples with posterior probabilities. The algorithm is theoretically robust as the kernel matrix is positive definite. The total computation complexity of the KIGP Gibbs sampler in each iteration is $O((M-1)pn^3)$ (Notes 3 and 4).

2.2. The Practice and Application

Various simulation studies and real data applications of the KIGP approach have been conducted and published. They showed that KIGPs performed very close to the Bayesian bound and consistently outperformed or performed among the best of a lot of state-of-the-art methods. Readers are referred to (16, 17) for more details. As an illustrative example, we show the application of KIGP approach for the acute leukemia microarray data analysis

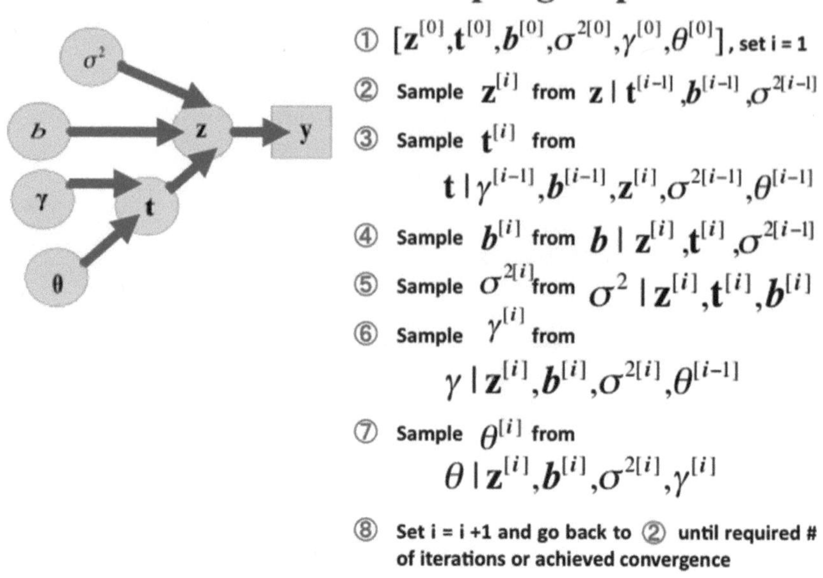

Fig. 3. Directed acyclic graph of the KIGP hierarchical Bayesian model and the KIGP Gibbs sampling algorithm.

below. The published acute leukemia data (1) consists of the bone marrow or peripheral blood samples taken from 72 patients with either acute myeloid leukemia (AML) or acute lymphoblastic leukemia (ALL). The training set has 38 samples, of which 27 are ALL and 11 are AML. The testing set has 34 samples, of which 20 are ALL and 14 are AML. Expression levels of 7,129 human genes were obtained from the Affymetrix high-density oligonucleotide microarrays.

The KIGPs with a PK, a GK, and a LK were applied to the training dataset. The prior parameter π_j for all j was uniformly set at 0.001. In both the "kernel parameter fitting phase" and the "gene selection phase," we ran 30,000 Gibbs sampling iterations and treated the first 15,000 iterations as the burn-in period; and in the "prediction phase," we ran 5,000 iterations and treated the first 1,000 iterations as the burn-in period. For the KIGP with a PK, the resulted posterior probability masses of the degree parameter d are Prob($d = 1$) = 0.985 and Prob($d = 2$) = 0.015. With the PK(1), 20 genes were identified as "significant" at 0.05 significance level. Using the PK(1) and the found significant genes, we made predictions for the 34 testing samples. We then ran a leave-one-out cross-validation (LOOCV) for the 38 training samples. This "loose" LOOCV procedure was however only involved in the "prediction phase." Since the fitted kernel parameter and the significant genes chosen from the first two phases had already contained the most information of the whole training dataset, it was not a proper validation measure for kernel type competition. More properly,

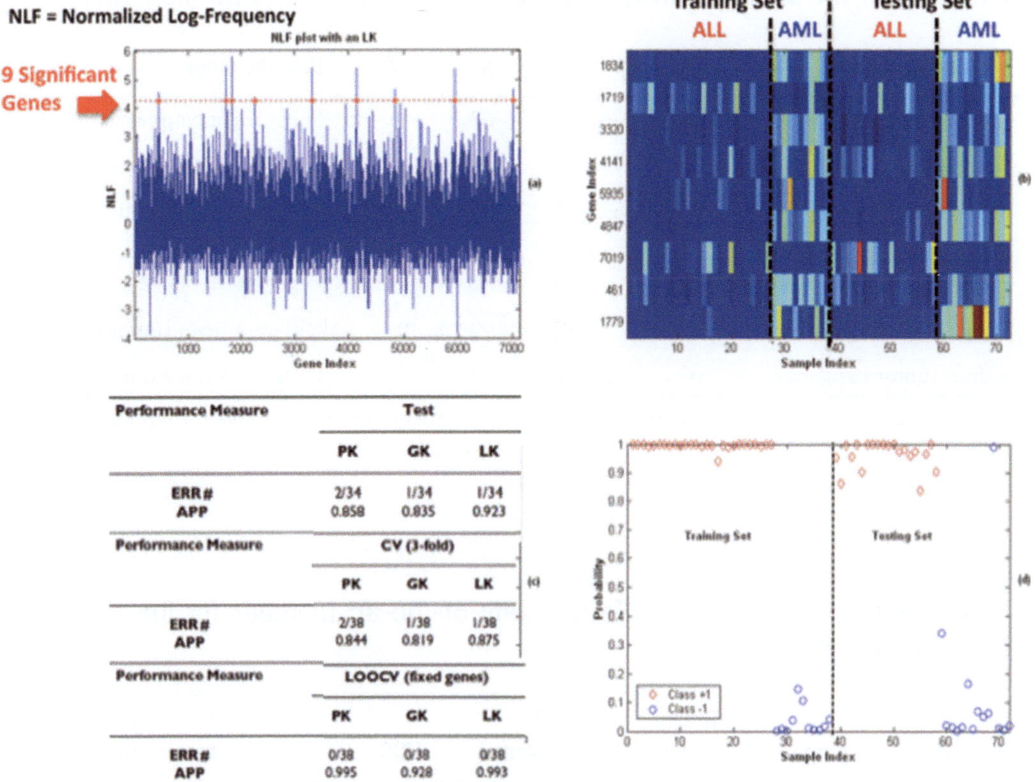

Performance Measure	Test		
	PK	GK	LK
ERR#	2/34	1/34	1/34
APP	0.858	0.835	0.923
Performance Measure	CV (3-fold)		
	PK	GK	LK
ERR#	2/38	1/38	1/38
APP	0.844	0.819	0.875
Performance Measure	LOOCV (fixed genes)		
	PK	GK	LK
ERR#	0/38	0/38	0/38
APP	0.995	0.928	0.993

Fig. 4. (**a**) Plot of the normalized log-frequency statistics. (**b**) Heat map of the nine highly significant genes for disease classification. (**c**) Performance summary of KIGP with a PK, a GK, and a LK. Test represents independent tests with the testing set. (**d**) Plot of posterior probabilities of class membership. *Red diamonds* (Class +1) represent ALL samples, *blue circles* (Class −1) represent AML samples.

we further did a rigorous threefold cross-validation (threefold CV) that included all three phases of the proposed algorithm (further details are described in refs. 16, 17). This whole procedure was then repeated for the KIGP with a GK and with a LK, respectively. As a result, the KIGP with a LK gave the best testing performance: only one misclassification error was found (same result from many other state-or-the-art analysis methods) and the average predictive probability (APP) of the true class labels was the largest. Nine highly significant genes were found by the KIGP with a LK (the normalized log-frequency (NLF) statistics were calculated for all genes and were used for gene selection). The KIGP analysis outputs are displayed in Fig. 4.

In addition, six more real microarray datasets (Table 1) were used for comparing the performance of KIGPs with other advanced methods. A summary is provided in Table 2 (Note 5).

Table 1
Summary of the six microarray datasets analyzed with the KIGP approach

Microarray dataset	*M*	*p*	*n*	*W*	Disease class
Lymphoma (32)	3	4,026	62	0	Subtypes of lymphoma
Breast cancer (33)	3	3,226	22	0	BRCA1/BRCA2/sporadic
MLL leukemia (34)	4	54,675	96	44	ALL/AML with/without MLL
Hepatocellular carcinoma (HCC) (35)	5	54,675	61	30	Different HCC classes
Brain tumor (36)	5	5,597	42	0	Different brain tumor types
Kidney tumor (37)	6	22,283	63	29	Different kidney tumor types

M number of classes, *p* number of investigated genes, *n* number of training samples, *W* number of testing samples

Table 2
Performance comparison of different state-of-the-art methods for the analysis of microarray data

Method	Lymphoma	Breast cancer	MLL leukemia	HCC	Brain tumor	Kidney tumor
KIGP/LK	0/62 (10)	0/22 (7)	0/44 (15)	3/30 (20)	0/42 (18)	0/29(20)
KIGP/GK	0/62 (4)	0/22 (6)	0/44 (11)	7/30 (10)	4/42 (22)	0/29 (15)
SVM/LK/UR	0/62 (41)	0/22 (50)	1/44 (8)	14/30 (42)	1/42 (40)	0/29 (19)
SVM/GK/UR	5/62 (17)	0/22 (12)	1/44 (8)	16/30 (46)	9/42 (50)	0/29 (31)
SVM/RFE	0/62 (15)	0/22 (6)	1/44 (8)	9/30 (103)	0/42 (20)	0/29 (6)
PLR/LK/UR	0/62 (11)	0/22 (10)	1/44 (4)	7/30 (38)	3/42 (22)	0/29 (11)
PLR/GK/UR	0/62 (11)	0/22 (10)	1/44 (13)	12/30 (98)	3/42 (16)	0/29 (27)
PLR/RFE	0/62 (8)	0/22 (6)	1/44 (12)	8/30 (71)	0/42 (20)	0/29 (6)
PAM	1/62 (1,987)	0/22 (48)	0/44 (2,331)	7/30 (4,401)	1/42 (5,521)	0/29 (8,339)

The format for each cell of the table is: "number of errors/number of testing samples (number of selected genes)." *Note*: If an independent testing set is not available, number of errors from leave-one-out cross-validation/number of training samples were reported

3. Notes

1. The most unique characteristic of the KIGP approach is its ability as a unifying framework to explore both the linear and the nonlinear underlying relationships between the target features of a given disease classification problem and the involved explanatory gene expression data.
2. Comparing to a regular SVM, the most popular kernel learning method, the KIGP has three key advantages. First, the probabilistic class prediction by the KIGP could be insightful for borderline cases in real applications. Second, the KIGP approach has implemented specific procedure for tuning the kernel parameter(s) (such as the width parameter of a GK) and the model parameters (such as the variance of the noise term). Tuning parameters has always been one of the key issues for nonlinear parametric learning methods. As the gene selection procedure is imbedded into the learner, the KIGP is also more consistent in identifying significant genes when comparing to regular UR or RFE method with a cross-validation procedure. In our simulated studies, the KIGP/GK significantly outperformed its SVM or PLR counterparts with either RFE or UR as gene selection strategy in the nonlinear example and in the example with mislabeled training samples. Third, the KIGP approach can provide more useful information, such as the posterior PDF of the parameters, for further statistical analysis and inference.
3. Computationally, the KIGP is robust and very amenable to be implemented to a Gibbs sampling system. Both the simulation studies and the real data studies have shown the effectiveness of the KIGP approach (16, 17). A major cost of using the KIGP is its computational complexity. With the prescreening procedure (17), we alleviated this cost, making computational complexity of the KIGP affordable in most real applications (e.g., 3.5 h for the MLL Leukemia dataset (17)). We found that the prescreening procedure dramatically decreased the computation intensity without losing predictive performance in both the simulated examples and the real case studies.
4. More recently, we have developed a new procedure of building a natural kernel, either a natural Gaussian Fisher kernel (NGFK) or a natural Student-*t* Fisher kernel (NTFK), which can address the issue of kernel selection for general kernel-induced learning methods. By implementing a natural kernel into the KIGP, we have also developed a Natural

kernel-imbedded Gaussian process (NKIGP) for microarray data analysis. Based on our simulated and real microarray data studies, the NKIGP can adaptively discover the underlying feature space in both linear and nonlinear cases with excellent results. Its performance was always very close to the theoretical Bayesian bound in all of our simulation studies. The NKIGP performed consistently very well without the need of tuning kernel parameters, even for datasets with multiple suspiciously mislabeled training samples. For nontrivial real datasets, such as the published colon tumor dataset (38), the NKIGP particularly showed its outstanding performance and demonstrated its promising potential for analyzing a dataset containing inconsistent information. This natural kernel-building procedure can be directly applied to other kernel-based learning algorithms (e.g., SVM) with minor or no changes. This work is currently in revision for publication. This line of research has also shed light on the broader usability of the KIGP approach for the analysis of other high-throughput omics data and omics data collected in time series fashion, especially when linear model based methods fail to work.

5. The code of the KIGP is available upon request. A user interface of the KIGP package is currently work in progress.

Acknowledgments

This work was partially supported by the Loyola University Medical Center Research Development Funds and the SUN Microsystems Academic Equipment Grant for Bioinformatics. The author would like to thank Dr. Xin Zhao at Sanjole Inc. for his involvement on the KIGP work.

References

1. Golub TR, Slonim D, Tamayo P et al (1999) Molecular classification of cancer: class discovery and class prediction by gene expression monitoring. Science 286:531–537.
2. Dudoit S, Fridlyand J, Speed T (2002) Comparison of discrimination methods for the classification of tumors using gene expression data. JASA 97:77–87.
3. Dudoit S, Shaffer J, Boldrick J (2003) Multiple hypothesis testing in microarray experiments. Statistical Science 18:71–103.
4. Efron B (2004) Large-scale simultaneous hypothesis testing: the choice of a null hypothesis. J. Amer. Statis. Assoc. 99:96–104.
5. Bair E, Hastie T, Paul D et al (2006) Prediction by supervised principal component. J. Amer. Statis. Assoc. 101:119–137.
6. Tibshirani R, Hastie T, Narasimhan B et al (2002) Diagnosis of multiple cancer types by shrunken centroids of gene expression. Proc. Natl Acad. Sci. USA 99:6567–6572.
7. Guyon I, Weston J, Barnhill S (2002) Gene selection for cancer classification using support vector machines. Machine Learning 46:389–422.
8. Zhu J, Hastie T (2004) Classification of gene microarrays by penalized logistic regression. Biostatistics 5:427–443.

9. Lönnstedt I, Britton T (2005) Hierarchical Bayes models for cDNA microarray gene expression. Biostatistics 6:279–291.
10. Chu W, Ghahramani Z, Falciani F et al (2005) Biomarker discovery in microarray gene expression data with Gaussian processes. Bioinformatics 21:3385–3393.
11. Lee KE, Sha N, Dougherty ER et al (2003) Gene selection: a Bayesian variable selection approach. Bioinformatics19:90–97.
12. Zhou X, Wang X, Dougherty ER (2004) Gene prediction using multinomial probit regression with Bayesian gene selection. EURASIP Journal on Applied Signal Processing 1: 115–124.
13. Zhou X, Liu K, Wong STC (2004) Cancer classification and prediction using logistic regression with Bayesian gene selection. Journal of Biomedical Informatics 37:249–259.
14. Pochet N, Smet FD, Suykens JAK et al (2004) Systematic benchmarking of microarray data classification: assessing the role of non-linearity and dimensionality reduction. Bioinformatics 20:3185–3195.
15. Zhou X, Wang X, Dougherty ER (2004) A Bayesian approach to nonlinear probit gene selection and classification. Journal of the Franklin Institute 341:137–156.
16. Zhao X, Cheung LWK (2007) A hierarchical Bayesian approach with kernel-imbedded Gaussian processes for micoarray gene expression data analysis. BMC Bioinformatics 8:67.
17. Zhao X, Cheung LWK (2011) Multi-class kernel-imbedded Gaussian processes for microarray data analysis. IEEE/ACM Transactions on Computational Biology and Bioinformatics 8(4):1041–1053.
18. Lin Y (2002) Support vector machines and the Bayes rule in classification. Data Mining and Knowledge Discovery 6:259–275.
19. MacKay DJC (1992) The evidence framework applied to classification networks. Neural Computation 4:720–736.
20. Kwok JT (2000) The evidence framework applied to support vector machines. IEEE Trans. on Neural Networks 11:1162–1173.
21. Gestel TV, Suykens JVK, Lanckriet G et al (2002) Bayesian framework for least-squares support vector machine classifiers, Gaussian processes, and kernel fisher discriminant analysis. Neural Computation 14:1115–1147.
22. Neal RM (1996) Bayesian learning for neural networks. Springer, New York.
23. Rasmussen CE, Williams CKI (2006) Gaussian processes for machine learning. The MIT Press, Cambridge, Massachusetts.
24. Cristianini N, Shawe-Tayer J (2000) An introduction to support vector machines. Cambridge University Press.
25. Kuh A (2004) Least Square Kernel Methods and Applications. In: Soft Computing in Communications. Wang L (ed) p:361–383. Springer, Berlin.
26. Müller K, Mika S, Rätsch G et al (2001) An Introduction to Kernel-Based Learning Algorithms. IEEE Trans. Neural Networks 12:181–202.
27. Diaz-Uriarte R, Andres SA (2006) Gene selection and classification of microarray data using random forest. BMC Bioinformatics 7:1–13.
28. Cheung LWK (2004) Use of runs statistics for pattern recognition in genomic DNA sequences. Journal of Computational Biology 11:107–124.
29. Nuel G (2006) Effective p-value computations using Finite Markov Chain Imbedding (FMCI): application to local score and to pattern statistics. Algorithms Mol Biol 1:5.
30. Aston J, Martin D (2007) Distributions associated with general runs and patterns in hidden Markov models. The Annals of Applied Statistics 1: 585–611.
31. Martin J, Regad L, Camproux A-C et al (2010) Finite Markov Chain Embedding for the Exact Distribution of Patterns in a Set of Random Sequences. In: Advances in Data Analysis- Statistics for Industry and Technology: Theory and Applications to Reliability and Inference, Data Mining, Bioinformatics, Lifetime Data, and Neural Networks. Skiadas C (ed). p.171-180. Springer.
32. Alizadeh AA, Eisen MB, Davis RE et al (2000) Distinct types of diffuse large B-Cell-lymphoma identified by gene expression profiling. Nature 403:503–511.
33. Hedenfalk I, Duggan D, Chen Y et al (2001) Gene expression profiles in hereditary breast cancer. The New England Journal of Medicine 344:539–548.
34. Zangrando A, Dell'orto MC, Te Kronnie G et al (2009) MLL rearrangements in pediatric acute lymphoblastic and myeloblastic leukemias: MLL specific and lineage specific signatures. BMC Med Genomics 2:36.
35. Chiang DY, Villanueva A, Hoshida Y et al (2008) Focal gains of VEGFA and molecular classification of hepatocellular carcinoma. Cancer Res 68:6779–6788.
36. Pomeroy S, Tamayo P, Gaasenbeek M et al (2002) Prediction of central nervous system embryonal tumoroutcome based on gene expression. Nature 415:436–442.
37. Jones J, Otu H, Spentzos D et al (2005) Gene signatures of progression and metastasis in renal cell cancer. Clin Cancer Res 11: 5730–5739.
38. Alon U, Barkai N, Notterman D et al (1999) Broad patterns of gene expression revealed by clustering analysis of tumor and normal colon tissues probed by oligonucleotide arrays. Proc. Natl Acad. Sci. USA 96:6745–6750.

Chapter 6

Biclustering of Time Series Microarray Data

Jia Meng and Yufei Huang

Abstract

Clustering is a popular data exploration technique widely used in microarray data analysis. In this chapter, we review ideas and algorithms of bicluster and its applications in time series microarray analysis. We introduce first the concept and importance of biclustering and its different variations. We then focus our discussion on the popular iterative signature algorithm (ISA) for searching biclusters in microarray dataset. Next, we discuss in detail the enrichment constraint time-dependent ISA (ECTDISA) for identifying biologically meaningful temporal transcription modules from time series microarray dataset. In the end, we provide an example of ECTDISA application to time series microarray data of Kaposi's Sarcoma-associated Herpesvirus (KSHV) infection.

Key words: Time series, Clustering, Bicluster, Iterative signature algorithm, Temporal module, Microarray, Time dependent, Enrichment constrained

1. Introduction

Biological processes including development, survival, replication, response to stimulus, and others are inherently dynamic. Understanding the temporal regulation of these processes comprises one of the most important aspects of biological research. At a molecular level, regulation of biological process can occur by controlling mRNA gene expression; examples include transcriptional regulation by transcription factors, posttranscriptional silencing by microRNAs, and epigenetic regulation such as DNA methylation. Microarray provides a powerful means to measure the dynamic regulation of a biological process at the gene expression level.

The so-called time series microarray experiments measure genome-wide expression at a consecutive series of time points over the course of a biological process of interest. So far, a large amount of genome-wide time series expression data measuring, for instance, yeast cell cycle (1) and Megakaryocytic differentiation (2), has been accumulated. These measurements can be

Junbai Wang et al. (eds.), *Next Generation Microarray Bioinformatics: Methods and Protocols*,
Methods in Molecular Biology, vol. 802, DOI 10.1007/978-1-61779-400-1_6, © Springer Science+Business Media, LLC 2012

considered as time series data samples, where expressions at two different time points are correlated. Time series analysis concerns the modeling and inference of the temporal patterns and correlation between genes embedded in expressions data. The temporal patterns are indicative of the regulations in the underlying biology processing of interest. For example, genes that have similar temporal expression patterns are likely to share similar functions. Also, genes regulated by the regulator genes often have a shared pattern at a time delay with the regulators gene expression and a gene regulatory network can be inferred by uncovering the delayed expression patterns among the genes.

Clustering plays a key role in time series microarray analysis. Many clustering algorithms have been developed including, most notably, hierarchical clustering (3), K-means clustering (4), self-organizing maps (5), and two-way clustering (6), and they have been applied to find transcriptional modules. These algorithms are less effective when applied to large and/or time series data sets due to two well-recognized limitations. First, standard clustering algorithms assign each gene to a single cluster, while many genes in fact belong to multiple transcriptional modules (7, 8); second, each transcriptional module may only be active in a few experiments (8–10) or a subperiod of entire time course. In fact, our general understanding of cellular processes leads us to expect transcriptional module to have shared gene components and be active at a specific period of time and/or under a specific experimental condition (11). Alternatively, biclustering algorithms have been proposed to address these problems of standard clustering algorithms. These algorithms can uncover temporal transcription modules (TTMs), or subsets of genes co-regulated under only certain time period. In this chapter, we discuss various biclustering algorithms applicable to analyzing time series microarray data.

2. Methods

In this chapter, we first introduce the concept of biclustering and the popular biclustering algorithm-iterative signature algorithm (ISA) (12); then, we show how to incorporate prior knowledge and time dependence into ISA to find biologically meaningful TTM from time series microarray data.

2.1. Biclustering and Its Interpretation

Biclustering or co-clustering is a data mining technique that allows simultaneous clustering of the rows and columns of a data matrix. It was recently introduced into the gene expression analysis by Cheng and Church (8). Different biclustering algorithms may have different definitions of bicluster, but in general, all biclustering algorithms seek to find patterns that are embedded into the whole

Table 1
Types of biclusters

(a) Constant value				
...	...	...	...	...
...	1	1	1	...
...	1	1	1	...
...	1	1	1	...
...	...	...	...	...
(b) Constant value on column				
...	...	...	...	...
...	1	3	2	...
...	1	3	2	...
...	1	3	2	...
...	...	...	...	...
(c) Constant value on rows				
...	...	...	...	...
...	1	1	1	...
...	2	2	2	...
...	3	3	3	...
...	...	...	...	...
(d) Constant difference on columns				
...	...	...	...	...
...	1	3	2	...
...	1.1	3.1	2.1	...
...	0.9	2.9	1.9	...
...	...	...	...	...

Since a bicluster is only a subset of the whole matrix, in this picture, we use "..." denotes the rest areas of the data matrix, which are not part of the bicluster

microarray dataset, where rows, representing genes in general, exhibit similar behavior across the columns, or conditions. Table 1 shows some popular bicluster definitions, each indicating a unique data pattern that may result from a particular underlying mechanism.

Under the context of microarray data analysis, the biological meaning of the above-mentioned bicluster types can be interpreted as well. In time series microarray data set, a row of the data matrix often represents the time series expression profile of a particular gene; each column represents a sample taken at a specific time. Then the biological meaning of the bicluster types in Table 1 can be explained as:

(a) A group of genes whose expressions level are similar and stay the same among several sample times.

(b) A group of genes whose expressions are the same at a particular time, and go up and down together across different time samples.

(c) A group of genes whose expressions stay unchanged across several time points.

(d) A group of genes whose expressions go up and down together.

Please note that different biclustering patterns can be highly related. In Table 1, (d) can be transformed into (b) by taking first-order difference along row dimension, and (b) can be transformed into (c) by taking transpose of the data matrix. In reality, by transforming the data matrix, a single biclustering algorithm can often be used to find many different biclustering structures (see Note 1). In practice, one would have to choose the bicluster type that best describes the research interest and seek to find it using a biclustering algorithm. We discuss in detail next a popular biclustering algorithm known as ISA (12).

2.2. Signature Algorithm

Proposed by Bergmann (12), the ISA has gained great success in gene expression analysis. Several extensions of ISA were developed, such as PISA (13), EDISA (14), and enrichment constraint time-dependent ISA (ECTDISA) (15), each focusing on a different aspect of biclustering of gene expression data. We briefly review the ISA in this section. Before starting to search for a bicluster, we have to first define a bicluster mathematically. A typical bicluster module can be defined as follows:

Let $\mathbf{Y} \in R^{G\times C}$ represents microarray data matrix that consists of expression of G genes sampled at C conditions. Given a pair of thresholds (τ_G, τ_C), bicluster or transcription module m can be defined by (1) as a group of genes $\mathcal{G}_m$ and a group of conditions $\mathcal{C}_m$ that satisfy

$$M(\tau_G, \tau_C) := \left\{ \{\mathcal{G}_m, \mathcal{C}_m\} \middle| \begin{matrix} \forall g \in \mathcal{G}_m & E(\mathbf{Y}_{g,\mathcal{C}_m}) > \tau_G \\ \forall c \in \mathcal{C}_m & E(\mathbf{Y}_{\mathcal{G}_m,c}) > \tau_C \end{matrix} \right\}, \quad (1)$$

where, $E(*)$ represents the average expression level of a vector of gene expressions; $\mathbf{Y}_{g,\mathcal{C}_m}$ represents the expression levels of the gth gene at the conditions defined condition set $\mathcal{C}_m$; $\mathbf{Y}_{\mathcal{G}_m,c}$ represents the expression levels of genes in sets $\mathcal{G}_m$ under the cth condition.

In this definition, the first term constrains the bicluster from the gene dimension, the second term constrains from the condition dimension. This equation essentially defines a rectangular area, of which the average expression level of each row is larger than τ_G, and the average expression level of each column is larger than τ_C. Biologically, the bicluster defines a group of genes that are upregulated under a group of conditions. If such pattern is identified from the data, presumably, we may conclude those genes may be positively related to those selected conditions. To find a bicluster, the ISA starts from an initial gene set and iteratively refines the condition and gene set by the following criteria,

1. Based on the previous gene set $\mathcal{G}_m$, find all the c that satisfy $E(\mathbf{Y}_{\mathcal{G}_m,c}) > \tau_C$ to form new $\mathcal{C}_m$.
2. Based on the previous condition set $\mathcal{C}_m$, find all the g that satisfy $E(\mathbf{Y}_{g,\mathcal{C}_m}) > \tau_G$ to form new $\mathcal{G}_m$.

The iteration stops until some convergence criterion is reached. The algorithm can be then restarted from another initial gene set to find another bicluster module. Let us consider an example. We want to find a bicluster module from the following microarray data:

$$\Upsilon = \begin{bmatrix} 2 & 5 & 2 & 2 & 1 \\ 2 & 5 & 2 & 4 & 1 \\ 1 & 1 & 1 & 1 & 1 \\ 1 & 5 & 2 & 5 & 1 \\ 1 & 1 & 1 & 1 & 3 \end{bmatrix}$$

with the setting $(\tau_G, \tau_C) = (2.5, 2.5)$ using ISA, and the detailed procedure is illustrated in Table 2.

As it can be seen from the example, the result converges very quickly, and it identifies the embedded module accurately. If this is a real microarray data, we will be able to further claim that the selected genes are upregulated under the selected conditions, or they could be positively related for some reason. However, ISA also suffers from the following limitations.

1. Given different values of the parameters (τ_G, τ_C), the identified biclusters could be significantly different; some of them may not be biologically meaningful.
2. When applied to time series data, ISA does not consider the dependence between samples, and thus could identify temporal modules that are discontinuous in time dimension, which would be hard to explain biologically.

In the next section, we introduce enrichment constrained time-dependent ISA (ECTDISA) (15), which was aimed to tackle the above-mentioned limitations.

2.3. Enrichment Constrained Time-Dependent Cluster Analysis

ECTDISA (15) consists of two main features:

1. An enrichment constrained framework that constrains the biological meaning of modules by choosing the optimal parameters of module defined based on prior knowledge.

Table 2
Procedure of ISA

1. $\begin{bmatrix} (2) & (5) & (2) & (2) & (1) \\ 2 & 5 & 2 & 4 & 1 \\ 1 & 1 & 1 & 1 & 1 \\ 1 & 5 & 2 & 5 & 1 \\ 1 & 1 & 1 & 1 & 3 \end{bmatrix}$	*Initial condition:* Initial gene set may contain any arbitrary genes, and in this example, it has only the first gene
2. $\begin{bmatrix} 2 & (5) & 2 & 2 & 1 \\ 2 & 5 & 2 & 4 & 1 \\ 1 & 1 & 1 & 1 & 1 \\ 1 & 5 & 2 & 5 & 1 \\ 1 & 1 & 1 & 1 & 3 \end{bmatrix}$	*First condition set refinement:* Given the initial gene set, the average expression levels of the conditions are [2, 5, 2, 2, 1]; since the condition parameter $\tau_C = 2.5$, only the second condition is selected
3. $\begin{bmatrix} 2 & (5) & 2 & 2 & 1 \\ 2 & (5) & 2 & 4 & 1 \\ 1 & 1 & 1 & 1 & 1 \\ 1 & (5) & 2 & 5 & 1 \\ 1 & 1 & 1 & 1 & 3 \end{bmatrix}$	*First gene set refinement:* Given the previous condition set, the average expression levels of the genes are [5, 5, 1, 5, 1]; since the gene parameter $\tau_G = 2.5$, the first, second and fourth genes are then selected
4. $\begin{bmatrix} 2 & (5) & 2 & (2) & 1 \\ 2 & (5) & 2 & (4) & 1 \\ 1 & 1 & 1 & 1 & 1 \\ 1 & (5) & 2 & (5) & 1 \\ 1 & 1 & 1 & 1 & 3 \end{bmatrix}$	*Second condition set refinement:* Given the previous gene set, the mean expression levels of the conditions are [5/3, 5, 2, 11/3, 1]; since the condition parameter $\tau_C = 2.5$, then the second and fourth conditions are selected
5. $\begin{bmatrix} 2 & (5) & 2 & (2) & 1 \\ 2 & (5) & 2 & (4) & 1 \\ 1 & 1 & 1 & 1 & 1 \\ 1 & (5) & 2 & (5) & 1 \\ 1 & 1 & 1 & 1 & 3 \end{bmatrix}$	*Second gene set refinement:* Given the previous gene set, the mean expression levels of genes are [7/2, 9/2, 1, 5, 1]; since the gene parameter $\tau_G = 2.5$, then the first, second, and fourth genes are selected. At the same time, compared with the result from step (4), we can see that the both results consist of genes 1, 2, 4 and condition 2, 4; in other words convergence of the algorithm is reached
6. $\begin{bmatrix} 2 & (5) & 2 & (2) & 1 \\ 2 & (5) & 2 & (4) & 1 \\ 1 & 1 & 1 & 1 & 1 \\ 1 & (5) & 2 & (5) & 1 \\ 1 & 1 & 1 & 1 & 3 \end{bmatrix}$	*Result:* The identified bicluster consists of the first, second, fourth genes at the second and fourth conditions

2. A time dependence module that constrains the continuity of the modules in time domain by incorporating the time dependence between samples of time series microarray data.

We introduce the two features of ECTDISA in detail next.

2.3.1. Enrichment Constrained Optimal Cluster

As mentioned in the previous section, ISA cannot determine the optimal parameters (τ_G, τ_C) that lead to the most biologically meaningful biclusters. In this section, we demonstrate how to seek the optimal clustering using enrichment analysis of gene

Table 3
Gene function

Gene ID	Gene annotation
First gene	Cell division
Second gene	Immunology
Third gene	Cell division
Fourth gene	Immunology
Fifth gene	Oncogene

ontology (GO) (16), which is a major gene annotation database. The same concept can be applied to different functional databases as well, such as KEGG pathway (17), NCI pathway interaction database (18), Molecular Signatures Database (MSigDB) (19), etc. (see Note 2). In functional analysis, enrichment of a gene function directly reveals the biological meaning of underlying data. To illustrate the concept of enrichment, suppose that in the genome, 1% genes are related to "cell cycle"; now, if 10% genes of a bicluster are related to "cell cycle," then the function "cell cycle" is clearly over-represented in the cluster and it is thus reasonable to infer that "cell cycle" would be a biological function possessed by the genes in it. Let us still look at the example in Table 2. Consider that, after we query GO database, we retrieve the functions of the five genes, which are listed in Table 3.

When different parameters are used, starting from the same initial gene set (first gene), the ISA may end up with different bicluster results (refer to Table 4). It can be seen that, when smaller parameters are used, larger modules will be identified with redundant genes that may not be related to the module function (see results 1–4 in Table 4); while when the parameters are too large, none or only a part of the module can be recovered (see results 6–7 in Table 4). Only when a parameter is properly chosen, can the result be biological most consistent, meaningful, and easy to interpret (see result 5 in Table 4).

In practice, multiple biologically functions can be enriched in a cluster with different degree of significance. The significance of enrichment can be evaluated by statistical tests, such as Fisher's exact test, which provide significance of enrichment in the form of p-values. Then, the concept of enrichment constraint cluster is to choose the biclustering parameters that generate the most significantly enriched result, which can be also considered as biologically most meaningful.

Table 4
Clustering results when using different parameters

Result index	(τ_G, τ_C)	Genes in the cluster	Gene function and count	Our interpretation
1	(0, 0)	First, second, third, fourth, fifth	Cell division 2 Immunology 2 Oncogene 1	Difficult
2	(1, 1)	First, second, third, fourth	Cell division 2 Immunology 2	Difficult
3	(2, 2)	First, second, fourth	Cell division 1 Immunology 2	Immunology module with redundancy
4	(3, 3)	First, second, fourth	Cell division 1 Immunology 2	Immunology module with redundancy
5	(4, 4)	Second, fourth	Immunology 2	Immunology module
6	(4.5, 3)	Fourth	Immunology 1	Part of the immunology module
7	(5, 5)	Empty	N/A	N/A

2.3.2. Time-Dependent Definition of Temporal Module

Different from independent data set, the samples of time series dataset are dependent on each other, i.e., the state of the previous sample is also likely to influence the state of the next sample. In Markov chain, the same idea is mathematically described in the state transition matrix, which defines the frequency of the state transitions. Let us review the most enriched cluster, i.e., result 5 that we obtained from Table 2.

$$\begin{bmatrix} 2 & (5) & 2 & (2) & 1 \\ 2 & (5) & 2 & (4) & 1 \\ 1 & 1 & 1 & 1 & 1 \\ 1 & (5) & 2 & (5) & 1 \\ 1 & 1 & 1 & 1 & 3 \end{bmatrix}.$$

It can be seen that the module are upregulated at time 2 and 4, but are not upregulated at time 1, 3, and 5; in other word, the latter state is always different from the previous state. However, in time series data, since the latter state is also likely to be correlated with the previous states, the frequent state transition as in result 5 would be hard to explain. This discrepancy of result is due to that time dependence between samples is considered in clustering.

To add the dependence between samples of time series dataset, we can redefine the definition of temporal modules as follows:

$$M(\tau_G, \tau_T) := \left\{ \{\mathcal{G}_m, \mathcal{T}_m\} \middle| \begin{array}{ll} \forall g \in \mathcal{G}_m & E(\mathbf{Y}_{g,\mathcal{T}_m}) > \tau_G \\ \forall t \in \mathcal{T}_m & E(\mathbf{Y}_{\mathcal{G}_m,[t-L:t+L]}, \mathbf{W}) > \tau_T \end{array} \right\}, \tag{2}$$

where $\mathbf{Y}_{g,\mathcal{T}_m}$ represents the expression levels of the gth gene that is also covered by the bicluster defined time set $\mathcal{T}_m$; $\mathbf{Y}_{\mathcal{G}_m,[t-L:t+L]}$ represents the expression levels of the genes in the bicluster m, or $\mathcal{G}_m$, from time $(t-L)$ to $(t+L)$; $E(*)$ represents the mean expression level of a vector of gene expressions; and $E(*,\mathbf{W})$ donates the weighted mean. The variable L defines the length of a time window, indicating how many adjacent samples should be included when deciding the state of a specific sample. Specifically, when $L=1$ and the weight vector $\mathbf{W}=[0.5\quad 1\quad 0.5]$ is applied, we have

$$E\left(\mathbf{Y}_{\mathcal{G}_m,[t-L:t+L]},\mathbf{W}\right)=E\left(\mathbf{Y}_{\mathcal{G}_m,[t-1:t+1]},[0.5\quad 1\quad 0.5]\right),$$
$$=\frac{0.5E\left(\mathbf{Y}_{\mathcal{G}_m,t-L}\right)+1E\left(\mathbf{Y}_{\mathcal{G}_m,t}\right)+0.5E\left(\mathbf{Y}_{\mathcal{G}_m,t+L}\right)}{0.5+1+0.5}. \quad (3)$$

Smaller weights for adjacent samples are used here to damp their influence. Correspondingly, the ISA includes the following iterations (Table 5)

As it can be seen in Table 5, after incorporating the dependency between samples, the resulting module is continuous in time domain. A more reasonable explanation can be reached: Genes 1, 2, and 4 are upregulated from time points 1–4, but not upregulated after time 4. Please note, for simplicity, in this particular example, we choose windows of length L equal to 1, and weight vector [0.5, 1, 0.5]. The choice of these two parameters should depend on the characteristics of microarray experiments that generate the data. In general, when the sampling interval is small, a larger window with more even weight vector can be used, and otherwise for larger sampling intervals (see Note 3).

2.3.3. ECTDISA for Finding Meaningful Temporal Modules

The enrichment constrained framework and time-dependent definition of bicluster can be thus combined to identify TTMs that are continuous in time domain and biologically meaningful. The resulted algorithm is known as enrichment constrained and time-dependent ISA (ECTDISA).

The goal of ECTDISA is to find co-regulated genes including upregulated gene sets. Accordingly, a more flexible bicluster definition is used:

$$M(\tau_G,\tau_T):=\left\{\{\mathcal{G}_m,\mathcal{T}_m\}\left|\begin{array}{ll}\forall g\in\mathcal{G}_m & \rho\left(\mathbf{Y}_{g,\mathcal{T}_m},\left\langle\mathbf{Y}_{\mathcal{G}_m,\mathcal{T}_m}\right\rangle\right)<\tau_G\\ \forall t\in\mathcal{T}_m & \frac{1}{|\mathcal{G}_m|}\sum_{g\in\mathcal{G}_m}\rho\left(\mathbf{Y}_{g,[t-1:t+1]},\left\langle\mathbf{Y}_{\mathcal{G}_m,[t-1:t+1]}\right\rangle\right)<\tau_T\end{array}\right.\right\}, \quad (4)$$

where ρ represents a distance measurement, such as Pearson's correlation or Euclidean distance, etc; we use Euclidean distance in this example; $\langle*\rangle$ represents the mean expression of module;

Table 5
Procedures of time-dependent ISA

1. $\begin{bmatrix} (2) & (5) & (2) & (2) & (1) \\ 2 & 5 & 2 & 4 & 1 \\ 1 & 1 & 1 & 1 & 1 \\ 1 & 5 & 2 & 5 & 1 \\ 1 & 1 & 1 & 1 & 3 \end{bmatrix}$	*Initial condition:* Initial gene set may contain any arbitrary gene, and in this example, we use only the first gene
2. $\begin{bmatrix} 2 & (5) & 2 & 2 & 1 \\ 2 & 5 & 2 & 4 & 1 \\ 1 & 1 & 1 & 1 & 1 \\ 1 & 5 & 2 & 5 & 1 \\ 1 & 1 & 1 & 1 & 3 \end{bmatrix}$	*First condition set refinement:* Given the initial gene set, the average expression levels of the conditions after incorporating adjacent samples are [3, 3.5, 2.75, 1.75, 1.33]; since the condition parameter $\tau_T = 2.5$, only the second condition is selected
3. $\begin{bmatrix} 2 & (5) & 2 & 2 & 1 \\ 2 & (5) & 2 & 4 & 1 \\ 1 & 1 & 1 & 1 & 1 \\ 1 & (5) & 2 & 5 & 1 \\ 1 & 1 & 1 & 1 & 3 \end{bmatrix}$	*First gene set refinement:* Given the previous condition set, the average expression levels of genes are [5, 5, 1, 5, 1]; since the gene parameter $\tau_G = 2.5$, the first, second, and fourth genes are then selected
4. $\begin{bmatrix} (2) & (5) & (2) & (2) & 1 \\ (2) & (5) & (2) & (4) & 1 \\ 1 & 1 & 1 & 1 & 1 \\ (1) & (5) & (2) & (5) & 1 \\ 1 & 1 & 1 & 1 & 3 \end{bmatrix}$	*Second condition set refinement:* Given the previous gene set, the average expression levels of conditions after incorporating adjacent samples are [2.8, 3.4, 3.2, 2.6, 1.9]; since the gene parameter $\tau_T = 2.5$, then the first, second, third, and fourth conditions are selected
5. $\begin{bmatrix} (2) & (5) & (2) & (2) & 1 \\ (2) & (5) & (2) & (4) & 1 \\ 1 & 1 & 1 & 1 & 1 \\ (1) & (5) & (2) & (5) & 1 \\ 1 & 1 & 1 & 1 & 3 \end{bmatrix}$	*Second gene set refinement:* Given the previous gene set, the average expression levels of mean heights of genes are [3.6, 4.3, 1, 4.3, 1]; since the gene parameter $\tau_G = 2.5$, then the first, second, and fourth genes are selected; Compared with the result from step (4), we can see that both results consist of genes 1, 2, 4 and conditions 2, 4; in other words convergence of the algorithm is reached
6. $\begin{bmatrix} (2) & (5) & (2) & (2) & 1 \\ (2) & (5) & (2) & (4) & 1 \\ 1 & 1 & 1 & 1 & 1 \\ (1) & (5) & (2) & (5) & 1 \\ 1 & 1 & 1 & 1 & 3 \end{bmatrix}$	*Final result:* The bicluster we identified consists of the first, second, fourth genes and the first, second, third, fourth conditions

$|\mathcal{G}_m|$ denotes the number of genes in $\mathcal{G}_m$. Moreover, the biological significance of a retrieved module is defined by the following score

$$S(M) = \frac{\sum_{j=1}^{|C|} \left(-\log P_{C_j,M}\right)}{\log(|\mathcal{G}_m|)}, \tag{5}$$

where $P_{C_j,M}$ is the significant p-value of the enrichment of a functional gene set C_j of a functional database in the gene set $\mathcal{G}_m$

of module $M(\tau_G, \tau_T)$ and can be calculated by Fisher's exact test, and $|\mathcal{G}_m|$ is the number of genes in module, which is used to penalize the module size. Note $S(M)$ is a function of (τ_G, τ_T). In ECTDISA, we search the optimal parameters that lead the bicluster that carries the largest significance score, hence the biological most meaningful result. Such search can be carried out by searching the 2-D grids (τ_G, τ_T) for $\tau_G = [0.05 : 0.05 : 0.5]$ and $\tau_T = [0.05 : 0.05 : 0.5]$.

2.4. Application of ECTDISA to Microarray Data of Virus Infection (See Note 4)

We show in this section the result of ECTDISA applied to time series microarray data of Kaposi's Sarcoma-associated Herpesvirus (KSHV) infection. The human time series microarray data were obtained from KSHV infection of human primary endothelial cells (20). The data were produced with Affymetrix Human Genome U133A Chips, consisting of the expression sample at time 0, 1, 3, 6, 10, 16, 24, 36, 54, 78 (h) after infection. Since priority was given to earlier states, sample times were unevenly chosen.

For the Affymetrix HGU133A Chip, 19,142 features (Probe set ID) of total 22,383 have corresponding official gene symbol; 19,142 features with corresponding gene symbols are further merged into 11,945 genes by taking the maximum value of all corresponding probe set IDs. An intensity filter (the intensity of a gene should be above 100 in at least one sample) and a variance filter (the interquartile range of log 2–intensities should be at least 0.2) were then applied to select 3,825 differentially expressed genes along with their expression profile in original scale. To make all remaining genes contributing equally to the algorithm, their expression profiles are further rescaled to standard normal distribution by subtracting mean and divided by standard deviation (see Note 5).

To apply ECTDISA, the initial gene sets are chosen in this way: A first gene is randomly selected and 30 genes that have the largest Pearson correlation with the first gene were added to form the initial gene set. To avoid repeated modules and cover a larger initial state, a gene can only appear once in initial gene sets, and must appear once. After ECTDISA, postprocessing was also applied to merge the modules with similar biological meaning and genes. In the end, 99 modules were obtained, among which there are both constant modules and TTMs (please see Fig. 1 for examples).

It can be seen from Fig. 1 that the 51st and 52nd modules are constant modules, which lasts for the entire period of the experiments, while the 53rd and 54th modules are temporal modules, which start from the fifth sample time. Associated with each module, we have a list of the most enriched pathways, some of which are listed in Table 6 as an example.

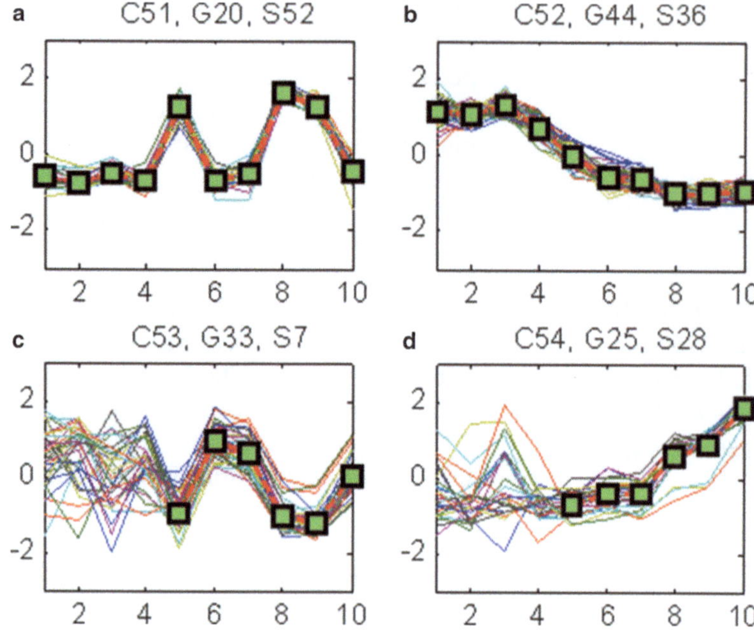

Fig. 1. Temporal transcription modules identified by ECTDISA.

Table 6
Biological meaning of the 51st module

Pathway name	Pathway annotation	−lg(*p*)
HIFPATHWAY	Under normal conditions, hypoxia inducible factor HIF-1 is degraded; under hypoxic conditions, it activates transcription of genes controlled by hypoxic response elements (HREs)	3.81
DREAMPATHWAY	The transcription factor DREAM blocks expression of the prodynorphin gene, which encodes the ligand of an opioid receptor that blocks pain signaling	3.6
BLADDER_CANCER	Genes involved in bladder cancer	3.40

3. Notes

1. Bicluster types are related. After some transformation of the data, a bicluster approach can often be used to discover some other bicluster types.
2. Incorporate prior knowledge could be very helpful. There have been enormous databases established for various kinds of biological information.

3. The dependency between samples is a very important feature of time series microarray data. When dealing with time series data set, it is important to model the dependency between samples; failing to do so may produce unreasonable result.
4. The complete data and MATLAB code are available at ref. 21. Please refer to ref. 15 for all details regarding this chapter and how ECTDISA is applied to other datasets.
5. Preprocessing is a very important step for clustering analysis. This step normally includes feature selection and data normalization.

Acknowledgments

This work is supported by an NSF Grant CCF-0546345.

References

1. Spellman PT, Sherlock G, Zhang MQ et al (1998) Comprehensive identification of cell cycle-regulated genes of the yeast *Saccharomyces cerevisiae* by microarray hybridization. Mol Biol Cell 9:3273–3297.
2. Fuhrken PG, Chen C, Miller WM et al (2007) Comparative, genome-scale transcriptional analysis of CHRF-288-11 and primary human megakaryocytic cell cultures provides novel insights into lineage-specific differentiation. Exp Hematol 35:476–489.
3. Eisen MB, Spellman PT, Brown PO et al (1998) Cluster analysis and display of genome-wide expression patterns. Proc Natl Acad Sci U S A 95:14863–14868.
4. MacQueen J (1967) Some methods for classification and analysis of multivariate observations. p 14. California, USA.
5. Tamayo P, Slonim D, Mesirov J et al (1999) Interpreting patterns of gene expression with self-organizing maps: methods and application to hematopoietic differentiation. Proc Natl Acad Sci U S A 96:2907–2912.
6. Alon U, Barkai N, Notterman DA et al (1999) Broad patterns of gene expression revealed by clustering analysis of tumor and normal colon tissues probed by oligonucleotide arrays. Proc Natl Acad Sci U S A 96:6745–6750.
7. Bittner M, Meltzer P, Trent J (1999) Data analysis and integration: of steps and arrows. Nat Genet 22:213–215.
8. Cheng Y, Church GM (2000) Biclustering of expression data. Proc Int Conf Intell Syst Mol Biol 8:93–103.
9. Getz G, Levine E, Domany E (2000) Coupled two-way clustering analysis of gene microarray data. Proc Natl Acad Sci U S A 97:12079–12084.
10. Ihmels J, Friedlander G, Bergmann S et al (2002) Revealing modular organization in the yeast transcriptional network. Nat Genet 31:370–377.
11. Madeira SC, Oliveira AL (2004) Biclustering algorithms for biological data analysis: a survey. IEEE/ACM Trans Comput Biol Bioinform 1:24–45.
12. Bergmann S, Ihmels J, Barkai N (2003) Iterative signature algorithm for the analysis of large-scale gene expression data. Phys Rev E Stat Nonlin Soft Matter Phys 67:031902.
13. Kloster M (2004) Self-organized criticality, competitive evolution and analysis of gene-expression data. Ph.D. Dissertation. Department of Physics, Princeton University.
14. Supper J, Strauch M, Wanke D et al (2007) EDISA: extracting biclusters from multiple time-series of gene expression profiles. BMC Bioinformatics 8:334.
15. Meng J, Gao S, Huang Y (2009) Enrichment constrained time-dependent clustering analysis for finding meaningful temporal transcription modules. Bioinformatics 25:1521–1527.
16. Ashburner M, Ball C, Blake J et al (2000) Gene ontology: tool for the unification of biology. The Gene Ontology Consortium. Nature genetics 25:25–29.
17. Kanehisa M, Araki M, Goto S et al (2008) KEGG for linking genomes to life and the

environment. Nucleic Acids Res 36: D480–484.

18. Krupa S, Anthony K, Buchoff J et al (2007) The NCI-Nature Pathway Interaction Database: A cell signaling resource. Nature Preceedings. http://dx.doi.org/10.1038/npre.2007.1311.1.
19. Subramanian A, Tamayo P, Mootha VK et al (2005) Gene set enrichment analysis: a knowledge-based approach for interpreting genome-wide expression profiles. Proc Natl Acad Sci U S A 102:15545–15550.
20. Gao SJ, Deng JH, Zhou FC (2003) Productive lytic replication of a recombinant Kaposi's sarcoma-associated herpesvirus in efficient primary infection of primary human endothelial cells. J Virol 77:9738–9749.
21. http://engineering.utsa.edu/yfhuang/ECTDISA.html.

Chapter 7

Using the Bioconductor GeneAnswers Package to Interpret Gene Lists

Gang Feng, Pamela Shaw, Steven T. Rosen, Simon M. Lin, and Warren A. Kibbe

Abstract

Use of microarray data to generate expression profiles of genes associated with disease can aid in identification of markers of disease and potential therapeutic targets. Pathway analysis methods further extend expression profiling by creating inferred networks that provide an interpretable structure of the gene list and visualize gene interactions. This chapter describes GeneAnswers, a novel gene-concept network analysis tool available as an open source Bioconductor package. GeneAnswers creates a gene-concept network and also can be used to build protein–protein interaction networks. The package includes an example multiple myeloma cell line dataset and tutorial. Several network analysis methods are included in GeneAnswers, and the tutorial highlights the conditions under which each type of analysis is most beneficial and provides sample code.

Key words: Network, Disease ontology, Gene ontology, Pathway analysis, GeneAnswers, Bioconductor

1. Introduction

Expression profiling, the practice of identifying the pattern of genes expressed at the level of genetic transcription under specific circumstances or in specific types of cells, has been practiced since the 1990s (1). Profiles of specific disease expression in breast cancer neoplasms and other tumor cells have indicated that patterns of expression may be useful markers of disease or for identifying targets for therapeutic intervention. Microarray analysis, and more recently, next-generation sequencing-based RNA-Seq (2), usually result in a list of genes. Besides the ranking by statistical properties (fold change and *p*-value) derived from the analysis of the expression profiles, there is no ordering of the gene list in terms of biological importance or network structure. Many commercial and open source packages exist to aid the researcher in

Junbai Wang et al. (eds.), *Next Generation Microarray Bioinformatics: Methods and Protocols*, Methods in Molecular Biology, vol. 802, DOI 10.1007/978-1-61779-400-1_7,

pathway analysis, in particular, offering inferential computation combined with graphical displays of this inferred network interactivity. This chapter describes the use of GeneAnswers, a package available from Bioconductor (3), an open source library of packages written in the statistical programming language R (4). GeneAnswers creates a gene-concept network and also can be used to build and analyze protein–protein interaction (PPI) networks. The conditions under which each type of analysis is most beneficial are described, and sample code is provided.

2. Materials

Software: We assume that the readers already have the R and Bioconductor installed. If not, R can be downloaded from ref. 5, and can be installed on Linux, Mac, or Windows machines. For Bioconductor packages, one can refer to ref. 6. For help in using GeneAnswers or any other Bioconductor packages, see Note 1.

Data set: To illustrate the methods discussed in the chapter, we utilize an example data set included with the GeneAnswers package. It is a subset of genes (86 genes only) from an Affymetrix microarray experiment of a multiple myeloma cell line treated with dexamethasone for 24 h (three biological replicates under each condition) as described previously (7).

3. Methods

When mapping genes to functional categories of gene ontology (GO) (8) annotations using tools such as the Web tool DAVID (database for annotation, visualization, and integrated discovery) (9, 10), many genes can map to several GO terms, but the output tables created by tools such as DAVID do not clearly illustrate how many genes map to multiple functional terms. GeneAnswers solves this problem by creating a network of genes-to-concepts and visually highlights genes that are involved in several functions (7).

To analyze the gene list in the context of biological functions, such as "cell cycle," gene ontology (GO) annotations can be used; to identify the disease involvements of a gene list, disease ontology (DO) annotations can be used (11). Note that a gene can be connected to a certain concept of interest (either GO or DO) indirectly via PPI networks. For instance, gene TLE1 can indirectly participates in the "protein kinase cascade" activity via its interaction with IL6ST (Fig. 1). As such, the PPI network can be used to augment the network inference.

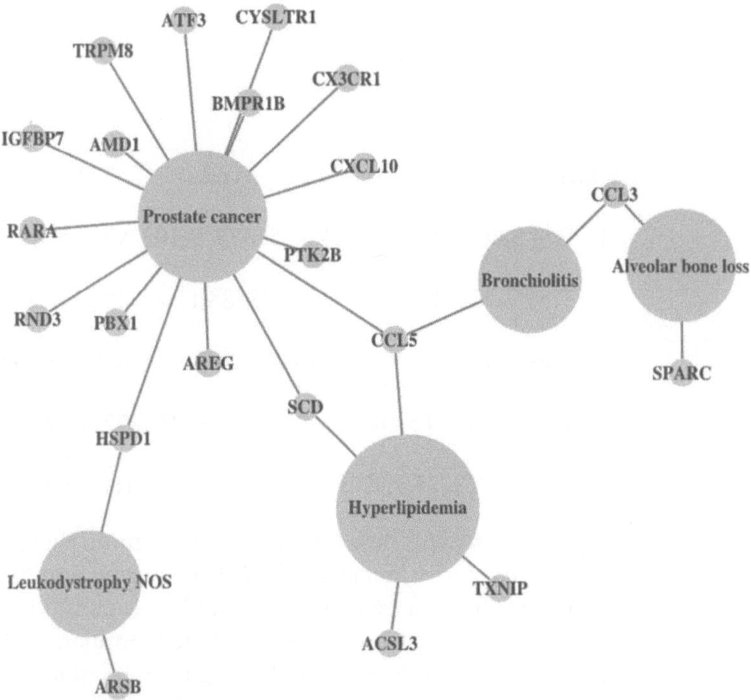

Fig. 1. A gene-concept network augmented by protein–protein interaction network. Colors and labels of nodes as in Fig. 4. Self–self interactions are indicated by looping back. Note that now gene interaction relationships are illustrated by the lines connecting the nodes, and thus network connectivity is increased (cf. Fig. 4).

3.1. Load the Library and Data Set

To install the geneAnswers package, please type the following in R (see Note 2):

```
source("http://bioconductor.org/biocLite.R")
biocLite("GeneAnswers")
```

Load the GeneAnswers package:

```
library (GeneAnswers)
```

We assume that readers have already analyzed the microarray raw data and derived a statistically significant list of genes. If not, see Note 3 for packages available from Bioconductor for the analysis of microarray data. As an example, the data set included with the GeneAnswers package, "humanExpr" is a data matrix of normalized, $\log_2$-transformed intensity of six microarray experiments (three controls and three treatments).

```
data('humanExpr')

> humanExpr [1:5,]

  GeneID    Ctrl1    Ctrl2    Ctrl3    Trt241    Trt242    Trt243
1    374 5.068725 4.833424 5.251337  9.105219  9.384213  9.065194
2   2289 7.764840 7.842467 7.973488 11.285194 11.189364 11.303396
3  10563 4.824265 4.780033 4.417243  8.362088  7.825695  8.087761
4    687 4.607949 5.169052 4.947950  7.463272  7.370329  7.726753
5  80149 7.729777 7.472989 7.617576 10.281806  9.541200  9.910506
```

"humanExpr" was analyzed by the limma package in Bioconductor to derive a table of 86 statistically significant genes as shown in the data frame of "humanGeneInput." "humanGeneInput" contains columns of Entrez gene identifier, fold change, and *p*-value statistics. Note that although other columns are optional, the human Entrez gene identifier column is necessary and always the first column. *See* Note 4 for information on conversion from other gene identifiers to Entrez gene IDs.

```
data('humanGeneInput')

humanGeneInput [1:5,]

      GeneID foldChange       pValue
374      374  17.553825 1.241073e-11
2289    2289  10.549139 5.513686e-14
10563  10563  10.688601 1.073456e-10
687      687   6.112663 1.351898e-09
80149  80149   4.939585 6.473841e-09
```

3.2. Identify the Gene-Concept Network

The GeneAnswers package will interpret a list of genes in the context of biological concepts. Relevant concepts come from the following gene annotation databases (Table 1).

For each concept, GeneAnswers will test its "enrichment" in the gene list versus the genome using the well-defined hypergeometric

Table 1
Annotation libraries supported by current GeneAnswers

CategoryType	Purpose	Species	Example concepts
"GO.BP," "GO.MF," "GO.CC," and "GO"	Biological process, molecular functions, cellular component, and all of them as defined by Gene Ontology	Human, mouse, rat, and fly	"Protein kinase cascade" ("GO:0006259")
"KEGG"	Biological pathway as defined by the KEGG database	Human, mouse, rat, and fly	"Butanoate metabolism" ("00650")
"DOLITE"	Disease as defined by the lite version of disease ontology	Human	"Prostate cancer " (DOLite:447)
"REACTOME. PATHWAY"	Biological pathway as defined by the REACTOME developed by EBI	Human, mouse, rat, and fly	"DNA Damage Reversal" ("73942")
"CABIO.PATHWAY"	Biological pathway as defined by the caBio developed by NCI	Human and mouse	"Nongenotropic Androgen signaling" ("7465")

statistical test (12). For example, the following code will test the gene list of humanGeneInput in the context of diseases (DOLITE).

```
data('humanGeneInput')

z <- geneAnswersBuilder(humanGeneInput,

   'org.Hs.eg.db', categoryType='DOLITE',

   pvalueT=0.1)

zz <- geneAnswersReadable(z)

zzz<-  geneAnswersSort (zz)

topDOLite(z, orderby='pvalue', top=8)
```

Table 2
Enrichment test results

Category	Number of genes	*p*-Value
Hyperlipidemia::DOLite:261	4	0.001982
Prostate cancer::DOLite:447	16	0.003963
Alveolar bone loss::DOLite:43	2	0.008729
Leukodystrophy NOS::DOLite:307	2	0.01148
Bronchiolitis::DOLite:93	2	0.01148
Macular degeneration::DOLite:330	3	0.01355
Esophagus cancer::DOLite:184	4	0.01455
HTLV-I infection::DOLite:229	2	0.01456

The result suggests that of the 86 genes in the list, 16 of them are associated with "prostate cancer" (DOLite: 447). The enrichment of prostate cancer-related genes cannot be interpreted by random chance ($p = 0.003963$); thus "prostate cancer" is statistically significantly associated with the gene list.

The first column in Table 2 is the names of categories. The default setting will print names with annotation library IDs separated by "::". This can be turned off by setting keepID to FALSE. In some cases, not all of top categories are interesting, so users can pick up some relative categories based on this table for further analysis. To show all categories, set "top" to a large number, like 1,000. All categories with statistical significance will then be printed on screen. But if you set top to "ALL," only the top 30 categories will show on the screen although all categories can be saved in user-specified file.

The genes associated with each category in Table 2 can be visualized as a gene-concept network by the following code. Results are shown in Fig. 2.

```
geneAnswersConceptNet(zzz, centroidSize = 'pvalue')
```

By default, geneAnswersConceptNet will draw the top five categories. For concept type of "GO.BP," sometimes it can result in an illegible drawing because of the complexity of the network. In other cases, one might only want to show a certain category of biological interest. The problems above can be solved by specifying which concepts to draw in the network. The following code illustrates how to select the following category in the drawing: "response

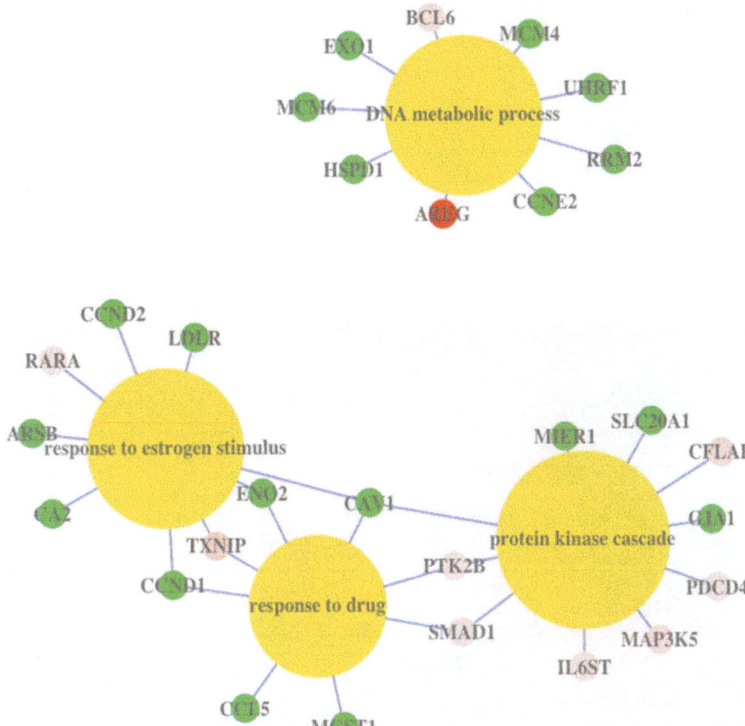

Fig. 2. Gene-concept network by disease ontology (DO) analysis. *Yellow nodes* are DO concepts and *gray nodes* represent genes. The sizes of the centroid nodes reflect *p*-values of the disease–gene associations as calculated by GeneAnswers.

to estrogen stimulus," "response to drug," "protein kinase cascade," and "DNA metabolic process." The corresponding GO IDs of "GO:0007049," "GO:0042592," "GO:0006259," and "GO:0007243" were collected from the topGO outputs.

```
GOBPIDs <- c("GO:0043627", "GO:0042493", "GO:0006259", "GO:0007243")

x <- geneAnswersBuilder(humanGeneInput, 'org.Hs.eg.db',

   categoryType='GO.BP')

topGO(x)

geneAnswersConceptNet(x, colorValueColumn='foldChange',

centroidSize='geneNum', showCats=GOBPIDs, catTerm=TRUE,

geneSymbol=TRUE)
```

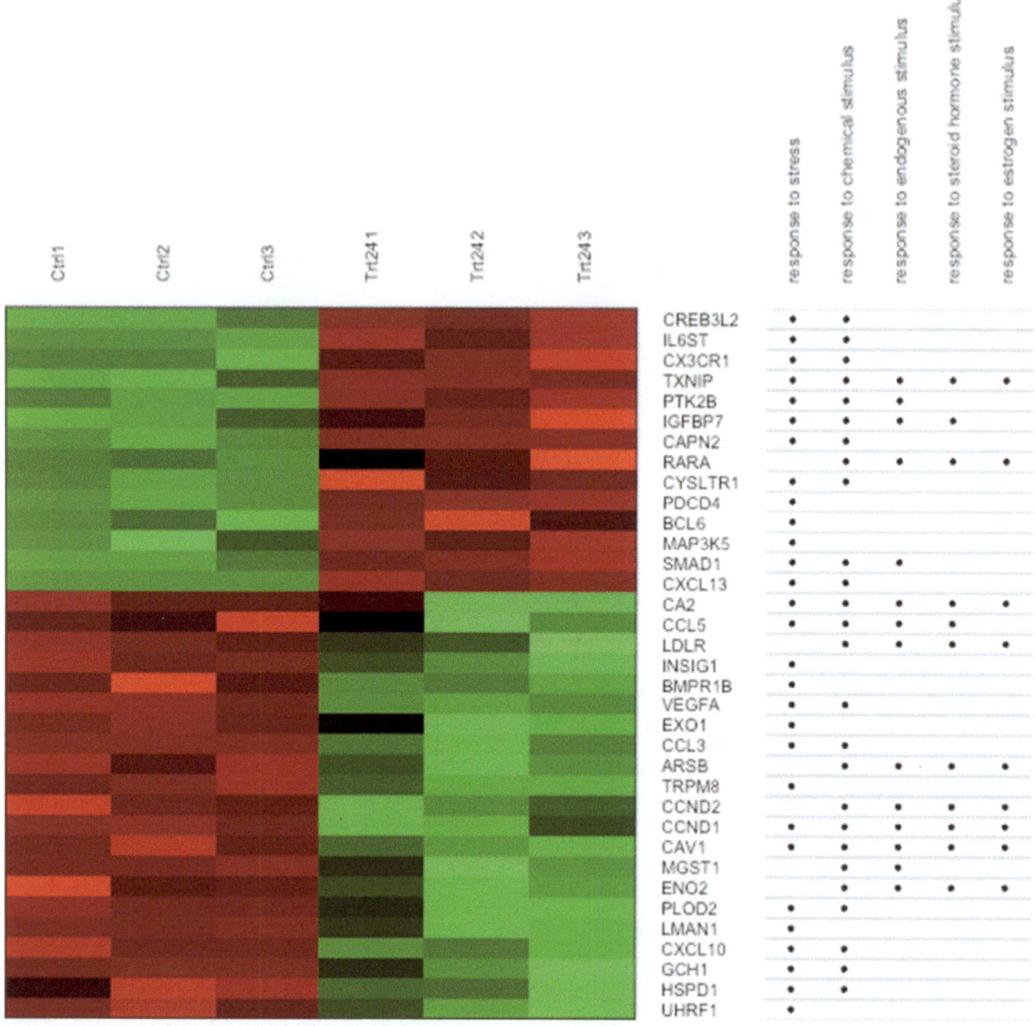

Fig. 3. Cross tabulation of the gene-concept network with a heatmap. The *left panel* shows the heatmap of the genes listed at the middle of the table in different experiments, while the *right panel* shows the relationships between these genes and KEGG pathways.

3.3. Cross Tabulate the Gene-Concept Network with Heatmaps

An expression profile can be more interpretable when cross-tabulated with functional annotations of each gene. As shown in Fig. 3, the CCND1 and CCND2 genes, which both involve in "p53 signaling," "focal adhesion," and "cell cycle" pathways from KEGG, were downregulated after dexamethasone treatment. By changing the categoryType to "GO.BP" or "DOLITE," the expression profile can be explored in the context of biological processes or diseases.

```
data('humanGeneInput')

data('humanExpr')

x <- geneAnswersBuilder(humanGeneInput, 'org.Hs.eg.db',

    categoryType='KEGG',

    geneExpressionProfile=humanExpr)

geneAnswersHeatmap(x, catTerm = TRUE, geneSymbol = TRUE)
```

3.4. Enhance the Interpretation with PPI Network

The gene-concept network discussed in Subheadings 2 and 3 can be further extended using the PPI network. With concepts or GO terms alone, genes that map to specific concepts may not be sufficiently connected to each other. The addition of PPI relationships can highlight functional links between genes that were previously unnoticed. The PPI network works especially well with gene knock-down experimental data, since the potential impact of dysregulation of one protein product can be spread throughout the entire inferred network. The following code will enhance results in Fig. 4 by taking PPI into consideration.

```
geneAnswersConceptNet(x, colorValueColumn = 'foldChange',
    centroidSize = 'geneNum', showCats = GOBPIDs, cat-
    Term = TRUE, geneSymbol = TRUE, geneLayer = 2)
```

Parameter geneLayer is set to one in default, which means no PPI information is included in the gene-concept network. When one wants to check how the PPI network is involved in the current gene-concept network, geneLayer should be set to an integer greater than one. For example, if geneLayer = 3, then two more levels of search for each given gene will be performed. Empirically, we find that more than six geneLayer searches will not make a difference in the gene-concept network, which coincides with "six degrees of separation" often described in social networks. In both cases, this is likely explained by the small world property of the networks.

3.5. Bias and Potential Misinterpretation

Results of the computational inference methods discussed in this chapter should be interpreted with caution. First, functional annotations of genes are far from complete. Current annotations are highly biased toward well-funded research areas, such as cancer. Second, there are often wrong annotations in the databases.

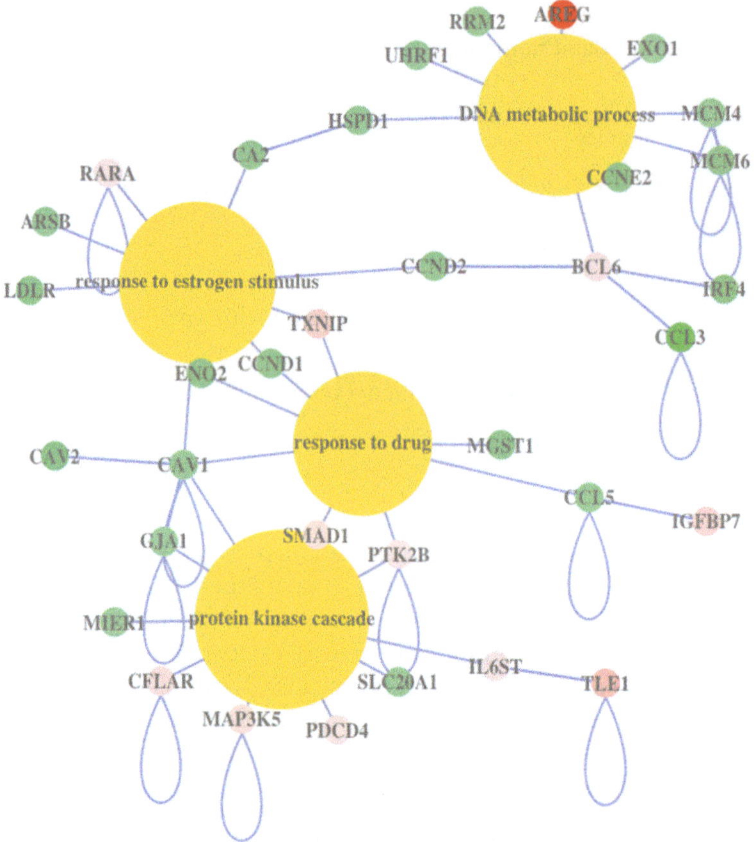

Fig. 4. Gene-concept network by gene ontology (GO) analysis. *Yellow nodes* are now gene ontology (GO) terms. *Green* and *red nodes* correspond to fold change values from the microarray data, with *green* representing downregulation and *red* representing upregulation, with intensity of color reflecting intensity of fold change in this case.

As such, the results should not be interpreted as a confirmation of the underlying biology but as a starting point for more biological investigation (see Notes 5 and 6).

4. Notes

1. As a community-supported software package, Bioconductor has a very active mailing list to answer all kinds of questions from users. Readers are encouraged to post questions related to GeneAnswers to the mailing list, or search the mailing list for previously discussed questions.
2. The GeneAnswers only needs to be installed once. When GeneAnswers is installed, it will also install any other Bioconductor packages it requires.

3. Bioconductor provides a number of packages for normalization of microarray data. The "limma" package (linear models for microarray data) is a popular package which provides a GUI for processing data. Output from LIMMA is shown in Subheading 1 in the "humanGeneInput" table.
4. To convert identifiers from other formats (Affymetrix array IDs, Ensembl IDs, gene symbols, etc.) to Entrez gene IDs, readers are encouraged to use the DAVID Gene ID Conversion Tool (13) available at ref. 14. The tool is easy to use and is able to convert identifiers from many major array platforms.
5. For users who wish to see disease–gene associations without expression values overlaid onto the network, the FunDO Web server is a simple tool that will convert a gene list to an interactive table and network diagram of disease–gene interactions. FunDO is based on disease ontology lite annotations and is useful for discovering unexpected disease associations from a gene list and can be valuable for initiation of new literature searches and disease–gene interaction investigation (15). The FunDO server can be found at ref. 16.
6. Many free and licensed packages are available for pathway analysis, in addition to GeneAnswers. Using GeneAnswers in combination with one or more of these other packages can create a more complete picture of interactivity and functional significance of a gene list. Combining GeneAnswers with the clustered GO annotation categories generated by DAVID (17) adds additional information about cell function and morphological features that may be enriched in a gene list. Using GeneAnswers in combination with licensed products such as ingenuity pathways analysis (18) or MetaCore by GeneGo (19) is also advantageous. Both of these licensed products feature user-friendly Web interfaces and functionality that allows users to upload array data and create networks of genes from the dataset, with canonical pathway overlays available.

References

1. Jordan B (2002) Historical background and anticipated developments. Ann N Y Acad Sci. 975:24–32.
2. Wang Z, Gerstein M, Snyder M (2009) RNA-Seq: a revolutionary tool for transcriptomics. Nat Rev Genet 10:57–63.
3. Reimers M, Carey VJ (2006) Bioconductor: an open source framework for bioinformatics and computational biology. Methods Enzymol. 411:119–134.
4. R Development Core Team (2010) R: A Language and Environment for Statistical Computing. Vienna, Austria: R Foundation for Statistical Computing.
5. http://www.r-project.org.
6. http://www.bioconductor.org.
7. Feng G, Du P, Krett NL et al (2010) A collection of Bioconductor methods to visualize gene-list annotations. BMC Res Notes 3:10.
8. Ashburner M, Ball CA, Blake JA et al (2000) Gene Ontology: Tool for the unification of biology. The Gene Ontology Consortium. Nat Genet. 25:25–29.
9. Dennis G Jr, Sherman BT, Hosack DA et al (2003) DAVID: Database for Annotation, Visualization, and Integrated Discovery. Genome Biol. 4:P3.
10. Huang da W, Sherman BT, Lempicki RA (2009) Systematic and integrative analysis of large gene lists using DAVID bioinformatics resources. Nat Protoc. 4:44–57.
11. Osborne JD, Flatow J, Holko M et al (2009) Annotating the human genome with Disease Ontology. BMC Genomics 10:S6.
12. Osborne JD, Zhu LJ, Lin SM et al (2007) Interpreting microarray results with Gene Ontology and MeSH. Methods Mol Biol. 377:223–242.
13. Huang da W, Sherman BT, Stephens R et al (2008) DAVID gene ID conversion tool. Bioinformation 2:428–430.
14. http://david.abcc.ncifcrf.gov/conversion.jsp.
15. Du P, Feng G, Flatow J et al (2009) From Disease Ontology to Disease-Ontology Lite: Statistical methods to adapt a general-purpose ontology for the test of gene-ontology associations. Bioinformatics 25:i63-i68.
16. http://fundo.nubic.northwestern.edu.
17. http://david.abcc.ncifcrf.gov/home.jsp.
18. http://www.ingenuity.com.
19. http://www.genego.com/metacore.

Chapter 8

Analysis of Isoform Expression from Splicing Array Using Multiple Comparisons

T. Murlidharan Nair

Abstract

There is a high prevalence of alternatively spliced genes (isoforms) in the human genome. Studies toward understanding aberrantly spliced genes and their association with diseases have lead researchers to profile the expression of alternatively spliced products. High-throughput profiling of isoforms has been done using microarray technology. Expression of isoforms reflects regulation both at transcriptional and posttranscriptional levels. This chapter details the methods to perform exhaustive comparison of isoforms using the R statistical framework.

Key words: mRNA isoforms, Multiple comparisons

1. Introduction

Alternative pre-mRNA splicing (AS) responsible for generating multiple transcripts from a single gene plays a central role in generating complex proteomes (1). It is estimated that more than 90% of the human genes have alternatively spliced products. Over the years, studies directed toward understanding alternative splicing using computational approaches have gained increased attention (2–5). Several studies have used microarray technology to quantify isoform expression levels either directly or indirectly (6–9). Quantifying isoform expression levels has the advantage in that it reflects the integrated outcome of the regulations at transcriptional and posttranscriptional levels. There is evidence that points to the functional integration of processes involved in transcription and RNA processing (10).

There are several disparate microarray platforms that have been used for expression analysis (11, 12); however, most platforms are not designed to specifically query isoforms. Multiplex mRNA isoform detection assays known as RASAL or DASL

Junbai Wang et al. (eds.), *Next Generation Microarray Bioinformatics: Methods and Protocols*,
Methods in Molecular Biology, vol. 802, DOI 10.1007/978-1-61779-400-1_8,

(RNA/DNA-mediated annealing, selection, and ligation), coupled with microarray were designed to uniquely profile mRNA isoforms in a high-throughput manner (13, 14). This chapter provides the computational methods for analyzing and extracting biological information from isoform expression data. For the purpose of this chapter, we have used data from Illumina BeadArray technology; however, the method described here can be easily extended to data collected from other high-throughput technologies, with some preprocessing of the data.

2. Materials

2.1. Hardware and Software Requirements

The computational protocol that is described here requires the following:

R is an open-source statistical computing environment available under the GNU Public License for different platforms (Windows/Linux/Unix/Mac) (15). R was developed by Robert Gentleman and Ross Ihaka. It has quickly become the language of choice for most large-scale computational analyses in Biostatistics and Bioinformatics. R has a command line interface where R commands are typed in. R has a rich library of add-on packages that has been developed for specific types of analyses. All the packages are available free to the user.

R can be downloaded from ref. 16. Binary versions are easy and straightforward to install. The analysis described in this chapter makes use of the multcomp package to carry out multiple-hypothesis testing (17). The multcomp may be installed using the R interface. It can be done by clicking on "packages" from the main menu and choosing "Install package(s)." Choose a mirror site closest to you geographically, and then choose the required package, in this case "multcomp", to be installed.

2.2. Dataset

The methods described here use the data generated using Illumina BeadArray (6, 18). For details of how the data was generated, the reader may refer to the original article by Li et al. (6). While there are several technologies that have been used for gathering information on expression of isoforms, the methods described in this chapter are not specific to any particular type of data set. However, some preprocessing of the data may be required so as to map the data obtained using other technologies to the one obtained using the BeadArray. For instance, the Affymetrix approach uses multiple probes to query a transcript; thus, care should be taken to combine the expression values from probes that query the same exon. This can then be used to compare expression levels of different exons within the same transcript using the method described here.

Table 1
Example of normalized isoform expression data: columns represent cell lines and rows represent isoforms/splicing event associated with the ATP-binding cassette, subfamily G, member 1 gene (ABCG1)

Isoform	HCE-7	HCE-7	MDA.MB-468	MDA.MB-468	PC3-E	PC3-E	PC3-E	DU145-E	DU145-E	DU145-E
ABCG1-0489	461.76	488.89	391.04	380.10	1,088.46	999.51	1,153.58	403.81	373.71	394.07
ABCG1-0490	507.13	479.91	541.21	676.18	275.79	272.72	260.40	277.59	258.48	257.83
ABCG1-0491	329.69	316.55	375.43	369.25	272.33	256.72	271.71	269.83	265.45	255.15
ABCG1-0492	337.54	338.47	441.79	456.05	248.00	248.04	260.30	246.47	245.60	252.56
ABCG1-0494	197.15	195.35	210.37	193.22	215.94	215.83	207.31	204.67	194.74	216.36
ABCG1-0495	279.85	326.66	491.95	601.24	1,132.31	1,207.01	1,260.44	429.28	421.67	389.16
ABCG1-1482	257.93	286.40	308.56	378.81	632.61	664.51	657.83	341.80	323.43	325.45
ABCG1-1483	212.34	200.27	203.42	219.46	188.74	214.67	209.19	220.03	196.99	207.44

The numbers following the gene name ABCG1 correspond to the different splicing events and are assigned at the time of experimental design.

2.2.1. Isoform Expression Data

The isoform expression data is read from a comma-separated value file (csv): each column represents a biological sample (cell line/tissue) and each row represents a different isoform or splicing event (see Table 1).

3. Methods

3.1. Experimental Design and Normalization

When profiling expression of isoforms/splicing events from biological samples, it is important to ensure that one takes the necessary steps to process the samples in batches and have biological and technical replicates. Careful attention should be paid when designing probes to minimize interference with hybridization due to secondary structure. Expression data from RASL/DASL assay used here has high specificity and sensitivity in querying isoform expression. The ligation step contributes to the specificity and the PCR step significantly enhances the sensitivity (6) in the assay. When extracting isoform expression information from other technologies like Affymetrix that use multiple probes, appropriate care should be taken to assign expression values to isoforms/splicing events (see Note 1) (19, 20).

Microarray data needs to be normalized before different data sets can be cross compared. Normalization enhances meaningful data characteristics and accounts for systematic differences across data sets. There are several methods that may be used to normalize expression data (21–23). The data used here was normalized against a synthetic average using locally weighted polynomial regression (LOWESS) (24). LOWESS uses a polynomial of degree 1 or 2, thus avoiding over-fitting. The procedure divides the data domain into several windows and uses the polynomial only to approximate over a narrow interval. Since normalization is not a one-size-fits-all solution, the user should decide, based on the data they have, which method is most suitable for their data. It is assumed here that data has been normalized.

3.2. Multiple Comparisons of Isoform Expression

In analyzing isoform expression data, we are confronted with the problem of testing the differences in expression between many means. This can be conveniently tackled using multiple comparisons. Differential analysis of isoform expression involves all possible comparisons and can be conveniently done using the R multcomp package (25). It is noteworthy to mention that such comparisons are compute intensive and it is advisable to use parallel processing (see Note 2). The output is in the form of confidence intervals, significant comparisons are those that do not intersect the zero line. We demonstrate the exhaustive comparisons using the data given in Table 1. R-code given in Table 2 can be used to carry out the analysis.

Table 2
R-code for carrying out the exhaustive comparisons using the multcomp package

```
library(mvtnorm)
library(multcomp)
par(mfrow = c(1,1),cex = 0.7, mai = c(3,2,1,2), ask = T)
complete.data < −read.csv("isoformSubset.csv",header = T)
lgth < −length(complete.data[1,])-1
complete.data.mat < −as.matrix(complete.data[,1:lgth + 1])
cell.line < −0
complete.data.frame < −as.data.frame(complete.data)
filename < −as.vector(complete.data.frame$Isoform)
cell.line < −colnames(complete.data.mat)
cell.line < −as.factor(substr(cell.line, 1,c(5,5,8,8,5,5,5,7,7,7)))
number.rows < −nrow(complete.data.mat)
i < −0
Expression < −0
mult.comp < −0
for(i in 1:number.rows)
{
  cat("Now computing::- > ", filename[i],"\n")
  for(j in 1:(lgth)){
    Expression[j] < −complete.data.mat[i,j]
}
  Expression < −as.numeric(Expression)
  isoform.expression < −data.frame(cell.line,Expression)
  isoform.expression$cell.line < −factor(isoform.expression$cell.line)
  amod < −aov(Expression ~ cell.line, data = isoform.expression)
  mult.comp < −glht(amod,linfct = mcp(cell.line = "Tukey"))
  conf.int < −confint(mult.comp,level = 0.99)
  plot(conf.int, main = filename[i],xlab = "99% Confidence interval")
  p.value < −summary(mult.comp)$test$pvalues
  out.data.mat < −data.frame(conf.int$confint[,1:3],p.value)
  filename.csv < −paste(filename[i], "csv",sep = ".")
  write.table(out.data.mat, file = filename.csv, sep = ",", qmethod = "double", col.name = NA)
  rm(amod,mult.comp,conf.int,p.value,out.data.mat,filename.csv)
}
```

Table 3
Output of the comparison of isoform ABCG1-0490

	Estimate	lwr	upr	*p*-Value
HCE.7-DU145.E	228.88653	44.68726	413.0858	0.003281
MDA.MB.4-DU145.E	344.06312	159.8638	528.2624	0.000266
PC3.E-DU145.E	4.9996982	−159.753	169.7525	0.998626
MDA.MB.4-HCE.7	115.17659	−86.6036	316.9568	0.10403
PC3.E-HCE.7	−223.8868	−408.086	−39.6876	0.003638
PC3.E-MDA.MB.4	−339.0634	−523.263	−154.864	0.000376

The preceding code may be written using any ASCII editor and saved as an R file. Lines 1 and 2 ensure that the two libraries are loaded. Line 3 sets the parameter for plotting. You may change these according to your requirements. Reading the isoform expression data is achieved in line 4. It is assumed here that the name of the file is "isoformSubset.csv." You should substitute your isoform expression data file name. Line 9 uses the isoform name from the expression data to create a file name to store the results of the analysis for a particular isoform. Line 11 creates a factor, in this case using the cell line names from the expression data. The substr function in line 11 is used to eliminate any additional differentiators that R introduces when the file header contains duplicate names. You may need to make changes to the substr function to reflect the size of the headers you have used. Lines 25 through 27 help achieve the multiple comparison. Confidence level used in computing the confidence intervals is set to 0.99 in line 27 to ensure low probability of type I error. Line 32 writes the output of each comparison to a file that has the isoform name as its filename. Table 3 shows a typical output that is written to the file created in line 32. In the interest of brevity, data contained in only one output file is shown.

3.3. Interpretation and Further Processing of the Output

The plots obtained from execution of line 28 are shown in Fig. 1. These plots are the graphical representation of the confidence intervals for the comparisons. The significant comparisons are those that do not intersect the zero line. Only comparisons for four of the isoforms are shown. The plots clearly show that there is a significant difference in expression of the isoform ABCG1-0490 between HCE.7 and DU145.E, and between MDA.MB.4 and DU145.E. The isoform ABCG1-0495 does not show a significant difference in expression between HCE.7 and DU145.E, and between MDA.MB.4 and DU145.E. Further, the isoform ABCG1-0494 does not show any significant difference in expression in any of the comparisons, as in all cases we see an intersection of the zero line.

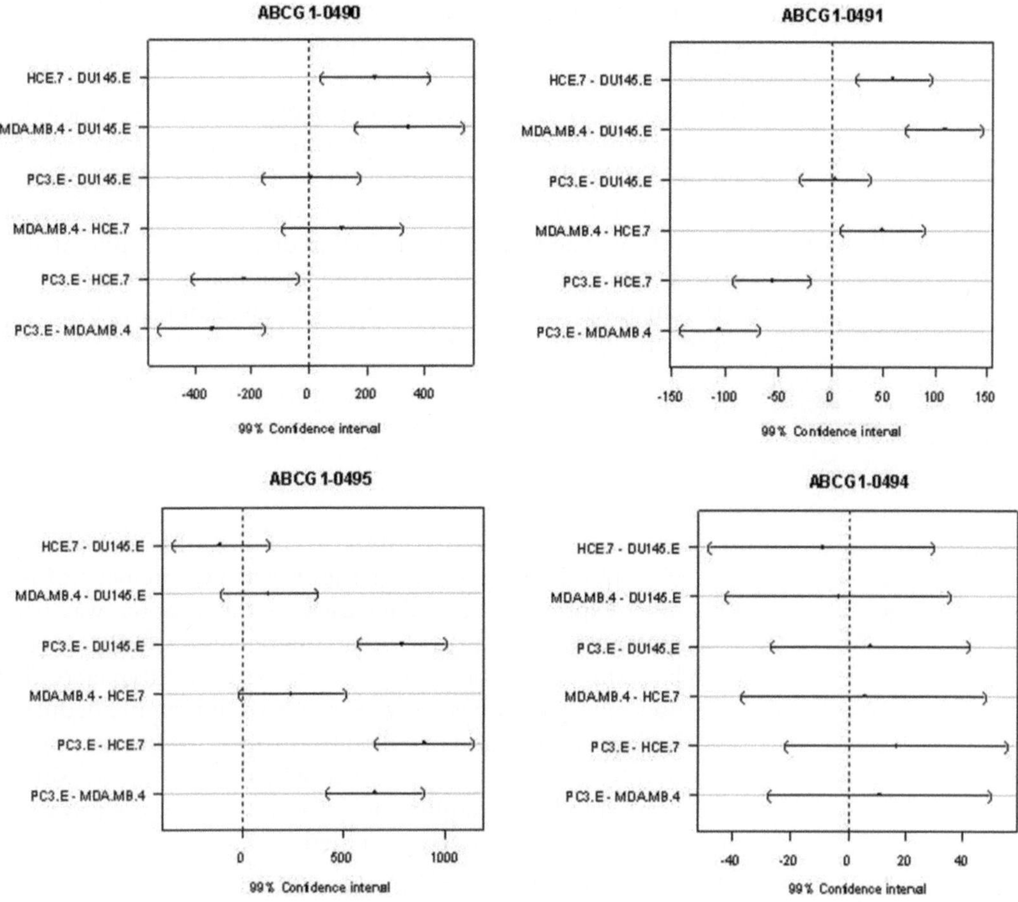

Fig. 1. Multiple comparisons on expression level of four different isoforms of the gene ABCG1. Comparisons that show significant difference in expression level are the ones that do not intersect the zero line.

The subset of data used here was part of a study to identify differential expression of isoforms in prostate cancer cell lines and nonprostate cancer cell lines (6, 18). The data generated as a result of this study consisted of isoform expression from cell lines. The cell lines for which expression data were collected included five prostate cancer cell lines, viz., LNCap, LAPC4, RWPE2, PC3, and DU145, and twelve nonprostate cancer cell lines, viz., colon cancer line (HT29, SW480, HCT116, LS174, Fet), breast cancer line (MCF7, MDA.MB-468), kidney cancer line (Caki-2), lung epidumoid carcinoma line (CALU1), and esophageal cancer lines (HCE-7, EC17 and TE3). Isoforms that exhibit differential expression between two classes of samples can be delineated from the output generated using multiple comparisons. Each isoform is given a unit score for every significant difference it showed in a comparison. The sum of the scores can be used to rank the isoform. In the example that we are using here, the isoforms ABCG1-0490 and ABCG1-0491 each have a sore of 4.

Even though the comparison between HCE.7 and MDA.MB.4 is significant, it is not considered, as both are nonprostate cancer cell lines. Isoform ABCG1-0495 has a score of 3, while ABCG1-0494 has a score of 0. Assignment of scores may be decided depending upon the question you are trying to answer, that is, whether you are doing a within-class comparison or a between-class comparison. Top ranking isoforms may be used as features for class separation or may be further studied to understand their potential to serve as biomarkers. Further, isoform levels may also reflect on the different levels of control that may be teased out in a problem-specific manner (see Note 3).

4. Notes

1. *Processing of expression data from disparate microarrays.* Not all microarrays permit the direct measurement of isoform expression. The data used in this chapter was from specially designed arrays that queried for splicing events. Isoform expression may be derived from Affymetrix that uses multiple probes. However, this would require deducing isoform information based on the probes that query the gene of interest. Care must be taken when such preprocessing is done and would require careful annotation of the probes to reflect the isoform being queried.
2. *Computational capacity issues.* Multiple comparisons are compute intensive, especially when one handles large datasets. It is advisable to use a cluster and process the data in parallel. The R/Parallel package helps to conveniently achieve this (26). In addition to this, computing efficiency may be improved by processing subsets of data and avoiding redundant comparisons.
3. *Deconvoluting controls at levels of transcription and splicing.* Controls of mRNA expression may be regulated at levels of transcription, RNA stability, and splicing. Depending on the type of data collected, it may be possible to tease this information from the data. For instance, multiple isoforms that are similarly elevated or depressed would indicate coordinated changes in transcription and/or RNA stability (6). The transcript change may be computed as the sum of the weighted fold change of the isoforms involved. The splicing change may be computed as the difference in fold change of the two isoforms. Thus, for isoforms that are similarly up- or down-regulated, the splicing change would be close to zero. These computations are data dependent and the reader is referred to an earlier work by the author for details of a specific case (6).

Acknowledgement

TMN would like to thank IUSB for research funding.

References

1. Matlin AJ, Clark F, Smith CW (2005) Understanding alternative splicing: towards a cellular code. Nat Rev Mol Cell Biol 6:386–398.
2. Kim N, Lee C (2008) Bioinformatics detection of alternative splicing. Methods Mol Biol 452:179–197.
3. Ferreira EN, Galante PA, Carraro DM et al (2007) Alternative splicing: a bioinformatics perspective. Mol Biosyst 3:473–477.
4. Chacko E, Ranganathan S (2009) Comprehensive splicing graph analysis of alternative splicing patterns in chicken, compared to human and mouse. BMC Genomics 10:S5.
5. Lee C, Wang Q (2005) Bioinformatics analysis of alternative splicing. Brief Bioinform 6:23–33.
6. Li HR, Wang-Rodriguez J, Nair TM et al (2006) Two-dimensional transcriptome profiling: identification of messenger RNA isoform signatures in prostate cancer from archived paraffin-embedded cancer specimens. Cancer Res 66:4079–4088.
7. Blencowe BJ (2006) Alternative splicing: new insights from global analyses. Cell 126:37–47.
8. Johnson JM, Castle J, Garrett-Engele P et al (2003) Genome-wide survey of human alternative pre-mRNA splicing with exon junction microarrays. Science 302:2141–2144.
9. Pando MP, Kotraiah V, McGowan K et al (2006) Alternative isoform discrimination by the next generation of expression profiling microarrays. Expert Opin Ther Targets 10:613–625.
10. Pandit S, Wang D, Fu XD (2008) Functional integration of transcriptional and RNA processing machineries. Curr Opin Cell Biol 20:260–265.
11. Hardiman G (2004) Microarray platforms – comparisons and contrasts. Pharmacogenomics 5:487–502.
12. Lee NH, Saeed AI (2007) Microarrays: an overview. Methods Mol Biol 353:265–300.
13. Yeakley JM, Fan JB, Doucet D et al (2002) Profiling alternative splicing on fiber-optic arrays. Nat Biotechnol 20:353–358.
14. Fan JB, Yeakley JM, Bibikova M et al (2004) A versatile assay for high-throughput gene expression profiling on universal array matrices. Genome Res 14:878–885.
15. http://www.r-project.org.
16. http://cran.r-project.org.
17. Hothorn T, Bretz F, Westfall P (2008) Simultaneous inference in general parametric models. Biom J 50:346–363.
18. Nair TM (2009) On selecting mRNA isoform features for profiling prostate cancer. Comput Biol Chem 33:421–428.
19. Bemmo A, Benovoy D, Kwan T et al (2008) Gene expression and isoform variation analysis using Affymetrix Exon Arrays. BMC Genomics 9:529.
20. Bemmo A, Dias C, Rose AA et al (2010) Exon-level transcriptome profiling in murine breast cancer reveals splicing changes specific to tumors with different metastatic abilities. PLoS ONE 5: e11981.
21. Bolstad BM, Irizarry RA, Astrand M et al (2003) A comparison of normalization methods for high density oligonucleotide array data based on variance and bias. Bioinformatics 19:185–193.
22. Zeller G, Henz SR, Laubinger S et al (2008) Transcript normalization and segmentation of tiling array data. Pac Symp Biocomput: 527–538.
23. Haldermans P, Shkedy Z, Van Sanden S et al (2007) Using linear mixed models for normalization of cDNA microarrays. Stat Appl Genet Mol Biol 6:Article 19.
24. Cleveland WS (1979) Robust Locally Weighted Regression and Smoothing Scatterplots. Journal of the American Statistical Association 74:829–836.
25. Hothorn T, Bretz F, Westfall P et al (2008) Multcomp: Simultaneous Inference for General Linear Hypotheses. URL http://CRAN.R-project.org.
26. Vera G, Jansen RC, Suppi RL (2008) R/parallel – speeding up bioinformatics analysis with R. BMC Bioinformatics 9:390.

Chapter 9

Functional Comparison of Microarray Data Across Multiple Platforms Using the Method of Percentage of Overlapping Functions

Zhiguang Li, Joshua C. Kwekel, and Tao Chen

Abstract

Functional comparison across microarray platforms is used to assess the comparability or similarity of the biological relevance associated with the gene expression data generated by multiple microarray platforms. Comparisons at the functional level are very important considering that the ultimate purpose of microarray technology is to determine the biological meaning behind the gene expression changes under a specific condition, not just to generate a list of genes. Herein, we present a method named percentage of overlapping functions (POF) and illustrate how it is used to perform the functional comparison of microarray data generated across multiple platforms. This method facilitates the determination of functional differences or similarities in microarray data generated from multiple array platforms across all the functions that are presented on these platforms. This method can also be used to compare the functional differences or similarities between experiments, projects, or laboratories.

Key words: Microarray, Biological pathway database, Functional comparison, Percentage of overlapping functions, R, Gene expression

1. Introduction

Several microarray platforms are currently available to measure gene expression on a genome-wide scale (1–3). These platforms differ in probe content, design, deposition technology, as well as labeling and hybridizing protocols. The types of probes usually include spotted cDNA sequences or PCR products (hundreds to thousands of base pairs in length), short (25–30-mer) or longer (60–70-mer) oligonucleotides. These probes can be either contact-spotted by pins, deposited by ink jet or synthesized directly on the arrays (4). Dye labeling, array hybridization, image acquisition, feature extraction, and signal data generation are often sources of variability across different microarray providers, and experiments are usually performed by using provider-specific kits

Junbai Wang et al. (eds.), *Next Generation Microarray Bioinformatics: Methods and Protocols*,
Methods in Molecular Biology, vol. 802, DOI 10.1007/978-1-61779-400-1_9, © Springer Science+Business Media, LLC 2012

and protocols (3–5). Therefore, gene expression measured by different microarray platforms might yield variable results even in the case where identical samples are used.

Comparisons can be performed across multiple platforms for various purposes. For example, concerns have been raised regarding whether concordance exists between gene expression data generated using different microarray platforms. In such cases, cross-platform comparisons are performed in order to address these concerns (3–5). Furthermore, a laboratory may employ multiple microarray platforms to conduct its experiments in order to increase the credibility of their experimental results by incorporating the advantages inherent to each platform (5). Moreover, microarray core facilities may need to compare gene expression data generated locally with data measured at contract facilities, or compare gene expression data generated by their custom microarrays vs. commercial microarrays. Cross-platform comparisons are also required for researchers who want to make use of the wealth of gene expression data available in public repositories, such as Gene Expression Omnibus (GEO) and Array Express Archive. The GEO, according to the statistics of the National Center for Biotechnology Information (NCBI) in 2008, holds over 10,000 experiments, 300,000 samples, and 16 billion individual abundance measurements (6). However, these microarray data were obtained by different laboratories and using different platforms (6, 7). Comparison across experiments, laboratories, and platforms is an important step to understand and mine these data in a robust manner.

Functional comparison means the comparison of microarray data at the level of biological functions derived from gene expression data. Functional comparison is an essential part of microarray evaluation since the ultimate purpose of microarray technology is to determine the differences in biological functions between samples of interest (8). Herein, we introduce a method called percentage of overlapping functions (POF) that performs functional comparisons across platforms, experiments, or laboratories. This method utilizes the biological functions generated by various biological pathway databases and enables a thorough analysis of the degree of similarity between multiple experiments.

2. Materials

This section provides materials of a test case example to illustrate how functional comparison might be performed in one scenario. In this test case, RNA samples were collected from the kidney tissue of rats treated with carcinogen aristolochic acid (AA) at a dose of 10 mg/kg body weight by gavage for 12 weeks (9, 10).

This treatment regimen induced kidney tumors in the rats (11). The rats receiving the vehicle, 0.9% sodium chloride, were used as controls. An aliquot of the RNA samples was sent to four microarray providers, Applied Biosystems (ABI), Affymetrix (AFX), Agilent (AG), and GE Healthcare (GEH) for assaying gene expression levels (12) (see Note 1). One list of differentially expressed genes (DEGs) was generated for each platform. These DEG lists were then analyzed using Ingenuity Pathway Analysis (IPA) to generate function lists for each platform. These function lists were then used to produce function tables which were input into an R-based program to perform POF.

Sixteen function tables in total were generated for the analysis, including 4 true function tables named "ABI_True.txt," "AFX_True.txt," "AG_True.txt," and "GEH_True.txt," and 12 random function tables named "ABI_Random1.txt," "ABI_Random2.txt," "ABI_Random3.txt," "AFX_Random1.txt," "AFX_Random2.txt," "AFX_Random3.txt," "AG_Random1.txt," "AG_Random2.txt," "AG_Random3.txt," "GEH_Random1.txt," "GEH_Random2.txt," and "GEH_Random3.txt." The four true function tables were generated from four function lists retrieved from IPA (13) according to the DEGs obtained from the gene expression analysis by four microarray platforms using the same set of RNA samples. The 12 random function tables, with 3 tables per platform, were generated using the same procedure as the true function tables from IPA-derived function lists. However, the function lists used to generate random function tables were retrieved from IPA using randomly selected genes from the corresponding platform (see Note 2). These random function tables were used to determine the background concordance between platforms as stated in Subheading 3.

R scripts used for performing functional comparisons are provided with this book chapter. R is an integrated suite of software functions used for data manipulation, calculation and graphical display. R can be freely downloaded at the R home website (14). Manuals and documentation about R can also be found on the website. All of the 16 example files and the R scripts are available for downloading at the Methods in Molecular Biology website.

3. Methods

3.1. Function Lists

A function list is a list of biological functions derived from a biological pathway database for a given set of genes and is used to describe what kinds of biological functions, cellular processes, molecular pathways, and/or disease/disorders are associated with the genes that are analyzed. The gene list can be derived by various means, but is typically composed of DEGs selected according to specific threshold criteria (*p*-values, *q*-values, fold changes, or other statistical values) (see Note 3) from an experiment using microarray, PCR array, or next-generation sequencing (NGS). Some biological pathway databases, such as IPA, Gene Ontology (GO) and Database for Annotation, Visualization, and Integrated Discovery (DAVID), can be used to functionally annotate a list of genes and generate function lists. There are a number of ways to determine biological functions and generate such lists from microarray data. It is important that whatever a database or method is chosen for functional annotation, it should be consistent across all analyses so that comparisons are always made based upon the same source. Readers are encouraged to explore these methods in the other related chapters of this book.

For most functional annotation or biological pathway databases, the functional annotation of a gene is assessed at several different levels in a hierarchical structure, each level giving some information according to its degree of confidence or specificity (see Note 4). The high levels, such as level 1 or 2, signify high confidence with low specificity and are used to convey large-scale characteristics. Conversely, the lower levels suggest higher specificity of functions but generally lower confidence for that particular designations. The level of specificity of functions increases as the levels decrease. Usually, one "parent" function at a higher level includes multiple "daughter" functions at the lower levels. For example, a level 1 function includes multiple level 2 functions, and a level 2 function includes multiple level 3 functions, and so on. Besides function names and levels, a function list generally also includes *p*-values. The *p*-values are usually calculated using Fisher's exact test (15, 16). The test measures the likelihood that a list of genes significantly represents a functional group relative to the total number of genes in that functional group. Smaller *p*-values mean a higher confidence; however, the number of genes usually decreases with *p*-value. Sometimes other information is also included, such as the number or names of genes involved in a specified function, or E values that represent a relative enrichment factor used to assess the significance of a function for a given list of genes (16). Whether or not such information is included in a function list depends on which functional annotation database is used. Figures 1 and 2 show two typical function lists generated from IPA and GO, respectively.

3.2. Function Tables

To make comparisons between datasets, function tables have to be made from the function lists. A function table includes two columns to display function names and *p*-values, respectively. Table 1 shows a typical function table for functional comparison (see Note 5). As mentioned above, a *p*-value is used to measure the likelihood that a function is significantly associated with a group of genes investigated. The smaller a *p*-value, the more

Category	Function	Function Annotation	P-value	Molecules	# Molecules
Cancer	G2/M phase	arrest in G2/M phase of pr	3.95E-04	AMACR, IL24, LIM	3
Tissue Development	aggregation	aggregation of hematopoie	3.95E-04	CXCL9, CXCL10, C	3
Hematological System Deve	aggregation	aggregation of hematopoie	3.95E-04	CXCL9, CXCL10, C	3
Cell Cycle	G2/M phase	arrest in G2/M phase of pr	3.95E-04	AMACR, IL24, LIM	3
Reproductive System Diseas	G2/M phase	arrest in G2/M phase of pr	3.95E-04	AMACR, IL24, LIM	3
Cellular Assembly and Orga	assembly	assembly of hemidesmosc	3.95E-04	EGF, LAMA3, LAN	3
Cellular Assembly and Orga	elongation	elongation of microvilli	3.95E-04	EGF, ESPN, ICAM	3
Hematopoiesis	aggregation	aggregation of hematopoie	3.95E-04	CXCL9, CXCL10, C	3
Cellular Function and Mainte	assembly	assembly of hemidesmosc	3.95E-04	EGF, LAMA3, LAN	3
Cell Morphology	elongation	elongation of microvilli	3.95E-04	EGF, ESPN, ICAM	3
Cancer	necrosis	necrosis of leukemia cell li	3.84E-04	FAS, S100A8, S10	4
Cancer	adhesion	adhesion of leukemia cells	8.44E-04	AMICA1, CD44, IC	4
Genetic Disorder	epidermolysis bullosa l	epidermolysis bullosa letal	1.36E-04	ITGB4, LAMA3, LA	4
Genetic Disorder	generalized atrophic be	generalized atrophic benig	1.36E-04	ITGB4, LAMA3, LA	4
Genetic Disorder	membranous glomerul	membranous glomerulone	3.84E-04	CLU, COL4A3 (incl	4
Inflammatory Disease	membranous glomerul	membranous glomerulone	3.84E-04	CLU, COL4A3 (incl	4
Immunological Disease	membranous glomerul	membranous glomerulone	3.84E-04	CLU, COL4A3 (incl	4
Cell Death	necrosis	necrosis of leukemia cell li	3.84E-04	FAS, S100A8, S10	4
Cellular Movement	cell movement	cell movement of lung cell	3.84E-04	EGF, FN1, FOSL1	4
Cellular Movement	chemotaxis	chemotaxis of hematopoie	3.84E-04	CCL19, CXCL9, CX	4
Cell-To-Cell Signaling and Int	adhesion	adhesion of hematopoietic	8.44E-04	CXCL9, CXCL10, C	4
Cell-To-Cell Signaling and Int	adhesion	adhesion of leukemia cells	8.44E-04	AMICA1, CD44, IC	4
Tissue Development	adhesion	adhesion of hematopoietic	8.44E-04	CXCL9, CXCL10, C	4
Hematological System Deve	adhesion	adhesion of hematopoietic	8.44E-04	CXCL9, CXCL10, C	4
Hematological System Deve	chemotaxis	chemotaxis of hematopoie	3.84E-04	CCL19, CXCL9, CX	4
Organismal Injury and Abnor	growth	growth of lesion	8.44E-04	CCL13, FLJ78302,	4
Hematological Disease	necrosis	necrosis of leukemia cell li	3.84E-04	FAS, S100A8, S10	4
Hematological Disease	adhesion	adhesion of leukemia cells	8.44E-04	AMICA1, CD44, IC	4
Cellular Assembly and Orga	binding	binding of plasma membra	3.84E-04	BDNF, CD44, LCK	4
Renal and Urological Diseas	membranous glomerul	membranous glomerulone	3.84E-04	CLU, COL4A3 (incl	4
Hematopoiesis	chemotaxis	chemotaxis of hematopoie	3.84E-04	CCL19, CXCL9, CX	4
Hematopoiesis	adhesion	adhesion of hematopoietic	8.44E-04	CXCL9, CXCL10, C	4
Tumor Morphology	adhesion	adhesion of leukemia cells	8.44E-04	AMICA1, CD44, IC	4
Dermatological Diseases an	epidermolysis bullosa l	epidermolysis bullosa letal	1.36E-04	ITGB4, LAMA3, LA	4
Dermatological Diseases an	generalized atrophic be	generalized atrophic benig	1.36E-04	ITGB4, LAMA3, LA	4
Organ Morphology	size	size of ovarian follicle	1.36E-04	FST, INHBA, INS1	4
Organ Morphology	morphology	morphology of mammary a	3.84E-04	AREG, EGF, NRG	4

Fig. 1. An example of a function list retrieved from IPA. This function list was generated by IPA based on 559 rat genes. The functional interpretation is arranged at three levels with Category (Level 1), Function (Level 2), and Function Annotation (Level 3). *p*-Values are calculated by right-tail Fisher's exact test. The column "Molecules" indicates which genes are involved in a specified Category/Function/Function Annotation. The column "# Molecules" indicates how many genes are involved in a specified Category/Function/Function Annotation (see Note 5).

closely a function is associated with the group of genes. To rank the functions according to the highest representation of genes, their *p*-values need to be increasingly sorted. After sorting, the functions with the smallest *p*-values will reflect the most represented biological functions of the gene list. Table 2 shows a function table that has been sorted by ascending *p*-values.

Functional level is an important factor that requires careful attention when making a function table. General functions, like functions at level 1, are usually not suitable for comparison as they are not detailed or descriptive enough to be meaningful. If their function levels are too low, the groups often contain only a few genes, thus they are not suitable for comparison either. The functions used for comparison should generally come from the same level. In many cases, the choice depends on the type of function pathway database used. For function lists from IPA,

Term	GO ID	Level (Average)	P value (Average)	Gene Hits	E value
positive regulation of biological process	GO:0048518	3.00	0.000000	135	1.73
regulation of developmental process	GO:0050793	3.33	0.000000	100	1.78
regulation of biological quality	GO:0065008	3.00	0.102783	121	1.91
positive regulation of cellular process	GO:0048522	4.00	0.102783	113	1.62
response to hormone stimulus	GO:0009725	4.00	0.103060	37	3.50
anatomical structure development	GO:0048856	3.00	0.103332	189	1.45
system development	GO:0048731	4.00	0.103372	170	1.49
response to steroid hormone stimulus	GO:0048545	5.00	0.103372	24	3.60
organ development	GO:0048513	4.75	0.103372	135	1.53
localization	GO:0051179	2.00	0.104129	236	1.40
regulation of nucleoside transport	GO:0032242	6.44	0.104729	1	24.47
low density lipoprotein mediated signalin	GO:0055096	6.60	0.105270	1	24.47
regulation of Rac GTPase activity	GO:0032314	9.71	0.108880	1	24.47
negative regulation of the force of heart c	GO:0001986	7.88	0.109755	1	24.47
negative regulation of histone H3-K9 met	GO:0051573	10.85	0.109755	1	24.47
negative regulation of protein binding	GO:0032091	6.00	0.040861	1	24.47
negative regulation of histone methylatior	GO:0031061	9.62	0.040861	1	24.47
arginine biosynthetic process via ornithin	GO:0042450	8.79	0.040861	1	24.47
negative regulation of chemokine product	GO:0032682	6.00	0.040861	1	24.47
protein kinase C deactivation	GO:0042313	6.80	0.040861	1	24.47
establishment of localization	GO:0051234	2.50	0.547478	206	1.41
organic acid metabolic process	GO:0006082	4.00	0.549235	66	2.17
nitrogen compound biosynthetic process	GO:0044271	4.75	0.551678	22	3.82
response to stress	GO:0006950	3.00	0.553897	124	1.81
response to external stimulus	GO:0009605	3.00	0.555415	106	2.50
developmental process	GO:0032502	2.00	0.557023	254	1.34
homeostatic process	GO:0042592	4.00	0.565981	64	1.93
amine biosynthetic process	GO:0009309	5.43	0.565981	20	4.94
regulation of response to stimulus	GO:0048583	3.33	0.565981	36	2.58
regulation of immune system process	GO:0002682	3.33	0.565981	35	2.56
response to wounding	GO:0009611	4.00	0.600750	70	2.53
response to drug	GO:0042493	4.00	0.603682	27	3.18
carboxylic acid metabolic process	GO:0019752	5.00	0.605314	66	2.18
wound healing	GO:0042060	5.00	0.605314	28	3.11
amino acid biosynthetic process	GO:0008652	6.33	0.610255	14	5.81
negative regulation of biological process	GO:0048519	3.00	0.611521	134	1.70
cell development	GO:0048468	4.33	0.617077	134	1.58

Fig. 2. An example of a function list retrieved from Gene Ontology (GO) via ArrayTrack (18). This function list was generated by the GO database based on 559 rat genes. In this function list, "Term" denotes the biological functions related to the input genes. "GO ID" is the identification number of the GO term. "Level" is the average level number of a term in the GO hierarchical tree. In GO, one term can belong to multiple daughter terms of the same parental term. They can also be located at different levels in the hierarchical tree. The level value here is the average of all the level numbers that a term can have when interpreting the group of genes. *p*-Values are calculated using right-tail Fisher's exact test. "Gene hits" indicate how many molecules are involved in a specified term. "E value" is a relative enrichment factor that is a direct measurement of the prevalence of a GO term among the input genes compared to the prevalence of the same term among all the genes in the GO database.

functions at level 2 (named "Functions" by IPA) or level 3 (named "Function Annotations" by IPA) are suitable for comparison (see Note 6). For function lists from GO, the functions at levels 1 and 2 or the functions with gene hits less than 4 (usually their levels are higher than 8) should be removed from the comparison. The GO functions at different levels can be mixed together for the comparison as the same functions may exist at different levels under different "parent" functions in a hierarchical structure.

Table 1
An example of a function table

Function	*p*-Value
Antibody response	1.79E – 04
Synthesis	3.28E – 06
Apoptosis	2.07E – 04
Biosynthesis	2.81E – 06
Metabolism	4.82E – 04
Transport	1.12E – 03
Cell growth	3.98E – 07
Cell proliferation	2.49E – 04
Tumor promotion	1.32E – 04

Table 2
An example of a function table sorted by ascending *p*-values

Function	*p*-Value
Cell growth	3.98E – 07
Biosynthesis	2.81E – 06
Synthesis	3.28E – 06
Tumor promotion	1.32E – 04
Antibody response	1.79E – 04
Apoptosis	2.07E – 04
Cell proliferation	2.49E – 04
Metabolism	4.82E – 04
Transport	1.12E – 03

3.3. POF Calculation

POF measures the similarity of two or more function lists. The POF is calculated using the following formula:

$$\mathrm{POF}(\%) = \frac{O_i}{T} \times 100\%,$$

where T is the number of the top-ranked functions and O_i ($i = 1, 2, 3\ldots$) is the number of overlapping functions between or among the T top functions that are being compared.

Table 3
An example of a percentage of overlapping function (POF) table

Rank	POF
1	0.01
2	50.00
3	66.66
4	100.00
5	80.00
6	83.33
7	85.71
8	75.00
9	66.67
10	60.00
11	63.64
12	66.67

For example, if there are 16 overlapping functions ($O_{20} = 16$) between the top-ranked 20 functions ($T = 20$) in two function tables, then

$$\text{POF}(\%) = \frac{16}{20} \times 100\% = 80\%.$$

POF can be calculated for each T retrieved from a pathway database (T can range into the hundreds depending on the pathway databases used). Generally, a POF generated by the top 20 or 50 functions is sufficient for comparison because these top functions are the most important in terms of biological meaning.

After POF calculation, a table with two columns (Rank and POF) will be generated. The Rank column consists of positive integers and each number indicates how many top-ranked functions are used for the comparison. Each number in the POF column is the percentage of overlapping functions for the given number of top functions shown in the rank column. Table 3 shows a typical POF table. This kind of tables demonstrates the level of similarity between two or among more function lists.

3.4. Visualization of POF

The POF vs. Rank can also be visualized to discern the similarity between two or among more function lists that are being compared. Various types of figures can be used for this purpose, but here we show a line-connected scatter plot, with the X and Y axes representing the rank number and POF value, respectively, using the example

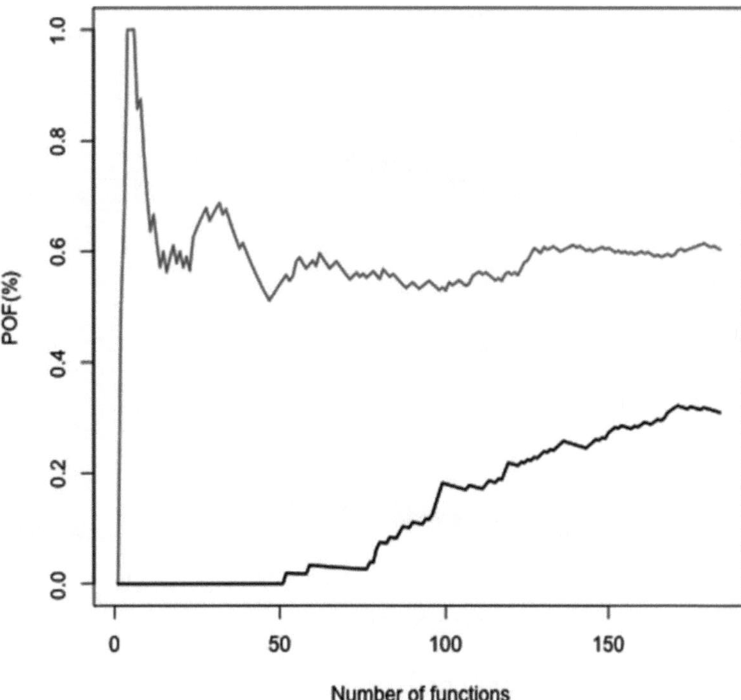

Fig. 3. An example of a POF graph . This figure is drawn using the R scripts provided with this chapter and based on the POF data calculated from one of the example function lists. The *grey line* indicates the POF between two true function lists while the *dark line* represents the POF between two random function lists.

data sets (Fig. 3). Usually, the POF values display dramatic variations for the first 10–20 rank numbers due to the small number of functions for comparison. As the rank number increases, the POF values become more stable because more functions are used for each comparison. Generally, rank numbers greater than 60 are required to stabilize the estimation of POF. If two function lists are identical, the POF will be 100% for all rank numbers. For the function lists generated from real DEGs, the POF values may vary between 0 and 100%, reflecting the degree of similarity among function lists at any given number of top-ranked functions.

3.5. Determination of Background Concordance Level

To determine whether the POF between function lists are significantly different from the background, it is necessary to determine the level of background concordance. As mentioned above, the function lists used for comparison are usually generated from a set of DEGs. To determine the background concordance, a list of randomly selected genes need to be generated. The set of randomly selected genes should be selected from the whole set of genes present on a microarray platform used for the gene expression experiment. The number of the randomly selected genes should be equal to that of the DEGs in the list. If you have two lists of DEGs generated from two different microarray

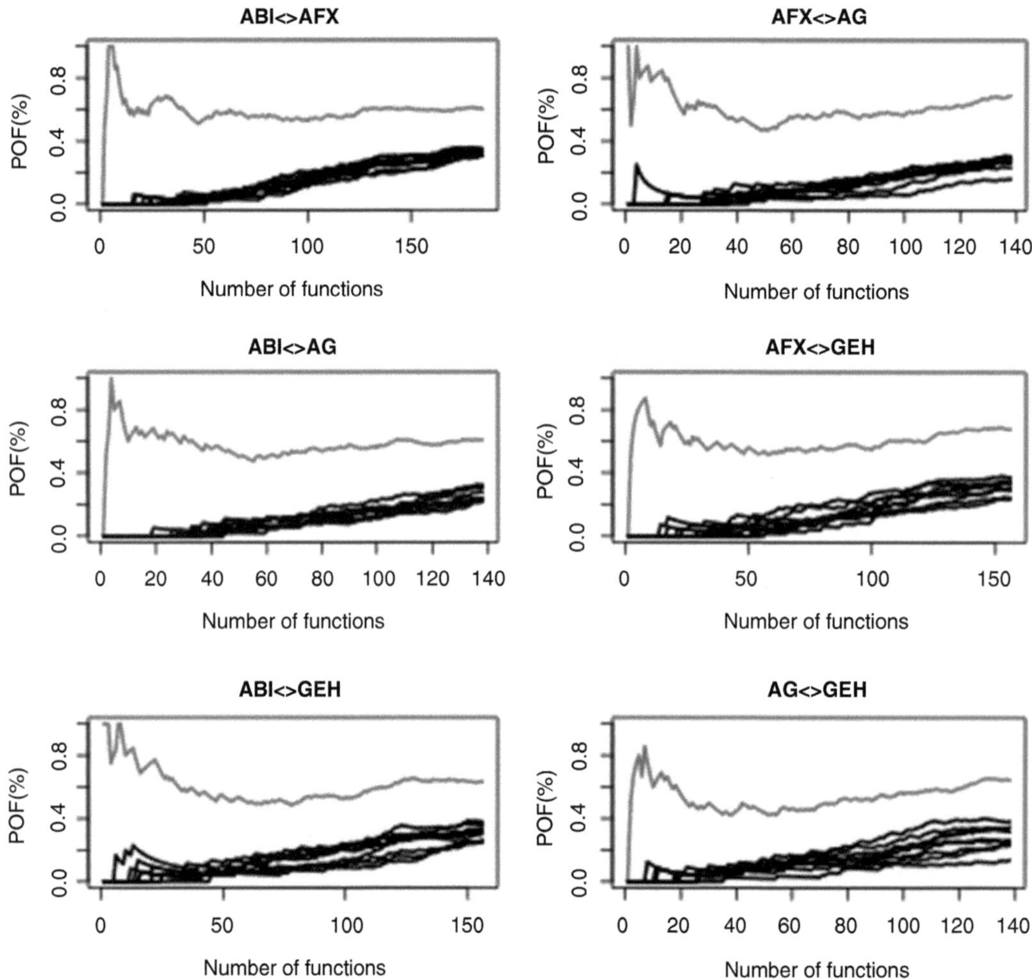

Fig. 4. Comparison between the six possible pairs of the four platforms based on the example function tables. The title for each graph indicates the two platforms that are being compared. The *X*-axis represents the number of top-ranked functions to be compared. *Y*-axis represents the POF values at any given number of the top-ranked functions. The *grey line* denotes the comparison between the two true function lists while the *dark lines* signify the comparisons between pairs of random function lists.

platforms, two sets of randomly selected genes should be produced from each platform with the same number of genes as those in their respective DEG lists. After obtaining the sets of the randomly selected genes, the random function lists can be generated using functional softwares as described previously. Accordingly, the POF values and graphs are then generated for the random function lists using the same method as described above for the true function lists. The POF from the random function lists serve as the comparator for evaluating the significance of the POF between the true function lists. To statistically determine significance between true POF lists and random POF lists, multiple random function lists (at least two) are expected to be generated for one true function list (see Fig. 4, for example).

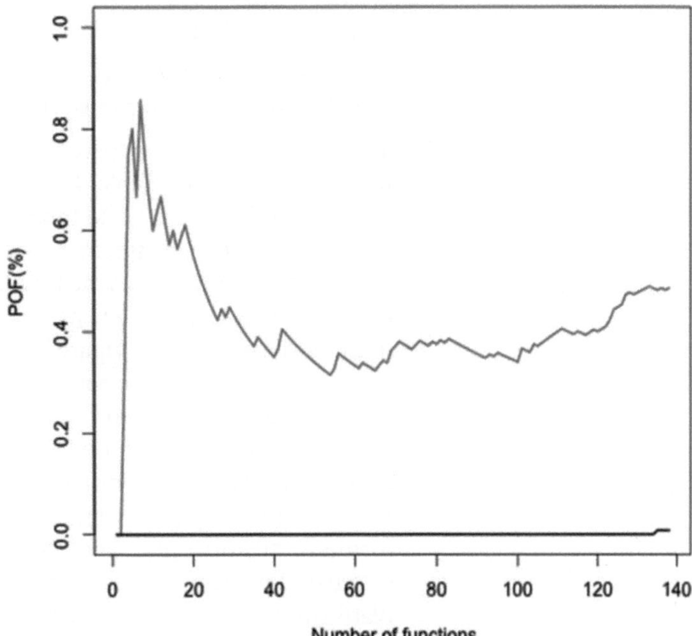

Fig. 5. Comparison across all the four platforms based on the example function tables. The *X*-axis represents the number of top functions compared. The *Y*-axis indicates the POF values at any given number of top functions. The *grey* and *dark lines* denote the comparisons between 4 true function lists and between 12 random function lists, respectively.

Then, multiple sets of POF values can be calculated. For example, four sets of POF values can be calculated if two random function lists are generated for each of the two true function lists. One sample *T*-test is used here to assess the significance of the difference between the true POF value and the multiple random POF values at each rank.

For the comparison of more than two function lists, the POF values obtained are generally lower than those between two lists. However, the background values are also usually very low. High concordance can still be obtained if a large difference exists between the true POF values and the background POF values (Fig. 5).

3.6. Visualization of the Comparison Using R

Functional comparisons presented in this chapter can be visualized by various programming software or bioinformatics tools depending on the preference of users. For those without programming experience, an executable program of R scripts is provided at the website of Methods in Molecular Biology. Brief instructions on how to use the program are provided below.

1. Install R onto your computer. R software is freely available at the R home web page (14). The installation can be completed automatically using the default settings at each step.

Readers are encouraged to read the reference documentation at the R web page FAQ and HOWTO documents to get familiar with R (17).

2. Once successfully installed, an R icon will appear on the desktop. Alternatively, you can start R at Start→Program→R.
3. The next step is to prepare your function tables. The table format is shown in Table 1. There are two columns named function and *p*-values. The function tables should be saved as Tab-separated txt documents for running the program. The file name is composed of two parts such as "PlatformB_Random2.txt" The first part denotes names of the microarray platforms used for the platform comparison. The name can also reflect the name of experiments, projects, laboratories, or other details of the functional comparison. "True" or "Random" designates the true or random function lists. The N in the "Platform_RandomN.txt" is applied to differentiate the random function lists. For example, if you compare three platforms named "PlatformA," "PlatformB," and "PlatformC" and you generate one true function list and two random function lists for each platform, the following file names could be used: "PlatformA_True.txt," "PlatformA_Random1.txt," "PlatformA_Random2.txt," "PlatformB_True.txt," "PlatformB_Random1.txt," "PlatformB_Random2.txt," "PlatformC_True.txt," "PlatformC_Random1.txt," and "PlatformC_Random2.txt" (see Note 7). All of the files need to be stored in a single folder, named, for example, as "FunctionTables."
4. The folder containing the function tables (in our case, "FunctionTables") will be used as the working directory of the current R session. Click "File" in the program, then press "Change dir." A dialogue will show up. Select the desired folder to use as working directory and click "OK." The computer will read data from or write data into this folder when you run the R program.
5. After downloading the file "FunctionComparison.txt" that contains the R scripts, open the file (see Note 8) and copy and paste the scripts into the R console to run the program. This will include sorting the *p*-values for each table, calculation of POF, and generation of POF graphs for any possible platform pairing between the true function tables and between the random function tables. In our example, the calculation of POF will be performed for the following pairs of true function tables: PlatformA_True vs. PlatformB_ True, PlatformA_True vs. PlatformC_True, and PlatformB_ True vs. PlatformC_True. The same calculation will be applied to the random tables such as PlatformA_Random1 vs. PlatformB_Random1, PlatformA_Random1 vs. PlatformB_Random2, PlatformA_Random2 vs. PlatformB_Random1, etc. A comparison across all the platforms will also be automatically performed.

6. After successfully running these R scripts, you will find a new folder named "Results" within the working directory that contains the TXT files of the POF values and a PDF file of figures. The TXT file names will indicate the comparisons performed. The resulting table in each file will have four columns for Rank, true POF, random POF, and *p*-values indicating the significance of a POF over the background. For our example, four TXT files will be generated, "PlatformA_PlatformB.txt," "PlatformA_PlatformC.txt," "PlatformB_PlatformC.txt," and "AcrossAllPlatforms.txt" in which there are only three columns without the *p*-values. The PDF file named "Figures_Functional Analysis.pdf" includes graphs for each TXT file (Figs. 4 and 5) (see Note 9).

3.7. Demonstration of this Analysis Using Example Data Sets

On the book web page, a folder named "ExampleFunctionTables" containing 16 function table TXT files can be found. These function tables were generated by platforms ABI, AFX, AG, and GEH from the same RNA samples (see Subheading 2 for details). One true function table and three random function tables were produced from each platform. These sample data sets are used here for demonstrating how to perform the functional comparison by the analysis of POF using the R scripts provided.

1. Create a new folder named "FunctionTables" and download the 16 function table files from the website and save it into this folder.
2. Run R. Select the "FunctionTables" folder as the working directory for the current R session as shown in **step 4** of Subheading 6.
3. Download the file "FunctionComparison.txt" that contains the R scripts from the web page.
4. Copy and paste all the R scripts into the R console. The R scripts will run automatically and complete the POF calculation and graph production.
5. Open the folder "Result" generated by the R scripts. All the resulting data can be viewed. There should be eight files in the "Result" folder, including "ABI_AFX.txt," "ABI_AG.txt," "ABI_AGH.txt," "AFX_AG.txt," "AFX_GEH.txt," and "AG_GEH.txt," "AcrossAllPlatforms.txt," and "Figures_-Functional Analysis.pdf." The seven TXT files contain the calculation results for POF. The PDF file includes two Figures. The first one displays six graphs for the comparisons between the possible pairs of the four platforms (Fig. 4). The second figure shows the comparison across all platforms (Fig. 5).

3.8. Interpretation of POF Data

One of the advantages of the POF method introduced here is to thoroughly evaluate the similarity of microarray data from two

Table 4
The top 20 functions related to the example microarray data of the four platforms

Rank	ABI	GEH	AG	AFX
1	Tumorigenesis	Tumorigenesis	Tumor	Tumor
2	Cancer	Cancer	Tumorigenesis	Cancer
3	Neoplasia	Neoplasia	Cancer	Neoplasia
4	Tumor	Metabolic disorder	Neoplasia	Tumorigenesis
5	Primary tumor	Tumor	Experimentally induced diabetes	Primary tumor
6	Malignant tumor	Malignant tumor	Primary tumor	Malignant tumor
7	Metabolic disorder	Primary tumor	Malignant tumor	Carcinoma
8	Carcinoma	Carcinoma	Diabetes	Experimentally induced diabetes
9	Infectious disorder	Endocrine system disorder	Rheumatoid arthritis	Genetic disorder
10	Colon cancer	Colorectal cancer	Inflammatory disorder	Diabetes
11	Colorectal cancer	Prostatic intraepithelial neoplasia	Carcinoma	Endocrine system disorder
12	Endocrine system disorder	Endometriosis	Endocrine system disorder	Digestive organ tumor
13	Autoimmune disease	Colon cancer	Colorectal cancer	Rheumatoid arthritis
14	Pathogenesis	Ovarian cancer	Connective tissue disorder	Genital tumor
15	Immunological disorder	Diabetes	Autoimmune disease	Colorectal cancer
16	Carcinoma in situ	Genital tumor	Rheumatic disease	Pancreatic cancer
17	Diabetes	Prostatic intraepithelial tumor	Inflammatory response	Prostatic intraepithelial tumor
18	Experimentally induced diabetes	Experimentally induced diabetes	Neovascularization	Immune response
19	Endometriosis	Cholestasis	Immunological disorder	Pancreatic adenocarcinoma
20	Prostatic intraepithelial tumor	Remodeling	Arthritis	Ovarian cancer

or more platforms in terms of the biological functions. The percentages generated from the comparison themselves indicate the similarity between or among the comparing datasets. The higher the percentages, the more comparable the datasets.

The p-values generated from one sample T-test using the R scripts may indicate whether the POF are different from those calculated from randomly selected gene lists although they cannot tell how similar two sets of data are. From Figs. 3 and 4, we can find that the background POF become higher and higher when the ranks increase, suggesting that the top functions, such as top 10 or 20 functions, are more meaningful for comparison. These top functions usually have smaller p-values determined by Fisher's exact test and are more reliably associated with the comparing gene lists. While the POF within the top functions reveal the similarity of the comparing datasets, the common functions between or among the different datasets may signify biological discoveries from the comparison. Therefore, more attention should be paid to the comparison of the top-ranked functions.

A comparison for the top 20 functions from the different microarray platforms was made using our sample data sets (Table 4). The comparison reveals similar ongoing biological processes in rat kidneys exposed to AA. The top functions reflect AA's carcinogenic characters in rat kidney; and the major functions from all the platforms were carcinogenesis-related, such as cancer, tumor, neoplasm, and tumorigenesis. Other functions like inflammatory disorder might reveal other toxicities of AA in rat kidney. Thus, the results demonstrate that the different platforms generated similar information that was related to the ongoing biological processes.

4. Notes

1. Our example data were generated using different microarray platforms. Your data sets, however, can be any other types of data from different experiments, projects, or laboratories, or different high-throughput technologies such as NGS and real-time PCR array.
2. The term "true" throughout this chapter refers to the gene lists, function lists, and function tables associated with DEGs determined by real microarray experiments while the term "random" refers to the gene lists, function lists, and function tables associated with sets of genes randomly selected from the entire gene pool present on microarrays. See step 6 in Subheading 3.6 and step 5 in Subheading 3.7 for examples about the two terms.
3. There are many different methods available for normalization of the raw data from microarray analysis and many different

criteria generally accepted for DEG list selection. The normalization and the gene selection methods can be the same or different for each platform. In the example provided, the normalization methods suggested by the microarray platform manufacturers were used. The DEG selection criteria, however, are expected to be the same across all the platforms.

4. The definition and number of function levels are usually dissimilar in different biological pathway databases. However, the common theme is that the functions are describing the related biological meaning from general to specific, depending on the levels.

5. At present, IPA can return up to 500 functional annotations after analysis of a set of input genes. A function list, however, may include more than 500 combinations of category, function, and function annotation. Even so, the number of unique functional annotations will not exceed 500.

6. In our experience, the comparison analyses based on IPA-derived "Function" or "Function annotation" generally result in very similar data.

7. No spaces are allowed in the file name. The underscore symbol "_" can only be used to separate the platform name and the word "True" or "Random," and is absolutely forbidden for other uses. All the letters in file names can be in either upper or lower case.

8. "FunctionComparison.txt" is a plain text file and can be opened by any text software like Windows Notepad. In this file, the lines starting with "#" are annotations and will not be executed by R. All of the other lines are commands and will be executed. An alternative way to run the scripts is using the command source. Input "source ('path/FunctionComparison.txt')" in R console has the same effects as pasting R scripts directly into R console. Here, "path" is the full file path to the file "FunctionComparison.txt". Do not save the script file in the directory that holds function tables.

9. For POF calculation and graph generation, the number of top functions for comparison will be determined by the function list with the lower number of functions if the lists are different lengths. For example, if a comparison is made between list A (140 functions) and list B (150 functions), only the top 140 functions in both the lists will be compared. This rule also applies for comparison across multiple function lists.

Acknowledgments

The authors would like to thank Drs. Minjun Chen and Zhihua Xu in Division of Systems Biology, National Center for Toxicological Research, U.S. Food and Drug Administration for their enlightening comments and hearty discussions in reviewing the manuscript, and Dr. Lin Xie in Department of Aquaculture and Fisheries, University of Arkansas at Pine Bluff for her advice on the statistical methods used in this manuscript. The views presented in this chapter do not necessarily reflect those of the Food and Drug Administration.

References

1. Barrett JC, Kawasaki ES (2003) Microarrays: the use of oligonucleotides and cDNA for the analysis of gene expression. Drug Discov Today 8:134–141.
2. Holloway AJ, van Laar RK, Tothill RW et al (2002) Options available–from start to finish–for obtaining data from DNA microarrays II. Nat Genet 32:481–489.
3. Shi L, Reid LH, Jones WD et al (2006) The MicroArray Quality Control (MAQC) project shows inter- and intraplatform reproducibility of gene expression measurements. Nat Biotechnol 24:1151–1161.
4. Yauk CL, Berndt ML, Williams A et al (2004) Comprehensive comparison of six microarray technologies. Nucleic Acids Res 32:e124.
5. Tan PK, Downey TJ, Spitznagel EL Jr et al (2003) Evaluation of gene expression measurements from commercial microarray platforms. Nucleic Acids Res 31:5676–5684.
6. Barrett T, Troup DB, Wilhite SE et al (2009) NCBI GEO: archive for high-throughput functional genomic data. Nucleic Acids Res 37:D885-890.
7. Barrett T, Suzek TO, Troup DB et al (2005) NCBI GEO: mining millions of expression profiles – database and tools. Nucleic Acids Res 33:D562-566.
8. Li Z, Su Z, Wen Z et al (2009) Microarray platform consistency is revealed by biologically functional analysis of gene expression profiles. BMC Bioinformatics 10:S12.
9. Chen L, Mei N, Yao L et al (2006) Mutations induced by carcinogenic doses of aristolochic acid in kidney of Big Blue transgenic rats. Toxicol Lett 165:250–256.
10. Mei N, Arlt VM, Phillips DH et al (2006) DNA adduct formation and mutation induction by aristolochic acid in rat kidney and liver. Mutat Res 602:83–91.
11. Mengs U, Lang W, Poch J-A (1982) The carcinogenic action of aristolochic acid in rats. Archives of Toxicology 51:107–119.
12. Guo L, Lobenhofer EK, Wang C et al (2006) Rat toxicogenomic study reveals analytical consistency across microarray platforms. Nat Biotechnol 24:1162–1169.
13. http://www.ingenuity.com/.
14. http://www.r-project.org/.
15. IPA. (2009) Calculating and Interpreting the p-values for Functions, Pathways, and Lists in Ingenuity Pathways Analysis. Ingenuity Systems, Redwood City, CA, 94063, USA.
16. Sun H, Fang H, Chen T et al (2006) GOFFA: gene ontology for functional analysis – a FDA gene ontology tool for analysis of genomic and proteomic data. BMC Bioinformatics 7:S23.
17. http://cran.r-project.org/faqs.html, FAQ R.
18. http://www.fda.gov/ScienceResearch/BioinformaticsTools/Arraytrack/default.htm, ArrayTrack.

Chapter 10

Performance Comparison of Multiple Microarray Platforms for Gene Expression Profiling

Fang Liu, Winston P. Kuo, Tor-Kristian Jenssen, and Eivind Hovig

Abstract

With genome-wide gene expression microarrays being increasingly applied in various areas of biomedical research, the diversity of platforms and analytical methods has made comparison of data from multiple platforms very challenging. In this chapter, we describe a generalized framework for systematic comparisons across gene expression profiling platforms, which could accommodate both the available commercial arrays and "in-house" platforms, with both one-dye and two-dye platforms. It includes experimental design, data preprocessing protocols, cross-platform gene matching approaches, measures of data consistency comparisons, and considerations in biological validation. In the design of this framework, we considered the variety of platforms available, the need for uniform quality control procedures, real-world practical limitations, statistical validity, and the need for flexibility and extensibility of the framework. Using this framework, we studied ten diverse microarray platforms, and we conclude that using probe sequences matched at the exon level is important to improve cross-platform data consistency compared to annotation-based matches. Generally, consistency was good for highly expressed genes, and variable for genes with lower expression values, as confirmed by QRT-PCR. After stringent preprocessing, commercial arrays were more consistent than "in-house" arrays, and by most measures, one-dye platforms were more consistent than two-dye platforms.

Key words: Microarray, Gene expression profiling, Bioinformatics, Data consistency, Probe matching

1. Introduction

Gene expression microarray technology has matured significantly over the past decade, and its role has been extended from an experimental tool for basic science research to clinical practice (see reviews 1–4). However, the diversity of platforms and microarray data raise the questions of whether and how data from different platforms can be compared and combined. The results of cross-platform comparisons have been mixed, and were much debated in initial investigations before 2004, whereas increasing knowledge and control of the

Junbai Wang et al. (eds.), *Next Generation Microarray Bioinformatics: Methods and Protocols*,
Methods in Molecular Biology, vol. 802, DOI 10.1007/978-1-61779-400-1_10, © Springer Science+Business Media, LLC 2012

factors that result in poor correlation among the technologies has led to much higher levels of correlation among publications after 2004 (see review 5). By analyzing the previously published studies, we summarized the following factors that may bias microarray cross-platform data comparison (6): (a) nonidentical samples on different platforms; (b) samples not being sufficiently distinct; (c) samples processed using different protocols; (d) lack of technical replicates; (e) data preprocessing steps not being standardized; (f) few types of platforms being directly compared; (g) measurements being matched using probe annotations; (h) "agreement" not unambiguously quantified, or (i) insufficient biological validation.

While some of the above conditions may be reflective of the anticipated use of these platforms in practice, they complicate assessing the magnitude of the disagreement attributable to the platforms. The user community of microarray technology has clearly expressed the need for well-controlled large-scale comparison studies. Recently, several large efforts to create standardized protocols for microarray experiments (from probe annotation to data analysis) were initiated, such as Minimum Information About a Microarray Experiment (MIAME) standards (7–9), the External RNA Controls Consortium (ERCC) (10, 11), and the Microarray Quality Control (MAQC) project (12, 13), aiming at quality improvement of microarray data through standardization (also see review 14).

We here present a comprehensive bioinformatics framework (see Fig. 1) for large-scale cross-platform comparison, and an example study which includes data from ten different mouse microarray platforms, as well as from different laboratories performing on the same microarray platform. This dataset is available in gene expression omnibus (GEO) (15) with accession number GSE4854. We considered the influencing factors as listed above and the best way to control each factor, when designing the biological experiments and the data analysis framework. The study included single- and dual-dye platforms, cDNA and oligonucleotide microarrays, and both commercial and "in-house" fabricated microarrays. Biological samples consisted of two pooled RNA samples of mouse retina (MR) and mouse cortex (MC), prepared by the same laboratory (16) and distributed to all participating laboratories. Following recent studies (17–19), we used probe sequence information to map probes both on levels of genes and exons to improve the stringency with which measurements are compared across platforms. For the data analyses, we combined well-described, commonly used and publicly available analytical approaches in a framework that can be used every time the reliability of a new platform needs to be assessed.

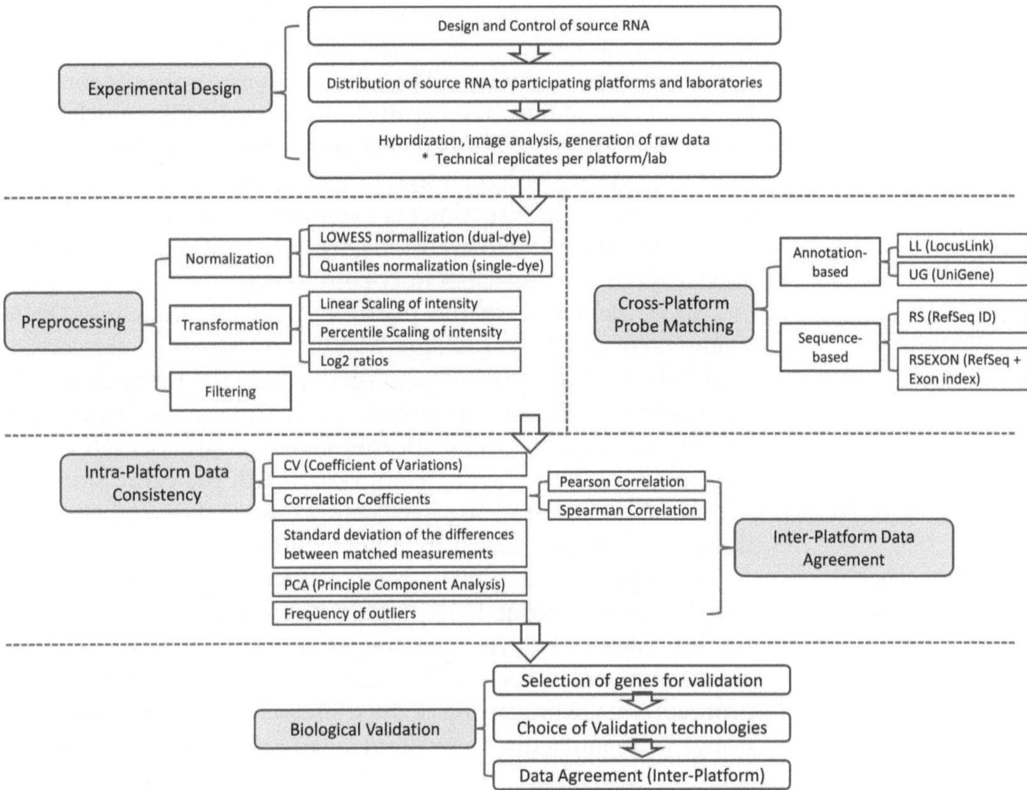

Fig. 1. Flowchart of the framework of microarray platform comparison study. The workflow contains, generally speaking, six modules: experimental design, data preprocessing, cross-platform probe-matching, intraplatform data consistency, and interplatform data agreement, and biological validation.

2. Methods

2.1. Quality Control on Biological Material

We minimized the biological bias by applying centralized preparation and quality control of biological sample in one laboratory. RNA samples used for all platforms were aliquoted from two pools of samples: C57/B6 adult mouse retina (MR) and Swiss-Webster postnatal day one (P1) mouse cortex (MC). MR and MC were chosen due to their availability and biological interest. MR samples were obtained from a pool of C57/B6 mice ($n = 350$) and MC were obtained from P1 Swiss-Webster mice ($n = 19$) (see Note 1). The mouse cortex was used as a reference sample for the dual-dye platforms. The total RNA from both samples was stored at −80°C.

2.2. Microarray Platforms and Dataset

Ten microarray platforms in this demonstrative study were: Affymetrix, Agilent, Applied Biosystems (ABI), Amersham (now GE Healthcare), Compugen (now Sigma-Genosys), Mergen, MWG BioTech (now Ocimum Biosolutions), Operon, "academic cDNA" arrays provided from the Cepko Laboratory, and "MGH long oligo" arrays long oligonucleotide arrays from Massachusetts General Hospital (MGH). The first eight platforms are commercially available, and the last two are custom-made. Oligonucleotides from both Compugen and Operon were printed together onto the same slide. A total of eight research laboratories were involved in this collaboration. The experiments on three platforms, Affymetrix, Amersham, and Mergen, were repeated at two different laboratories and analyzed for cross-laboratory consistency.

Six of the ten microarray platforms (Agilent, academic cDNA, Compugen, MGH long oligo, MWG, and Operon) are considered to be two-dye platforms, as they require the hybridization of two samples, whereas the others (ABI, Affymetrix, Amersham, and Mergen) are one-dye platforms. Five replicates of each sample were used to assess the degree of variation in the expression data within each platform (20) (see Note 2). A total of 91 hybridizations were completed and are reported in this study.

Each participating laboratory received aliquots of the RNA samples: mouse retina (MR) and mouse cortex (MC), from the Cepko laboratory. All labeling and hybridization methods were completed as specified by each manufacturer's hybridization protocol. Image processing of the scanned images were conducted using the manufacturer's recommended scanners and settings.

2.3. Data Quality Examination Using Visualization and Descriptive Statistics

We recommend using some data visualization techniques and descriptive statistics for preliminary investigation of the quality of microarray dataset. Descriptive statistics can include the mean, standard deviation, minimum and maximum of signal intensities, which shows if there is any obvious outlier. Some chosen percentiles, such as 5, 25, 75, 95% etc., are also good indicators showing whether the signal distribution was correct or abnormally skewed. To visualize microarray data, the commonly used methods are intensity scatter plot (i.e., intensities of sample 1 vs. sample 2), histogram or box-plot of intensity distributions, M-A plot, etc.

In our study, we examined our dataset using all these techniques as quality control tools, and ensured there was no artifact being introduced when conducting each microarray experiment. The R programming language/environment (21) was used for generating descriptive statistics and graphics. Unless specified otherwise, R was also used in the following data analysis work.

2.4. Data Preprocessing: Filtering

The filtering criteria chosen in this study were either recommended by the vendors or have been broadly adopted by the research community (see Note 3). For Affymetrix and Amersham, probe set and spot quality flags were referenced, where only "present" and "good" calls were adopted, respectively. The signal-to-noise ratio (SNR) threshold of 3 was used for ABI, in addition to removal of flagged spots as recommended by the vendor. A SNR threshold was set to 2 for Agilent, Compugen, Mergen, and Operon platforms. For academic cDNA, MGH long oligo, and MWG arrays, the images were scanned using GenePix software 3.0 (see Note 4). The software automatically generated flags at default settings for poor and missing spots, which were removed. We wrote Perl scripts to implement the above-mentioned filtering criteria in parsing the data files for each platform.

Stringent filtering for spot quality has been reported to improve consistency across different platforms (22, 23). This is also verified in our results (see Fig. 2).

		Affymetrix		Amersham		Mergen		ABI		cDNA Academic		MGH		MWG		Agilent		Compugen		Operon	
UG	Affymetrix	9341	1	0.84	0.84	0.85	0.87	0.84	0.85	0.23	0.29	0.79	0.80	0.66	0.70	0.77	0.79	0.66	0.67	0.79	0.81
	Amersham	6119	0.76	9750	1	0.82	0.84	0.83	0.84	0.24	0.30	0.74	0.75	0.62	0.67	0.74	0.75	0.60	0.61	0.77	0.80
	Mergen	5892	0.71	6575	0.73	8505	1	0.81	0.86	0.26	0.31	0.78	0.78	0.64	0.68	0.74	0.74	0.65	0.65	0.79	0.81
	ABI	4055	0.79	4438	0.79	4118	0.73	14310	1	0.24	0.29	0.76	0.77	0.60	0.68	0.73	0.74	0.62	0.65	0.77	0.80
	cDNA Academic	1682	0.31	1665	0.28	1341	0.32	4674	0.29	7517	1	0.25	0.31	0.24	0.36	0.22	0.26	0.08	0.22	0.24	0.30
	MGH	6496	0.68	7796	0.64	6715	0.58	5560	0.66	2060	0.30	12513	1	0.59	0.61	0.69	0.68	0.58	0.60	0.68	0.69
	MWG	5980	0.60	6764	0.59	7078	0.54	4322	0.62	1486	0.35	7469	0.53	8768	1	0.60	0.68	0.49	0.56	0.58	0.64
	Agilent	4972	0.67	4988	0.67	4169	0.65	3400	0.70	1765	0.25	5951	0.63	4534	0.67	8509	1	0.56	0.54	0.69	0.69
	Compugen	1593	0.39	1776	0.38	1931	0.35	1108	0.38	365	0.05	1678	0.34	1822	0.34	971	0.35	2022	1	0.65	0.69
	Operon	6064	0.64	7494	0.65	6600	0.59	4604	0.63	1634	0.26	8546	0.51	6647	0.49	4807	0.58	1877	0.35	11711	1
LL	Affymetrix	8384	1	0.84	0.84	0.85	0.86	0.84	0.87	0.26	0.30	0.80	0.81	0.66	0.70	0.77	0.79	0.65	0.66	0.79	0.80
	Amersham	6168	0.76	9769	1	0.81	0.83	0.83	0.86	0.25	0.29	0.74	0.76	0.61	0.67	0.74	0.75	0.60	0.61	0.77	0.79
	Mergen	5927	0.70	6615	0.73	8971	1	0.82	0.85	0.27	0.30	0.78	0.79	0.63	0.68	0.74	0.73	0.64	0.64	0.79	0.81
	ABI	6450	0.80	7825	0.80	7681	0.73	15142	1	0.24	0.29	0.78	0.79	0.61	0.68	0.74	0.76	0.62	0.67	0.78	0.80
	cDNA Academic	2842	0.31	3243	0.28	2967	0.31	4750	0.28	6453	1	0.25	0.32	0.23	0.33	0.21	0.24	0.17	0.31	0.24	0.28
	MGH	4986	0.69	6072	0.65	6037	0.58	7309	0.69	2818	0.30	7931	1	0.61	0.62	0.70	0.69	0.56	0.59	0.70	0.72
	MWG	5933	0.61	6818	0.59	7359	0.55	7514	0.63	2899	0.31	5809	0.54	8689	1	0.60	0.68	0.49	0.57	0.59	0.63
	Agilent	4526	0.68	5019	0.68	4387	0.65	6573	0.71	3476	0.24	4311	0.64	4522	0.67	8757	1	0.55	0.53	0.70	0.70
	Compugen	1633	0.39	1819	0.37	2005	0.35	1975	0.41	742	0.15	1620	0.34	1939	0.35	1039	0.33	2103	1	0.65	0.70
	Operon	6259	0.63	7757	0.64	7051	0.59	8942	0.65	3536	0.24	6320	0.54	7265	0.49	5288	0.58	2025	0.36	10976	1
RS	Affymetrix	4747	1	0.86	0.87	0.89	0.91	0.89	0.91	0.29	0.30	0.80	0.81	0.64	0.68	0.81	0.81	0.62	0.57	0.80	0.83
	Amersham	3267	0.78	7930	1	0.83	0.85	0.84	0.87	0.21	0.33	0.74	0.77	0.60	0.67	0.75	0.76	0.62	0.62	0.77	0.80
	Mergen	3243	0.74	5280	0.75	7051	1	0.86	0.89	0.30	0.34	0.78	0.78	0.64	0.71	0.80	0.79	0.63	0.64	0.80	0.82
	ABI	3058	0.84	4273	0.80	3928	0.79	9450	1	0.24	0.25	0.77	0.79	0.60	0.68	0.78	0.78	0.62	0.58	0.80	0.83
	cDNA Academic	468	0.35	528	0.31	493	0.37	638	0.29	964	1	0.31	0.34	0.19	0.32	0.18	0.23	0.19	0.18	0.26	0.31
	MGH	2661	0.67	4659	0.66	4136	0.59	4506	0.68	466	0.30	7699	1	0.59	0.62	0.70	0.68	0.58	0.63	0.70	0.72
	MWG	2036	0.59	3484	0.60	3452	0.61	2557	0.64	302	0.36	2990	0.55	4624	1	0.59	0.66	0.48	0.56	0.59	0.62
	Agilent	2712	0.72	3638	0.69	3115	0.72	3596	0.73	579	0.25	3146	0.64	1966	0.66	6251	1	0.59	0.53	0.72	0.72
	Compugen	874	0.36	1386	0.37	1421	0.37	935	0.39	118	0.07	1035	0.40	955	0.33	693	0.33	1704	1	0.66	0.63
	Operon	3658	0.66	5997	0.65	5368	0.61	4835	0.66	549	0.30	4990	0.55	3768	0.50	3713	0.61	1389	0.33	8313	1
RSEXON	Affymetrix	4869	1	0.88	0.89	0.90	0.91	0.90	0.92	0.27	0.28	0.81	0.82	0.67	0.71	0.82	0.82	0.69	0.65	0.81	0.83
	Amersham	2093	0.81	7996	1	0.85	0.87	0.86	0.88	0.22	0.38	0.74	0.79	0.60	0.66	0.76	0.77	0.63	0.62	0.79	0.82
	Mergen	2712	0.75	2771	0.78	7216	1	0.88	0.90	0.30	0.36	0.78	0.77	0.65	0.71	0.81	0.80	0.64	0.67	0.81	0.82
	ABI	2868	0.84	2788	0.83	3178	0.81	9736	1	0.25	0.28	0.76	0.79	0.61	0.68	0.79	0.79	0.64	0.55	0.81	0.84
	cDNA Academic	416	0.34	338	0.34	368	0.40	544	0.30	987	1	0.41	0.38	0.14	0.14	0.17	0.24	0.21	0.21	0.27	0.34
	MGH	748	0.65	1305	0.66	1016	0.62	1879	0.67	138	0.40	7861	1	0.54	0.63	0.73	0.72	0.62	0.68	0.68	0.74
	MWG	1441	0.61	1632	0.59	2070	0.62	1803	0.63	182	0.19	934	0.57	4656	1	0.59	0.65	0.48	0.53	0.58	0.63
	Agilent	2488	0.73	1955	0.68	2295	0.74	3165	0.74	494	0.26	646	0.68	1184	0.65	6529	1	0.64	0.61	0.73	0.73
	Compugen	682	0.38	711	0.37	903	0.38	721	0.39	89	0.08	297	0.44	583	0.30	444	0.40	1712	1	0.73	0.64
	Operon	3374	0.67	3516	0.67	4080	0.61	4177	0.67	449	0.34	1418	0.52	2483	0.49	2986	0.63	987	0.37	8532	1

Fig. 2. Summary of the interplatform performance measures (including probe-matching statistics, correlation coefficients with and without filtering). This figure lists the interplatform data correlation results when using various probe-matching approaches (including LL, UG, RS, and RSEXON-based matches). For a given probe-matching scheme, each pair of platforms corresponds to four numbers, two at the upper-triangle (above the diagonal) are the correlation coefficients (from left to right, Pearson and Spearman correlation, respectively) of the two platforms with filtered data; two at the lower-triangle (below the diagonal) are the probe-matching statistics (to the left) and the Pearson correlation coefficient without data filtering.

2.5. Data Preprocessing: Normalization

Normalization methods were chosen based on the past microarray studies that have indicated their maturity and potential advantages over other methods in single and dual-dye platforms (24–26). For single-dye platforms, quantile normalization (25) was applied, where ten arrays (five for MR and five for MC) were considered as one group; two-dye platforms were normalized using locally weighted scatterplot smoothing (LOWESS) normalization (24, 26) (see Notes 5 and 6). The "affy" and "marray" packages from Bioconductor were used, respectively.

2.6. Data Preprocessing: Scaling Transformation for Comparison of Raw Intensities

We suggest using two scaling transformations, linear scaling and percentile scaling, to compare raw intensities quantified by different software packages. These two methods were applied to all platforms for different purposes of comparison. Linear scaling was used when we measure intraplatform coefficient of variations of the intensities, whereas percentile transformation was mainly used in the interplatform comparisons (see Note 7).

Linear scaling was performed such that for each slide, within each channel, the minimum and maximum of intensities were transformed to 1 and 100, respectively. Then, all other intensity measurements were linearly mapped to an analogous number within the range of [1, 100]. Percentile transformation projected the data to a hundred discrete levels (i.e., 1–100) according to percentiles of the intensity values, i.e., for each slide, within each channel, 100‰ define 100 intervals of intensities, then the intensities falling in the interval between the $(N-1)$th percentile and the Nth percentile may be transformed to N.

2.7. Data Preprocessing: Calculating Log_2 Ratios

Log_2 ratios were computed to allow the comparison of single-dye and two-dye platforms. Five log_2 ratios were obtained from five technical replicates of each two-dye platform, and from five randomly paired arrays across samples without replacement for each single-dye platform. The averaged log_2 ratios of technical replicates for each platform were used to assess interplatform variation.

2.8. Probe Matching

We demonstrate two approaches of gene matching: annotation-based and sequence-based. For the annotation-based approach, MatchMiner (27) was used to map UniGene (UG) clusters and LocusLink (LL) identifiers by using GenBank accession numbers that were provided by each platform (see Note 8).

For the sequence-based approach, the probe sequences from each microarray platform were mapped to the mouse genome using the BLAT stand-alone program (28), based on the February 2003 version of the mouse reference sequences (mm3) downloaded from the UCSC Genome Site (29). The context sequences for Affymetrix, which is 255 base pairs correspond to the length of the sequences spanned by the 11 probe pairs of each gene, were obtained from their NetAffx analysis center (30). ABI provided us with 180 bps long sequences where the actual 60-mer probe for each gene lies within.

The probes from different platforms were matched both at the gene level by RefSeq identifiers (RS) and on the exon level by RefSeq exon (RSEXON). "Probe-to-exon" match meant only aligned sequences positioned completely within an exon were considered as a match. If multiple within-exon matches for a probe sequence occurred, the best match in terms of the length of "hit" was selected. If no match was found, that probe was excluded. If there was more than one probe that matched to a particular identifier, the expression values were averaged. In most instances, however, each gene was represented by only one probe on all platforms.

The probe-matching statistics is shown in Fig. 2.

2.9. Data Consistency Measurements

In terms of the measurements of data consistency in this framework, we applied the two commonly used indexes: coefficients of variations (CV) and correlation coefficients. In addition, a few other measurements, such as standard deviations of the difference between matched expressions, principal component analysis (PCA), plot of correspondence at the top (CAT), and the degree of deviations by defining outliers across various platforms' measurements for each gene, were used to help us corroborate our conclusion of the comparison.

The CV measures the reproducibility among multiple replicate experiments within each platform. Besides the use of CV on channel-specific intensities, we also defined a segmental function for the CV of $\log_2$ ratios (see Note 9). When the mean of $\log_2$ ratios was between -1 and $+1$, the CV equals the standard deviation, otherwise, the conventional definition of CV (the variation among multiple measurements in proportion to their mean) was applied. The CV of our dataset indicated very good within-platform data consistency for all platforms (data not shown).

Pearson and Spearman correlation coefficients were calculated for both intra- and interplatform comparisons. Intraplatform correlations consisted of computing the correlations for both linearly transformed intensities within each sample and their $\log_2$ ratios (data not shown). For interplatform comparisons, the correlations were calculated based on the averaged $\log_2$ ratios. In our data, the two correlation coefficients showed general agreement (see Fig. 2), indicating the distribution of experimental data as expected.

Standard deviation (SD) of the differences between matched measurements was applied to technical replicates in the case of intraplatform agreement and to cross-platform matched measurements. For sequence-based matching, we only considered the measurements being matched across at least six platforms, among which the four most widely used platforms ABI, Affymetrix, Agilent, and Amersham are present.

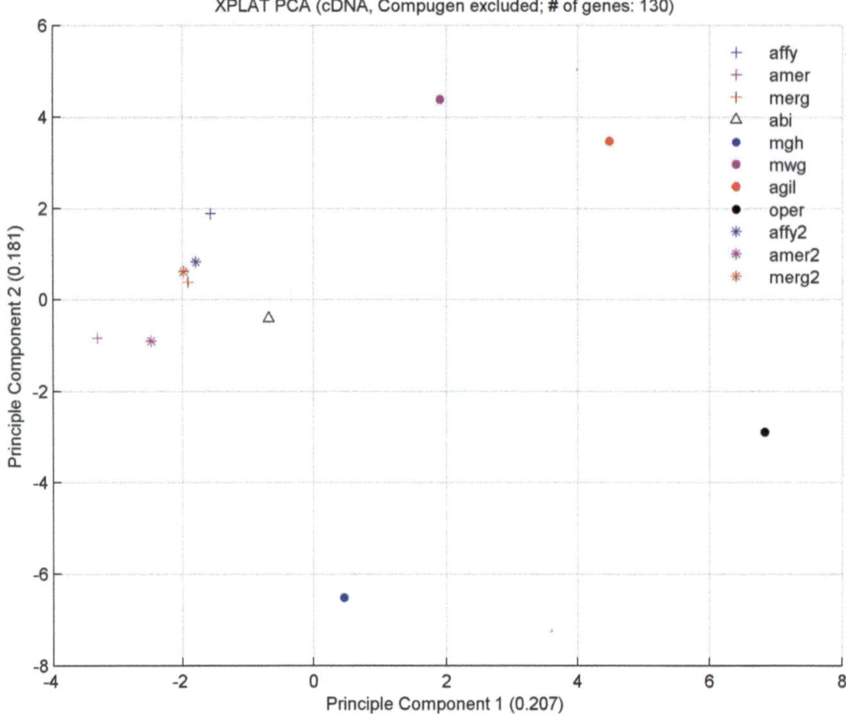

Fig. 3. Principal component plot of gene expression measurements from eight platforms. Principal component analysis (PCA) was used on eight microarray platforms including three in which measurements originated in two different laboratories (Affymetrix, Amersham, and Mergen), but excluding academic cDNA and Compugen as there were few matched probes for these platforms. The numbers in parenthesis on the *x*- and *y*-axis label give the percentage indicating the variance accounted for the first and second principal components, respectively. A total of RS-matched 130 probes were used in this analysis.

The frequency of outliers for each platform examines the degree of variation of each platform from the rest. For a given gene that has been measured in at least five platforms, if a platform's measurement lies outside of the range of the mean expression ratios ± one standard deviation, it was identified as an outlier.

We performed PCA on the probe-matched dataset of all platform-laboratory combinations, in order to identify, which platforms are closely correlated and which are more distant from the others. Figure 3 gives an intuitive indication of agreement between datasets. This analysis was conducted after standardization so that each gene has a zero mean and unit standard deviation.

Furthermore, we also found CAT plots (31) to assess cross-platform agreement useful. This method was proposed based on an educated and proved assumption that higher gene expression measurements tend to be more reliable and reproducible than lower ones. CAT plots were generated using the top 200 genes for up- and downregulated genes, as shown in Fig. 4a, b, respectively, using filtered normalized $\log_2$ ratios on the RSEXON-matched expression measurements.

2.10. Biological Validations Using QRT-PCR

In our study, we considered the following criteria for gene selection for biological validation: (1) genes, based on RSEXON match, present in at least six platforms (must include: ABI, Affymetrix, Agilent, and Amersham); (2) the expression of genes span the dynamic range, based on the percentile-transformed intensity, from the high expression group (67–100 percentiles), medium (34–66 percentiles), and low (1–33 percentiles); and (3) some genes were chosen for validation due to their disagreement of the microarray measurements. In total, 165 genes were selected based on this criterion.

Among these, 74 and 91 genes were conducted, using Roche LightCyclers® and TaqMan® Gene Expression Assays, respectively, on the identical samples used for the microarray experiments (see Note 10). Expression ratios measured by QRT-PCR were

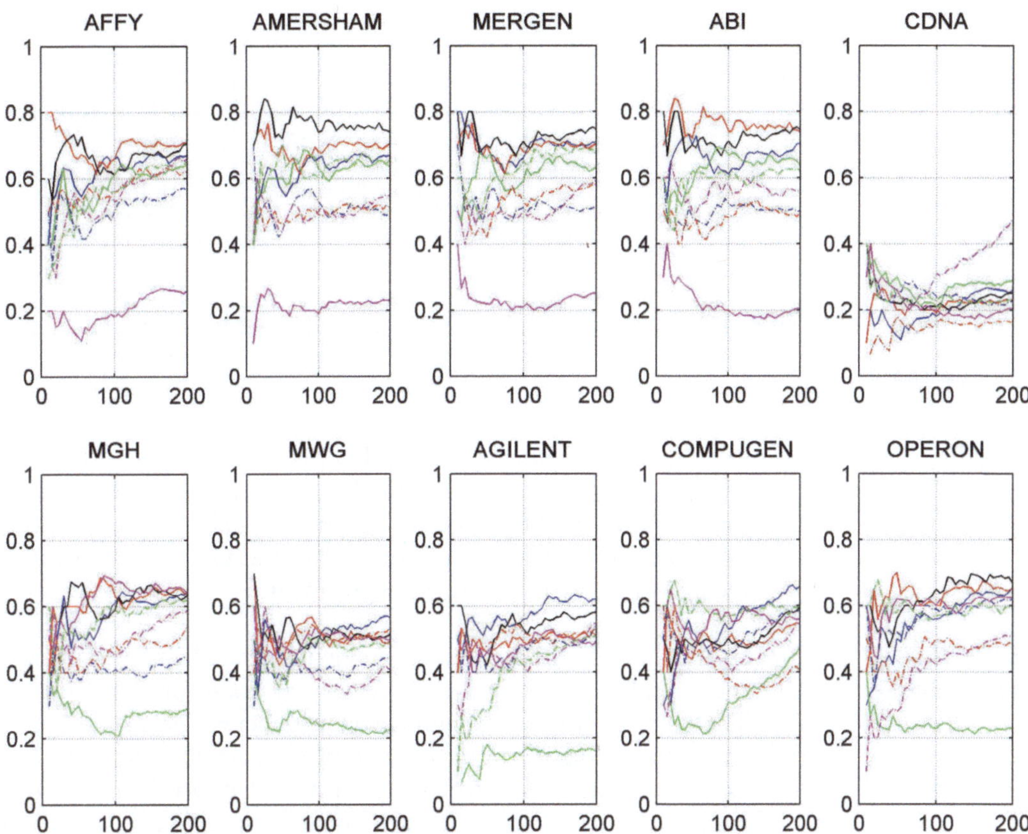

Fig. 4. Assessment of cross-platform agreement of RSEXON-matched data using CAT plots. CAT plots were generated using RSEXON-matched normalized $\log_2$ ratios (filtered) for (**a**) up- and (**b**) downregulated genes. The list sizes were chosen to be from 10 to 200, with an increment of 5. The platform used for reference is listed at the top of each plot. The color and each line type correspond to a particular platform. The "*blue solid*," "*red solid*," "*black solid*," "*magenta solid*," "*green solid*," "*blue dash*," "*red dash*," "*black dash*," "*magenta dash*," and "*green dash*" correspond to "Affymetrix," "Amersham," "Mergen," "ABI," "academic cDNA," "MGH long oligo," "MWG," "Agilent," "Compugen," and "Operon" platforms, respectively.

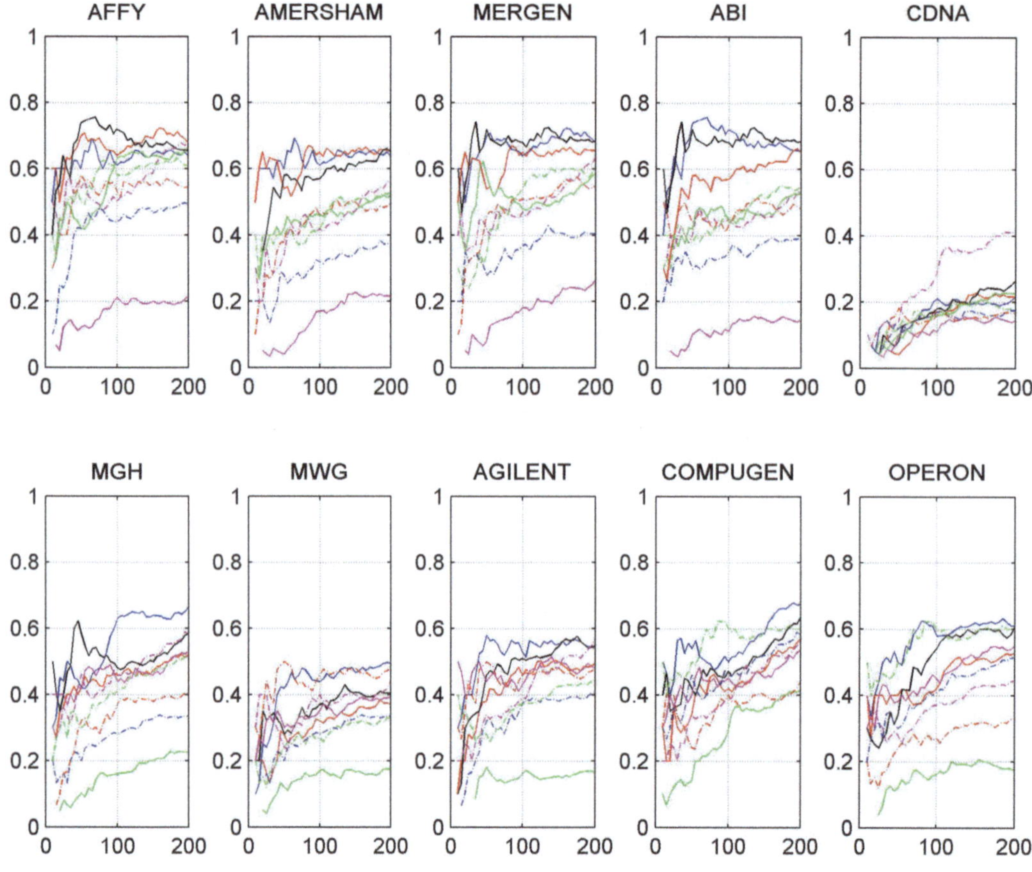

Fig. 4. (continued).

calculated as follows: $\log_2$ ratio(MR/MC) $= -(C_{t\,MR}{}' - C_{t\,MC}{}')$, where $C_{t\,MR}{}'$ and $C_{t\,MC}{}'$ correspond to the mean cycle thresholds for mouse retina and mouse cortex, respectively. The Pearson correlation coefficient between microarray data and QRT-PCR measurements was used to evaluate data agreement.

2.11. Results and Conclusion

In this example study, our results demonstrated that, first of all, the intraplatform data consistency is very good for all platforms. The cross-platform data agreement is generally good, especially when the biological sample is identical and data filtering is applied. One-dye platforms out-performed two-dye platforms in our study. And, when applying each vendor's specific protocols and image analysis methods, the commercial microarray vendors had better data consistency than the academic in-house arrays in general. We tested our four probe-matching strategies for pairing up the gene expression measurements between different platforms. Among them, the two sequence-based methods yielded better results than the annotation-based methods, and the most stringent approach based on RSEXON matching results in the best cross-platform data agreement. When validating the microarray measurements by QRT-PCR, the QRT-PCR results were in good agreement with most of the microarray platforms, except the academic cDNA arrays. We confirmed that the genes of higher expression have more reproducible measurements than those of lower expression.

In our opinion, the key factors to a successful microarray cross-platform comparison study is: (a) to minimize possible biological bias in sample preparation; (b) to follow each microarray vendor's recommended protocols in data generation (including experiments, image analysis, data preprocessing), but avoid using any specialized methods favoring a particular platform; (c) to utilize up-to-date sequence-based probe-matching strategy; (d) to apply as many as possible various measures of data agreement because different measure investigates on different aspects of data quality and characteristics, and (e) to draw conclusion by considering over all measures. We believe our proposed framework has all these attributes, as well as good flexibility to include new platforms as they emerge.

3. Notes

1. The choice of biological sample: For the general usefulness of the comparison, the RNA samples should be selected from a commonly used organism, and should have a diverse set of transcripts covering a wide expression range. Some commercial universal reference RNA sources (such as products from Ambion (32) and Stratagene (33)) may be a good choice. They were, however, not available at the time of our study. We extracted RNA from tissues of cortex and retina from the well-studied *Mus musculus*, because these tissues have broad gene expression profiles and some well-known tissue-specific transcripts (34, 35). Inbred mice were selected to eliminate genetic variability, and pooling tissue from many animals minimized the biological variations within tissue RNA preparations. Both tissue samples can be considered as replenishable sources of RNA with little variability, as observed by laser-based capillary electrophoresis of labeled samples.
2. Number of technical replicates: The number five was chosen as a reasonable compromise between the wish to reduce the effect of array-to-array variability and resource limitations.
3. Filtering criteria: Due to the diversity of the technical approaches of the various platforms, different scanners with their proprietary image analysis algorithms were used, and this

limited our ability to apply the same filtering criteria to all the platforms. In spot quality filtering procedures, we chose to prioritize the quality flags generated by image analysis software according to recommendations from the platform vendors. Our results demonstrated that stringent spot quality filtering can improve data consistency, confirming reports of previous studies (22, 23).

4. Scanner saturation and dual-scan procedure: Scanner saturation was observed for some experiments in some platforms. It is difficult to assess to what extent the limitations of scanner intensity ranges influenced the comparisons reported. However, a dual-scan procedure was tested for one platform having saturation, but did not result in better agreement (data not shown). Such observations emphasize the need for careful design of cross-platform protocols and performance tuning throughout the execution of the experimental procedures.
5. Lack of external spikes common across all platforms: Each platform may have a proprietary set of quality control features, including external spikes, alien probes, and positive and negative controls. Such features were not present in all platforms, thus affect their inclusion for comparison purposes. This reflects the current usage of these platforms in laboratory environments.
6. Normalization of Compugen and Operon: The oligonucleotide probes from Compugen and Operon were printed onto the same slide. LOWESS normalization was performed on the whole chip before the two sets of probe measurements were separated and analyzed in the study. However, we also examined and confirmed that when this normalization was performed for each platform independently, the results were similar (data not shown).
7. Linear scaling vs. percentile scaling: Differences in technical and instrumental choices among platforms, such as image analysis algorithms, make direct comparisons based on raw intensity signals impossible. The two scaling transformations aim to bring the signal ranges to a uniform scale to compensate for differences in signal intensity ranges between platforms. This was found useful in comparing intraplatform variations. Beyond this, percentile scaling can also correct the artifacts introduced by different distribution characteristics among various platforms, as well as purposefully neglect some minor fluctuations in expression levels.
8. Annotation-based probe match vs. sequence-based probe match: The agreement between platforms on matched data tended to increase with increasing mapping specificity, i.e., in

the following order: (annotation-based) UG, LL, (sequence-based) RS, RSEXON. A possible interpretation is that the RefSeq mapping eliminates biases due to splice variants, being on the transcript level, and that the RSEXON mapping possibly forces the probes of different platforms to be more similar, as they are confined to a limited region of each gene.

9. Segmental CV: This measure effectively avoid including small denominators to distort the CVs considerably when a large proportion of probes having a mean of $\log_2$ ratio close to zero are expected in microarray experiments.
10. Biological validation: Overall, the microarray results were in agreement with QRT-PCR for genes with medium and high expression, while there was little agreement for genes with lower or variable expression. We interpret this as stochastic variation appearing at low transcript numbers in both microarrays and validation procedures. We also found evidence for the importance of careful primer design when using QRT-PCR, as the results from TaqMan were more consistent than those from Universal ProbeLibrary. For the former, primers had been designed to be on the same exon as the microarray probes. This was not enforced for the latter, where the primers were designed to be optimal for their kit using proprietary software. The differences in measurements of the two QRT-PCR methods suggest that the use of QRT-PCR for biological validations must be carried out carefully.

Acknowledgments

The authors would like to thank all the microarray vendors and facilities/laboratories which have actively participated this large-scale study. The authors were supported by the functional genomics program (FUGE) in the Research council of Norway for this work.

References

1. Bauer JW, Bilgic H, Baechler EC (2009) Gene-expression profiling in rheumatic disease: tools and therapeutic potential. Nat Rev Rheumatol 5:257–265.
2. Cheang MC, van de Rijn M, Nielsen TO (2008) Gene expression profiling of breast cancer. Annu Rev Pathol 3:67–97.
3. Garcia-Escudero R, Paramio JM (2008) Gene expression profiling as a tool for basic analysis and clinical application of human cancer. Mol Carcinog 47:573–579.
4. Giordano TJ (2008) Transcriptome analysis of endocrine tumors: clinical perspectives. Ann Endocrinol (Paris) 69:130–134.

5. Yauk CL, Berndt ML (2007) Review of the literature examining the correlation among DNA microarray technologies. Environ Mol Mutagen 48:380–394.
6. Kuo WP, Liu F, Trimarchi J et al (2006) A sequence-oriented comparison of gene expression measurements across different hybridization-based technologies. Nat Biotechnol 24:832–840.
7. Brazma A (2009) Minimum Information About a Microarray Experiment (MIAME) – successes, failures, challenges. Scientific World Journal 9:420–423.
8. Brazma A, Hingamp P, Quackenbush J et al (2001) Minimum information about a microarray experiment (MIAME) – toward standards for microarray data. Nat Genet 29:365–371.
9. MIAME. (Minimum Information About a Microarray Experiment) http://www.mged.org/Workgroups/MIAME/miame.html.
10. Baker SC, Bauer SR, Beyer RP et al (2005) The External RNA Controls Consortium: a progress report. Nat Methods 2:731–734.
11. ERCC. (The External RNA Controls Consortium) http://www.cstl.nist.gov/biotech/Cell&TissueMeasurements/GeneExpression/ERCC.htm.
12. Shi L, Reid LH, Jones WD et al (2006) The MicroArray Quality Control (MAQC) project shows inter- and intraplatform reproducibility of gene expression measurements. Nat Biotechnol 24:1151–1161.
13. MAQC. (Microarray Quality Control) http://www.fda.gov/nctr/science/centers/toxicoinformatics/maqc/.
14. Enkemann SA (2010) Standards affecting the consistency of gene expression arrays in clinical applications. Cancer Epidemiol Biomarkers Prev 19:1000–1003.
15. GEO. (Gene Expression Omnibus) http://www.ncbi.nlm.nih.gov/geo/.
16. The Cepko Laboratory at Harvard Medical School (http://genetics.med.harvard.edu/~cepko/).
17. Carter SL, Eklund AC, Mecham BH et al (2005) Redefinition of Affymetrix probe sets by sequence overlap with cDNA microarray probes reduces cross-platform inconsistencies in cancer-associated gene expression measurements. BMC Bioinformatics 6:107.
18. Mecham BH, Klus GT, Strovel J et al (2004) Sequence-matched probes produce increased cross-platform consistency and more reproducible biological results in microarray-based gene expression measurements. Nucleic Acids Res 32:e74.
19. Mecham BH, Wetmore DZ, Szallasi Z et al (2004) Increased measurement accuracy for sequence-verified microarray probes. Physiol Genomics 18:308–315.
20. Lee ML, Kuo FC, Whitmore GA et al (2000) Importance of replication in microarray gene expression studies: statistical methods and evidence from repetitive cDNA hybridizations. Proc Natl Acad Sci U S A 97:9834–9839.
21. The R Project for Statistical Computing: http://www.r-project.org/.
22. Pounds S, Cheng C (2005) Statistical development and evaluation of microarray gene expression data filters. J Comput Biol 12:482–495.
23. Shippy R, Sendera TJ, Lockner R et al (2004) Performance evaluation of commercial short-oligonucleotide microarrays and the impact of noise in making cross-platform correlations. BMC Genomics 5:61.
24. Berger JA, Hautaniemi S, Jarvinen AK et al (2004) Optimized LOWESS normalization parameter selection for DNA microarray data. BMC Bioinformatics 5:194.
25. Bolstad BM, Irizarry RA, Astrand M et al (2003) A comparison of normalization methods for high density oligonucleotide array data based on variance and bias. Bioinformatics 19:185–193.
26. Workman C, Jensen LJ, Jarmer H et al (2002) A new non-linear normalization method for reducing variability in DNA microarray experiments. Genome Biol 3: research0048.
27. Bussey KJ, Kane D, Sunshine M et al (2003) MatchMiner: a tool for batch navigation among gene and gene product identifiers. Genome Biol 4:R27.
28. Kent WJ (2002) BLAT – the BLAST-like alignment tool. Genome Res 12:656–664.
29. UCSC Genome Site: http://www.genomearchive.cse.ucsc.edu/goldenPath/mmFeb2003/bigZips/.
30. Liu G, Loraine AE, Shigeta R et al (2003) NetAffx: Affymetrix probesets and annotations. Nucleic Acids Res 31:82–86.
31. Irizarry RA, Warren D, Spencer F et al (2005) Multiple-laboratory comparison of microarray platforms. Nat Methods 2:345–350.
32. Ambion: http://www.ambion.com/catalog/CatNum.php?6050.
33. Stratagene: http://www.stratagene.com/manuals/740000.pdf.

34. Blackshaw S, Fraioli RE, Furukawa T et al (2001) Comprehensive analysis of photoreceptor gene expression and the identification of candidate retinal disease genes. Cell 107:579–589.

35. Blackshaw S, Harpavat S, Trimarchi J et al (2004) Genomic analysis of mouse retinal development. PLoS Biol 2:E247.

Chapter 11

Integrative Approaches for Microarray Data Analysis

Levi Waldron, Hilary A. Coller, and Curtis Huttenhower

Abstract

Microarrays were one of the first technologies of the genomic revolution to gain widespread adoption, rapidly expanding from a cottage industry to the source of thousands of experimental results. They were one of the first assays for which data repositories and metadata were standardized and researchers were required by many journals to make published data publicly available. Microarrays provide high-throughput insights into the biological functions of genes and gene products; however, they also present a "curse of dimensionality," whereby the availability of many gene expression measurements in few samples make it challenging to distinguish noise from true biological signal. All of these factors argue for integrative approaches to microarray data analysis, which combine data from multiple experiments to increase sample size, avoid laboratory-specific bias, and enable new biological insights not possible from a single experiment. Here, we discuss several approaches to integrative microarray analysis for a diverse range of applications, including biomarker discovery, gene function and interaction prediction, and regulatory network inference. We also show how, by integrating large microarray compendia with diverse genomic data types, more nuanced biological hypotheses can be explored computationally. This chapter provides overviews and brief descriptions of each of these approaches to microarray integration.

Key words: Microarray, Meta-analysis, Bioinformatics, Coexpression, Functional interaction networks, Biomolecular networks, Bayesian networks, Regulatory networks, Protein function prediction, MEFIT, COALESCE

1. Introduction

A single microarray, like any experimental assay, takes place under a specific set of relevant environmental conditions: temperature, media, pH, strain, source tissue, growth protocol, and so forth. The power of genome-scale assays (see Note 1) is to capture a snapshot of molecular activity spanning many or all of a system's genes under one particular condition. The metadata describing these conditions can thus be considered as part of the experimental results themselves. This has driven the flurry of activity surrounding metadata standards such as MIAME (1) and MAGE-ML (2), which in turn has enabled the integration of independent experiments on a

Junbai Wang et al. (eds.), *Next Generation Microarray Bioinformatics: Methods and Protocols*,
Methods in Molecular Biology, vol. 802, DOI 10.1007/978-1-61779-400-1_11,

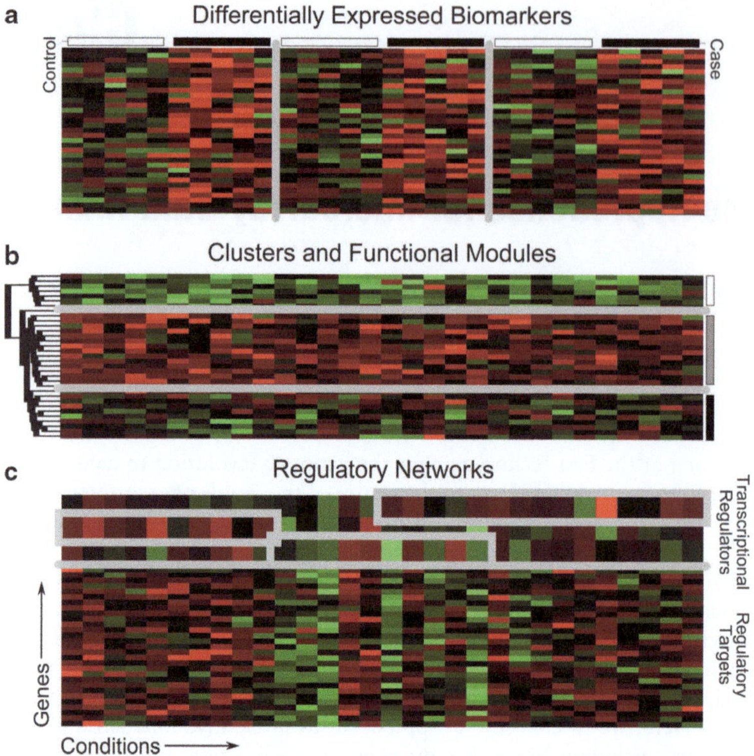

Fig. 1. Integrative approaches for microarray data analysis. While a carefully designed set of related microarray experiments can answer any number of interesting biological questions, three main areas are typically explored using large-scale integrative microarray analyses. These are questions where the added statistical power and diversity of experimental conditions offered by large microarray compendia can be particularly helpful. (**a**) Biomarker discovery – that is, the determination of differentially expressed genes – is one of the first and most widespread uses of microarray data integration. Statistical meta-analyses allow multiple experiments testing the same set of differential conditions (e. g., disease cases and controls, or cancer and normal tissue pairs) to be combined in order to more reliably determine genes whose expression is consistently differential in the biological condition of interest. (**b**) While genes (and conditions) can be clustered within any one microarray dataset in order to extract functionally related coexpression modules, this technique can be expanded to cluster or bicluster many microarray conditions. This approach can answer specific questions about the functional roles of as-yet-uncharacterized genes based on their coexpression partners and the experimental conditions where that coexpression occurs within a diverse compendium. (**c**) Similarly, by studying the relationships between transcriptional regulators and their potential regulatory targets over a wide range of integrated conditions, extensive regulatory networks can be derived. By performing this task in a sufficiently large data collection and by incorporating additional biological knowledge (e.g., binding sites or physical interactions), it is possible to begin teasing apart causation versus correlation within the regulatory network.

previously unrealizable scale. This chapter introduces integrative microarray analysis in the contexts of biomarker discovery, gene function and interaction prediction, and regulatory network inference (Fig. 1).

1.1. Biomarker Discovery

Biomarker discovery was one of the earliest applications of microarray integration (3–5), and it remains one of the primary uses for microarray compendia from large-scale human population cohorts. Meta-analysis (see Note 2) of multiple independent data sets has helped to reduce or overcome limitations otherwise intrinsic to biomarker discovery studies. The "p greater than n" problem of high-dimensional statistics (6) (see Note 3) is mitigated through the increase in sample size from combining multiple studies (see, for example, Note 4). Furthermore, the potential for bias due to batch effects (7) is reduced because independent experiments are unlikely to repeat the same relationships between batch and phenotype. Meta-analysis for biomarker discovery typically consists of three stages: a summary process where predictor and response variables are converted to effect sizes within each study, a regression (or comparable) procedure for combining multiple studies, and a corresponding inferential process for determining the significance of the combined result. We discuss the use of test statistics to integrate potentially incomparable response variables from different studies through unitless effect sizes, including Cohen's d for differential expression (see Notes 5 and 6). We also discuss the use of meta-regression to explicitly incorporate inter-study differences in the framework of linear modeling, and the use of rank products to combine studies without the need to combine variables *or* test statistics between studies.

1.2. Prediction of Gene Function and Interactions

The prediction of gene function and interaction (see Note 7) from coexpression not only benefit from integrative analysis due to increasing sample size, but additionally by incorporating a greater range of experimental conditions or treatments. In this context, predictions are based on the shared response of genes to a variety of experimental conditions or observed samples, so the consideration of additional samples creates more possibilities to observe coexpression. In a seminal early paper using gene coexpression observed by microarray to predict gene function (8), roughly 200 strains of yeast were constructed, each with a single gene removed from the genome. Expression arrays were used to profile the resulting changes in transcriptional activity, and uncharacterized genes could thus be assigned function if the transcriptional profile resulting from their deletion was similar to that of known pathways. For example, the *ERG28* gene was determined in this manuscript to be involved in ergosterol biosynthesis due to its clustering with a group of seven known ergosterol synthesis transcripts. While this study demonstrated the power of such an approach, only a single environmental context comprising standard rich medium growth conditions was assayed, and greater potential to discover novel gene function exists in using integrative approaches with large microarray compendia (see Note 8 for an example of time courses). We discuss the use of Microarray Experiment Functional Integration Technology (MEFIT) to gain global information about gene expression patterns, discover interactions too weak to detect in a single data set, and determine specific conditions in which the genes interact.

1.3. Regulatory Network Inference

It is essentially impossible to infer a fully accurate regulatory network using expression data alone, but large-scale integration of microarrays performed under many conditions with additional data types has shown promise in tackling this challenging problem (9–12). This highlights a final important biological motivation for integrative microarray analysis, which is that transcriptional activity is only one aspect of cellular biomolecular activity. Genetic and epigenetic variation, post-transcriptional and post-translational behavior, and intercellular signaling all come together to bridge the gap from genome to phenotype. We introduce an approach in which gene expression data and DNA sequence data are co-analyzed simultaneously, called the Combinatorial Algorithm for Expression and Sequence-based Cluster Extraction (COALESCE). Finally, we mention several methods for integrating microarray with other omics data such as protein interaction networks.

2. Methods

2.1. Biomarker Discovery

2.1.1. Data Collection and Normalization

The first step in a microarray meta-analysis is data collection, comparison, and normalization. While archives such as the Gene Expression Omnibus (GEO) (13) and ArrayExpress (14) have made it relatively easy to obtain large numbers of expression arrays, ensuring that the conditions or phenotypes assayed in multiple studies are minimally comparable is still very much a manual process. For example, the Gene Expression Atlas (GXA) (15) currently lists 16 microarray datasets in which one or more leukemia samples were assayed. Were these fresh samples or cell lines, blood or bone marrow, treated or new patients? Were the arrays all performed on the same platform, using the same protocol, with the same scanner, and with the same normalization and software postprocessing? While some of these factors can be corrected for during meta-regression (see below), many cannot, and an analyst must balance these issues when deciding which studies are biologically (as opposed to statistically) comparable.

Normalization of microarray measurements to effect sizes comparable between studies is, fortunately, a more straightforward process, and several procedures have become common (16). First, for either single or dual channel microarrays, the correspondence of individual probes with a phenotype can be converted to dimensionless test statistics such as t, z, P, or Q values within each study independently (17, 18). A useful test statistic for differential expression between groups I and J is Cohen's d (19), similar to a z-score using pooled deviation:

$$d = \frac{\mu_I - \mu_J}{\sqrt{((|I| - 1)\sigma_I + (|J| - 1)\sigma_J)/(|I| + |J|)}}, \tag{11.1}$$

for within-group means and standard deviations μ and σ, respectively. Methods exist for combining intrastudy test statistics or p-values between studies; for example, the metaMA Bioconductor package (20) extends Linear Models for Microarray Analysis (limma) (21) to meta-analysis.

Second, the preponderance of single channel Affymetrix arrays has lent itself to direct interstudy comparisons in units of raw transcript abundance after multiarray normalization of all arrays together with a procedure such as Robust Multichip Average (RMA) (22), GCRMA (23), or frozen RMA (24), which normalizes microarray probes to a distribution pre-determined from thousands of arrays, rather than determined from the arrays at hand. Multiple microarray datasets of the same platform, processed together using one of these methods, can generally be directly compared at least at the level of gene expression, keeping in mind the likelihood of batch effects (7) between experiments, and that no normalization will correct for gross differences in biological conditions. Coexpression meta-analysis (25–27) performs a similar normalization on Pearson, Spearman, or other correlation values between pairs of genes (rather than individual probes) within each study. One example of a successful coexpression effect size measure is the within-study Fisher transformation of correlation (2) followed by z-transformation $Z_{i,j}$ between genes i and j (3) (28):

$$z_{i,j} = \frac{1}{2}\ln\frac{1+\rho_{i,j}}{1-\rho_{i,j}}, \tag{11.2}$$

$$Z_{i,j} = \frac{z_{i,j} - \mu_z}{\sigma_z}, \tag{11.3}$$

for $\rho_{i,j}$ the Pearson correlation between genes i and j and μ_z and σ_z the mean and standard deviation of all z-transformed correlations within a dataset. Meta-analysis then produces a combined coexpression network, again weighting each study by sample size and noise characteristics, which can be analyzed directly or tested for differential coexpression biomarkers (29).

Finally, the rank product approach (30) provides a flexible alternative for combining studies on heterogeneous microarray platforms, to estimate the significance of effects for the union of genes present on any of the platforms. These authors found the rank product approach to perform especially well relative to classical test statistics in situations of small within-study sample size, since it does not rely on any estimate of expression variance. In the two-class case, fold-change is used to rank each gene in each pair of samples between the two classes within each study. In the RankProd Bioconductor package (31) (see Note 9) , the product of these ranks across all studies is used as a nonparametric test statistic:

$$RP_g = \left(\prod_i \frac{r_{gi}}{n_i} \right)^{1/K}, \tag{11.4}$$

where r_{gi} is the rank of gene g, n_i is the number of genes in the ith pairwise comparison, and K is the total number of pairwise comparisons. False discovery rate (FDR) is estimated by a permutation test.

2.1.2. Meta-regression

Meta-regression is an alternative approach to integrative biomarker discovery where differences between studies are explicitly incorporated in a regression model (32). These differences can include differences in sample size, systematic biases (e.g., the entire genome is more highly expressed in one study versus another), and differential responses (e.g., more or less effective treatment conditions) among studies. Statistically, this combination process is typically modeled as a regression in which each study's effect is modeled as a linear function of the unobserved "true" effect and of zero or more additional factors thought to impact the effect (sample size, experimenter, exposure, etc.) The simplest form of this regression assumes that intrastudy variation has been fully normalized, that interstudy variation is Gaussian and homoscedastic, and for each gene i solves a system of equations over studies s and factors t:

$$y_{i,s} = \beta_i + \sum_t \beta_t x_{s,t} + \varepsilon, \tag{11.5}$$

for observed effect size $y_{i,s}$, true effect β_i, unobserved coefficients β_t, and interstudy variance ε. A fixed effects model augments this by allowing each study to have its own intrastudy variance η_s (at the expense of not modeling interstudy variance ε). Finally, random-effects meta-analysis (33) fully models both intrastudy variance η_s and interstudy variance ε:

$$y_{i,s} = \beta_i + \sum_t \beta_t x_{s,t} + \eta_s + \varepsilon. \tag{11.6}$$

Estimators have been derived for each of these statistics, their p-values, and their confidence intervals, typically using maximum likelihood methods.

2.1.3. An Example: Rhodes et al. (34) Prostate Cancer Microarrays

As an illustrative example of biomarker discovery using microarray meta-analysis, we reproduce here the seminal study of Rhodes et al. (34), which examined four prostate cancer datasets comprising over 120 individual microarrays to determine a 153-gene marker of prostate cancer relative to benign tissue. For each gene in each study, a t-statistic of prostate/benign differential expression was calculated:

$$t_{i,s} = \frac{\mu_{i,s}(I) - \mu_{i,s}(J)}{\sqrt{\sigma_{i,s}(I)/|I| + \sigma_{i,s}(J)/|J|}}, \tag{11.7}$$

for gene i in study s containing groups I (prostate) and J (benign), and $|I|$ and $|J|$ are the numbers of samples in the two respective groups. Significance was not calculated using the parametric t-distribution; instead, an empirical p-value was determined using the bootstrap (35) with 10,000 random permutations (i.e., randomizations of I and J within s). These p-values were combined into a summary statistic S_i:

$$S_i = -2 \sum_s \log p_{i,s}. \tag{11.8}$$

The significance P_i of this statistic was also calculated empirically using a 100,000 sample bootstrap permutation test. Finally, a FDR correction for multiple hypothesis testing was applied (36), transforming each P_i into a q-value:

$$Q_i = \frac{P_i N}{N_i}, \tag{11.9}$$

where N is the total number of genes and N_i is the number with p-values less than or equal to P_i. Applying this process to each combination of studies resulted in 50 up- and 103 downregulated prostate cancer genes with $Q_i < 0.1$, and subsequent work (18) further developed this technique for broad meta-analysis of cancer microarrays.

2.1.4. Final Thoughts on Meta-analysis for Biomarker Discovery

Even when the statistical aspects of a meta-analysis have been fully addressed, there remain human elements that can bias results. Such bias occurs most often in an anticonservative direction, leading yet again to pitfalls that can impede the reproducibility of gene expression biomarkers (37, 38). A simple example is any systematic bias not explicitly modeled during meta-analysis. For example, if one combines five microarray studies and three are from the same laboratory, they will almost certainly share correlated technical artifacts; six studies performed on three different cell lines are likely to group into three pairs with lower-than-expected interstudy variation. Care must thus be taken when selecting which factors $x_{s,t}$ to model as described above. Second, the *file drawer problem* (39) is the tendency for negative results to go unpublished: an effect size truly distributed around zero might thus show up as strictly positive in the literature (Fig. 2a), and models have likewise been developed to take this into account (40, 41). Finally, even more unintuitive behaviors such as Simpson's paradox (42) can emerge in which studies show a trend in one direction individually, but in the opposite direction when combined (Fig. 2b). These diverse and very domain-specific pitfalls to meta-analyses have contributed to its associated controversy in the literature (43, 44), and as with any bioinformatic

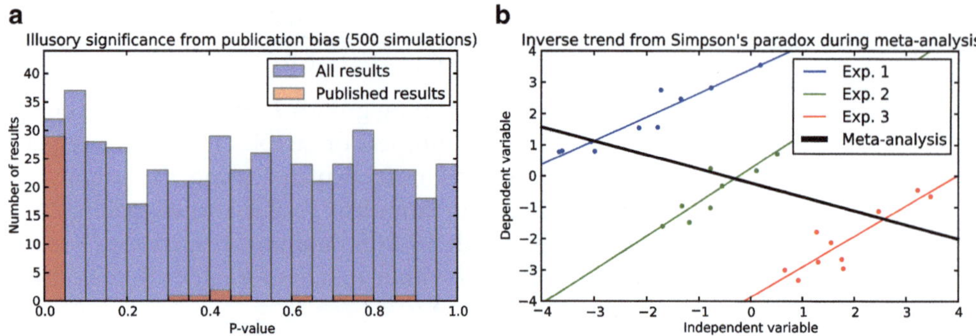

Fig. 2. Pitfalls of microarray meta-analysis. Any meta-analysis, including that of microarrays and expression biomarkers, is subject to a number of potential drawbacks. Many are obvious; a meta-analysis by definition attempts to combine a variety of experiments carried out by different laboratories under potentially different conditions and using different protocols. Even proper statistical normalization of these effects can lead to misleading conclusions. (**a**) Publication bias can produce illusory significance, since nonsignificant results will never reach the literature to be meta-analyzed. This is also known as the file drawer effect, since nonsignificant experiments are quietly "filed away." Here, 500 experimental results have been simulated in which the null hypothesis is true – there is no real biological effect. However, if we assume that significant results are published with a probability of 95% and nonsignificant results at only 1%, almost all published *p*-values are less than 0.05, and a biological effect appears artificially when the literature is meta-analyzed. (**b**) Simpson's paradox refers to the possibility of a correlative trend apparent in several experiments reversing itself when their results are combined during meta-analysis. This is most often the case when an unknown confounding variable is present. For example, suppose here that our independent variable is patient age and our dependent variable is survival. Each individual study would conclude that older patients have better outcomes, yet this is clearly untrue in meta-analysis. This counterintuitive effect might be observed if the actual determinant of survival is tumor size, and each study inadvertently sampled larger tumors from younger patients.

methods, care should be taken in microarray meta-analysis that the end result is biologically feasible and experimentally verifiable.

2.2. Prediction of Gene Function and Interaction

2.2.1. Integrating Multiple Microarray Datasets Using Coexpression Network Models

Meta-analysis provides a means to combine microarrays with a focus on experimental conditions and sample phenotypes; it is also possible to combine microarray data to focus on molecular mechanisms, gene function, or biomolecular networks. One example is the Microarray Experiment Functional Integration Technology (MEFIT) platform, which provides a supervised approach to leveraging information from multiple microarray datasets (28). MEFIT takes arbitrary microarray data as input and integrates it to predict functional relationships between specific genes. By integrating data from many different microarray analyses instead of focusing on a small set of results, MEFIT offers the opportunity to gain global information about gene expression patterns. Integrating data from many microarrays may also allow for the discovery of gene–gene interactions that are too subtle to be detected in a single dataset. MEFIT provides information on the specific conditions in which these genes interact, and these data may lead to the development of hypotheses that can be tested experimentally.

The MEFIT platform uses Bayesian networks to combine microarray data (45). Importantly, this allows microarray data generated on different platforms with different protocols and experimental conditions to be integrated. Data are analyzed within the context of different biological functions, and the probability of each gene–gene interaction is defined within this context. The primary output is thus one genome-wide functional interaction network per context, in addition to information on the importance of specific datasets in each context's specific biological process. As an example, a small number of microarrays may have been performed under conditions in which yeast sporulate; as a result, these microarrays may be particularly informative about functional interactions between genes involved in sporulation. Biological functions representing contexts can be provided by a scientist with an interest in particular biology, or they can be assigned automatically based on catalogs such as the Gene Ontology (46) or KEGG (47). Based on the genes in each of these contexts, all available microarray results are up- or downweighted so as to emphasize datasets active in each biological context. This results in more accurate context-specific functional interactome predictions, as well as quantifying how informative each of the input microarray datasets is for each biological function.

2.2.2. MEFIT Algorithm and Methodology

As shown in Fig. 3, the inputs for MEFIT are microarray datasets that have been preprocessed in order to ensure uniformity among platforms (26). Within each dataset, replicated genes are averaged so that each gene has a single gene expression vector, and missing values are imputed using KNNImpute (48). Biological contexts of interest are provided as input gene sets, lists of genes involved in, e.g., mitosis or fatty acid biosynthesis. These provide known positive gene–gene interactions, as any two genes in such a set are functionally related. Negative controls – gene pairs thought to be functionally unrelated – can be obtained by selecting pairs of genes from different contexts or by selecting random pairs (since most gene products perform unrelated biological functions). These contexts, whether defined manually or using curated functional annotations, serve as gold standards that define instances in which a gene–gene interaction is known to exist and instances in which gene pairs are known to be unrelated. In addition, the same gene lists are used to define the contexts in which different Bayesian networks should be constructed. In addition to describing pathways and processes, they can also be generalized to other categories such as tumor type, tissue of origin, or signaling pathways (49, 50).

Next, Pearson correlations are calculated between every pair of genes, and these are then normalized to generate *z*-scores with an average of zero and a standard deviation of one (see above). For each dataset, a collection of gene pair *z*-scores are generated,

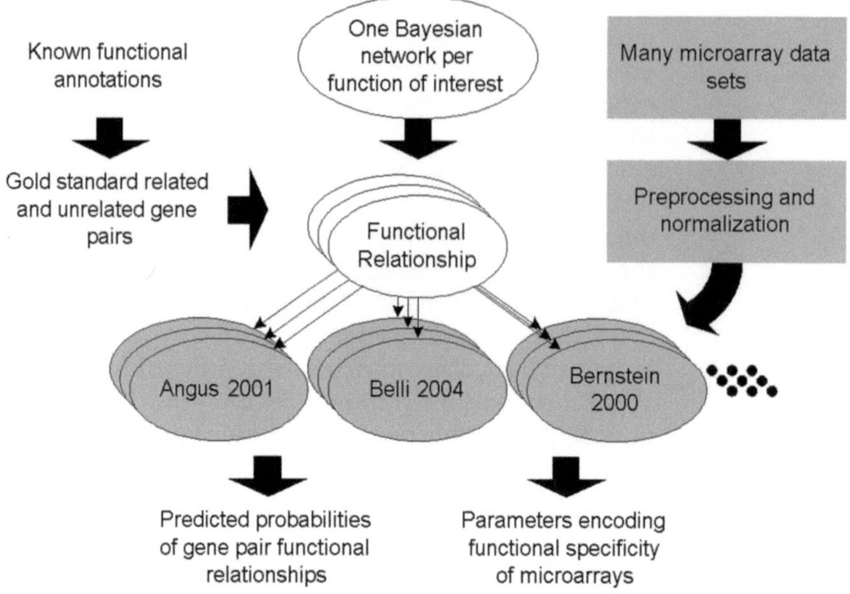

Fig. 3. Schematic overview of the MEFIT algorithm. Microarray data are provided as input, preprocessed, and normalized. This information is combined with prior knowledge regarding curated gene functions, and these together allow us to learn a set of Bayesian networks each representing a different biological context of interest. Functional relationships can be inferred from each network for its respective context, providing predicted probabilities of gene–gene functional interactions, as well as information about the specific microarrays that are most important for determining these interactions. Reproduced from Huttenhower, C., Hibbs, M., Myers, C., and Troyanskaya, O. G. A scalable method for integration and functional analysis of multiple microarray datasets. (2006) Bioinformatics 22 (23), 2890–7 by permission of Oxford University Press.

each representing the number of standard deviations their correlation lies from the dataset-specific mean. For each context, these data are used to learn a naive Bayesian classifier, such that the probability of observing a functional interaction FR within some context c and given some collection of datasets D_i is:

$$P_c(\mathrm{FR}|D) \propto P_c(D|\mathrm{FR})P_c(\mathrm{FR}) = P_c(\mathrm{FR})\prod_i P_c(D_i|\mathrm{FR}). \tag{11.10}$$

Using Bayes rule, we know that the probability of observing a functional relationship given some data is proportional to the probability of that data given that we have observed a relationship, times the prior $P_c(\mathrm{FR})$ of observing a functional relationship in the first place in context c. For example, many strong interactions occur among the components of the ribosome, so $P_c(\mathrm{FR})$ in the context of translation might be high; the prior probability of a functional relationship occurring in a sparse, specific biological process such as organelle fusion might be very low. A naive classifier assumes that each dataset (i.e., each observation) is independent, allowing us to separate all data D into a product over individual datasets D_i. The probability distribution $P_c(D_i|\mathrm{FR})$ over results from dataset D_i in context c is learned from the gold standard by picking out each pair of genes that are functionally related in that context. Finally, for these genes, we simply build a histogram by counting the number of times each result $D_i = d$ is observed, where d might be, e.g., high (>2), medium (−2 to 2), or low (<−2) z-scored correlation. This allows us to infer functional relationships $P_c(\mathrm{FR}|D)$ for other genes in the future for which we have experimental data but no prior knowledge in the gold standard.

The outputs from the MEFIT platform are thus the predicted probabilities that every pair of genes have a functional interaction within some context, modeled as a weighted undirected biological network. Each of these function-specific networks also learns during training how reliable each dataset will be for that function. These reliabilities allow a single confidence score to be assigned for each dataset for each context, by finding the difference between the prior and posterior probabilities of a functional relationship for each dataset and context independently.

2.2.3. MEFIT Results

To computationally evaluate such predictions, 20% of the genes were randomly selected as test genes. The remaining 80% of the genome was used for training, and performance on the test genes was determined by comparison with annotations in a GO-based gold standard. Compared to correlation alone, simple z-scoring, and several alternative methods, MEFIT resulted in increased areas under ROC curves and precision/recall for almost every biological context. The functions for which MEFIT provided the least improvement were functions already possessing higher AUC scores; since these functions are easily detected in a variety of data, they are by definition difficult to improve on. The other class of functions for which MEFIT provided little improvement (but still predicted accurately) were rarely observed in the available data (e.g., autophagy) such that there was not enough information for MEFIT to provide improved results. MEFIT thus provided the most benefit for relatively frequent functions that are poorly predicted by more traditional methods. One way in which MEFIT achieves this is by downweighting datasets in which the genes tend to show a high functional correlation nonspecifically, and the result is that, unlike other methods, it retains high precision even when recall is low. Gene expression-based predictions about gene function that result from MEFIT, in many circumstances, are likely to represent novel and accurate predictions when other types of data are considered. Thus, MEFIT represents one methodology for the simultaneous analysis of large numbers of microarray datasets using Bayesian integration on a function-by-function basis. MEFIT leverages both prior biological knowledge and the intrinsic condition-specificity of every microarray dataset to boost precision, sensitivity, and relevance to specific biological questions of interest. Further work has shown clearly the importance of establishing gene–gene interactions in a context-dependent way (49, 51, 52). Additional information regarding MEFIT can be found online at ref. 53.

2.3. Microarray Data Analysis for Regulatory Networks

Another goal of combining microarrays is to derive regulatory information, particularly when the microarray data are coupled with one or more complementary data sources. It has been shown that complete regulatory networks cannot be derived from microarrays alone (54), but progress in this area has been made by integrating additional sources of information. For example, the genomic sequences upstream and downstream of coding regions contain information about the situations in which gene products should be expressed. These include the binding sites for transcription factors (55), recognition sites for microRNAs (56) and RNA binding proteins (57), and chromatin remodeling signals (58). Microarray data can thus be analyzed for the purpose of discovering regulatory elements, that is, motifs that control when a gene is expressed. By analyzing the patterns in which genes are expressed using a large number of conditions and by incorporating the DNA sequences surrounding the genes, it becomes possible in some instances to identify the regulatory interactions controlling the expression of specific genes under specific conditions.

Several approaches to defining regulators of gene expression have been published, incorporating DNA sequence alone (10, 59), ChIP-chip (60, 61) or ChIP-seq results (62, 63), chromatin structure (64), and physical binding information (65). In unicellular organisms, DNA sequence motifs alone can be relatively informative about transcription factor binding sites (9). In mammalian systems, however, associating motifs with gene expression patterns is much more difficult. Regulatory motifs in higher organisms tend to be short and degenerate, making them difficult to identify clearly within a longer DNA sequence (50). Also, while regulatory motifs tend to be found close to the transcriptional start site in unicellular organisms, metazoan functional regulatory motifs can present a significant distance upstream of transcriptional start (66, 67). A final confounding factor is the complex integration of many transcription factors, both activators and repressors, into regulatory modules controlling gene expression. These factors together make it exceptionally difficult to identify the key functional components of regulatory networks in higher organisms, and integrative analysis is critical to the unraveling of these processes.

2.3.1. An Overview of COALESCE

An approach to this problem taken by several groups has been to first group genes together based on clustering in expression microarray data (68–70). Then, the DNA sequences upstream of the genes within each group is inspected for statistically enriched sequences (71, 72). We have developed an alternative approach in which gene expression data and DNA sequence data are co-analyzed simultaneously, called the Combinatorial Algorithm for Expression and Sequence-based Cluster Extraction (COALESCE, available online at ref. 73; Fig. 4). The advantage of this approach is that regulatory motifs associated with gene expression patterns can be identified even in the presence of noise in either data type individually, because clustering occurs based on both gene expression and DNA sequence information. To enable inclusion of as many diverse expression conditions as possible, COALESCE was designed to be extremely scalable; it can be used on datasets of >20,000 genes and has been applied to extremely large microarray compendia of 15,000 or more conditions.

The output is a set of clusters that contain coregulated genes, the specific conditions in which they display coordinate regulation, and any DNA sequence motifs that are enriched in the up- or downstream regions surrounding the clustered genes. The algorithm runs iteratively, with each cluster determined serially and initiated by identifying a small group of genes that have similar expression patterns. Features of this gene set are then defined, including the conditions in which the coexpression is strongest and any motifs enriched in the specific genes' DNA sequences. Based on this information, discordant genes are eliminated from the group and new genes are added based on a probabilistic model. The cluster is redefined for the next iteration, updating genes, conditions, and motifs, then the process is repeated until no more changes occur. When a stable group of genes is identified and the cluster has converged, this group is reported as a cluster, its signature is removed from the full data set, and a new cluster is initiated with another group of coexpressing genes. All clusters are then consolidated at the end of a complete COALESCE run.

2.3.2. COALESCE Algorithm and Methodology

The COALESCE algorithm is initiated with a set of expression datasets that serve as input. These microarrays are combined to create a single large matrix of gene expression values and conditions. The data are normalized so that the expression levels in each column have an average value of zero and a standard deviation of one; missing values do not affect the algorithm's performance and are left unchanged. Each iteration of module discovery begins with the identification of the two genes that are maximally correlated across all expression conditions. During the subsequent rounds of optimization, genes, conditions, and motifs are designated as "in" or "out" of the module. A condition is included in the module if the distribution of that condition's expression values for genes in the module differs from that of the genomic background (genes out of the module). A standard z-test is used for this analysis and requires the associated p-value to be below a user-defined cutoff p_e (typically

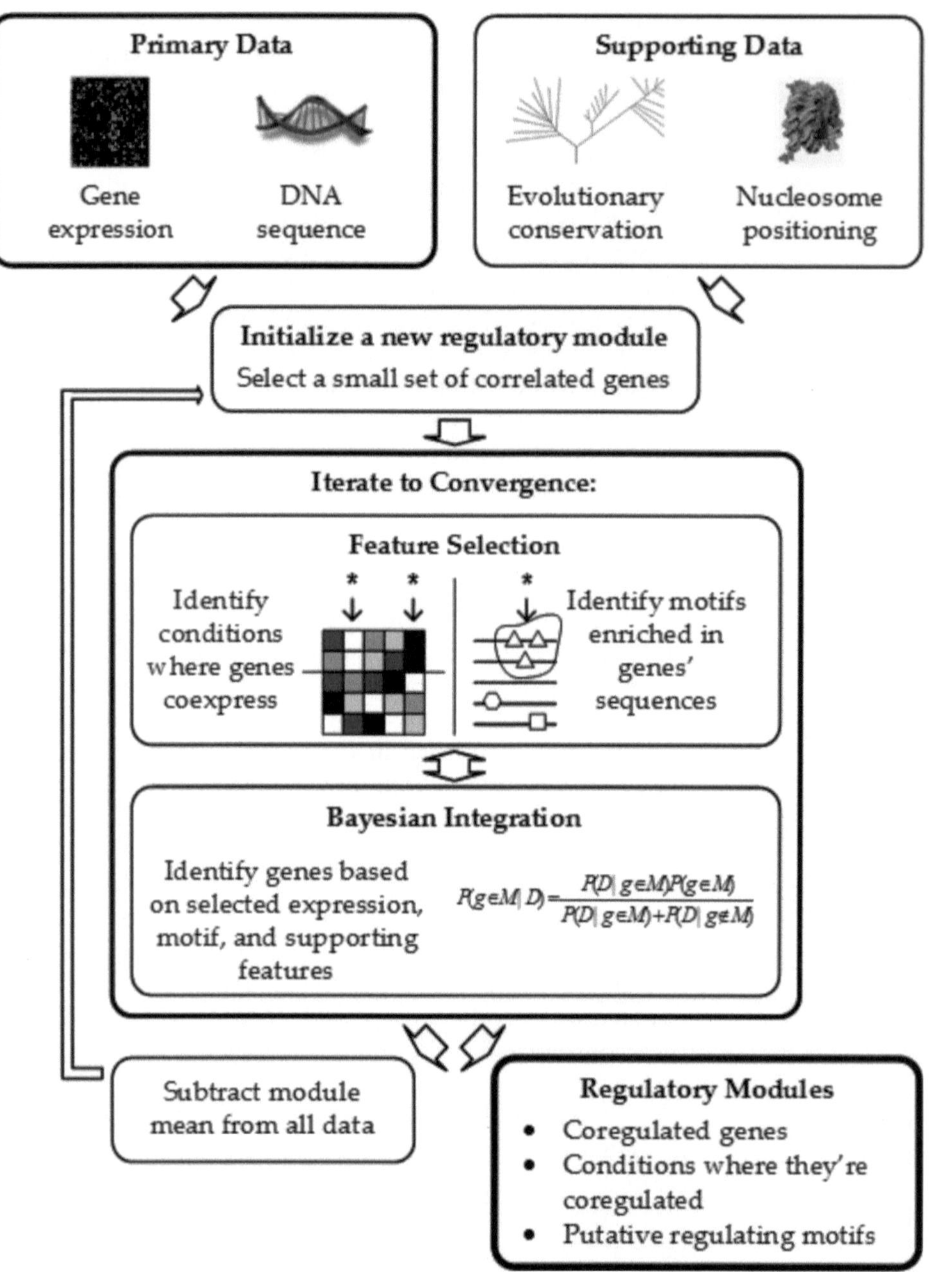

$$P(g \in M \mid D) = \frac{P(D \mid g \in M)P(g \in M)}{P(D \mid g \in M) + P(D \mid g \notin M)}$$

Fig. 4. Schematic of the COALESCE algorithm for regulatory module discovery. Gene expression and, optionally, DNA sequence data are provided as inputs; supporting data such as evolutionary conservation or nucleosome positions can also be included. The algorithm predicts regulatory modules in series, each initialized by selecting a small group of highly correlated genes. Conditions in which the genes are coexpressed are identified, as are motifs enriched in their surrounding sequences. Given this information, genes with similar expression patterns or motif occurrences are added to the module, and dissimilar genes are removed. Finally, given this new set of genes, conditions, and motifs are once again elaborated, and the process is iterated to convergence. At this point, the regulatory module (genes, conditions, and motifs) is reported, its mean subtracted from the remaining data, and the algorithm continues with a different set of starting genes. When no further significant modules are discovered, the predicted modules are merged into a minimum unique set describing predicted regulation in the input microarray conditions. Reproduced from Huttenhower, C., Mutungu, K. T., Indik, N., Yang, W., Schroeder, M., Forman, J. J., Troyanskaya, O. G., and Coller, H. A. Detailing regulatory networks through large-scale data integration. (2009) Bioinformatics 25 (24) 3267–74 by permission of Oxford University Press.

0.05). Similarly, motifs are considered significant if their frequency in gene sequences within the module likewise differs significantly from the background distribution (by some threshold p_m).

Based on the selected features (conditions and motifs), COALESCE calculates the probability that a gene is in the module using a Bayesian model. This calculation is performed based on a combination of the probabilities of observing the gene's expression data D (conditions) and sequence motifs M given the corresponding distributions of data from all other genes in and out of the cluster. Also included is a prior $P(g \in C)$ based on whether the gene was in the cluster during the previous iteration, which helps to stabilize module convergence. Thus:

$$\begin{aligned} P(g \in C|D, M) &\propto P(D, M|g \in C)P(g \in C) \\ &= P(g \in C)\prod_i P(D_i|g \in C)\prod_j P(M_j|g \in C), \end{aligned} \tag{11.11}$$

$$P(D_i|g \in C) = N(\mu_i(C), \sigma_i(C)), \tag{11.12}$$

where the probability of a motif $P(M_j|g \in C)$ is the relative number of times it occurs in any gene already in cluster C. Genes with a resulting probability $P(g \in C|D, M)$ above p_g, a user-defined input, are included in the cluster, and those below are excluded. The distribution of conditions and motifs in and out of the cluster are then redefined. After a sufficient number of iterations, the module converges, and the mean gene expression values and motif frequencies are subtracted from the remaining data. The entire process then begins again with a new pair of seed genes to determine the next module. Once no additional significant modules can be found, all identified clusters are merged based on simple overlap to form a minimal set of output clusters. Given the randomized nature of module initialization, the entire algorithm can then be run again if desired, and the results from multiple runs can be combined to define the most robustly discovered clusters.

2.3.3. Motifs, DNA Sequences, and Supporting Data Types

The basic type of binding motif identified by COALESCE is a simple string of DNA base pairs (of length defined by user input). It can also identify enriched motifs that are reverse complement pairs, e.g., AACG *or* CGTT. The algorithm can also identify probabilistic suffix trees (PSTs) that are overrepresented. These are trees with a node for each base to be matched, each representing the probability that specific base is present at a location corresponding to its depth in the tree. They represent degenerate motifs in a manner similar to position weight matrices (PWMs), but with the added benefit of allowing dependencies between motif sites. As COALESCE determines enriched motifs, if similar motifs are discovered, they are merged to a PST, and the algorithm tests whether the PST as a whole is enriched. For any of these three types of motifs – strings, reverse complements, or PSTs – the gene-specific motif score is determined by assuming each locus in the provided sequence is independent and determining the probability of observing that sequence, normalized by the probability of a match of identical length occurring by chance.

COALESCE has been designed so that it can be used to analyze any type of microarray data as well as supporting data including evolutionary conservation or nucleosome positions. Some of this supporting information can be included in a microarray-like manner; for instance, one can discover clusters in which both expression and the density of nucleosome occupancy within a group of genes is coordinately changed. More often, however, it is useful to include sequence-oriented supporting data such as the degree of site-specific conservation or ChIP-chip/-seq for transcription factors or nucleosomes. This is incorporated into the probability calculations as described above by indicating the relative weights given to each locus during motif matching. The incorporation of supporting data can, for instance, leverage information on nucleosome occupancy. Base pairs that are determined to be covered by histones are less likely to interact with transcription factors, and this provides weights for specific base pairs: their likelihood of being part of a regulatory motif is lower if they are occluded by a histone and higher if they are not. Evolutionary conservation is another example of data that can be incorporated in a similar manner, since conserved bases can be assigned higher weights. This weight information directly influences the amount by which each motif present in the sequence surrounding the gene affects the overall probability distributions used for cluster convergence.

2.3.4. COALESCE Results

Validation in synthetic data. The COALESCE method was validated on synthetic data with and without "spiked-in" regulatory modules. When significant coexpression or regulatory motifs were not spiked-in, no false-positive modules were identified by the algorithm. Conversely, COALESCE output on data with modules spiked-in resulted in precision and recall on the order of 95% for all of modules, motifs, genes, and conditions (50).

Recovery of known biological modules in yeast. To evaluate its ability to recover known biological modules, COALESCE has been applied to *Saccharomyces cerevisiae* expression data and the resulting clusters compared with coannotations in the Gene Ontology. Even without sequence information, COALESCE performs extremely well when clustering together genes with the same Gene Ontology annotations, outperforming earlier biclustering approaches such as SAMBA (74) and PISA (75), although the addition of information about nucleosome position and evolutionary conservation provided little improvement by this metric.

Identification of known transcription factors. In addition, COALESCE performed well in an analysis designed to determine whether targets of transcription factors were accurately identified. A comparison of COALESCE results with Yeastract (76), a database of experimentally verified binding sites, determined that COALESCE consistently provides reliable data on targets of yeast transcription factors (performing comparably to, e.g., cMonkey (77) and FIRE (78)). Further analysis of COALESCE's ability to recover transcription factor targets was performed in *Escherichia coli* and demonstrated comparably high accuracy (recovering known targets for ~50% of the TFs covered comprehensively by RegulonDB (79)).

Application to metazoan systems. However, COALESCE was initially designed to tackle the much more challenging problem of discovering regulatory motifs within metazoan systems. Correspondingly COALESCE reported coherent clusters when applied to data from *Caenorhabditis elegans*, *Drosophila melanogaster*, *Mus musculus*, and *Homo sapiens*. Each of these analyses identified regulatory modules with genes and transcription factors (motifs) that both reproduce existing information and extend our knowledge. Still, it should also be recognized that transcriptional regulation in metazoans is complex. While COALESCE represents a powerful approach to identifying regulatory modules, it does not model the full complexity of the regulation of transcript activity in these systems, which likely involves a summation of proximal, distal, inducing, inhibitory, insulating, posttranscriptional and post translational, and epigenetic factors. Fully understanding the mechanisms of regulation of transcript abundance in mammalian systems will require both richer models and even more extensive data integration.

2.4. Combining Microarrays with Other Genomic Data Types

Every assay, be it of gene expression or of another biomolecular activity, provides a snapshot of the cell under some specific environmental condition. Most microarrays measure mRNA transcript abundance alone, and they do so for a controlled population of cells with a defined medium, temperature, genetic background, and chemical environment. We have discussed above the advantages of integratively inspecting many such conditions simultaneously; we now consider the additional benefits provided by integrating microarrays with other genomic data types (see Note 10). For example, if two transcripts are coordinately upregulated when the cell is provided with specific carbon sources, this provides evidence that they may be functionally linked to each other and to carbon metabolism. If additional data is considered in which they physically interact, one contains an extracellular receptor domain, the other a kinase domain, and they both colocalize to the cellular membrane, a clearer composite picture of their function in nutrient sensing and signaling can be inferred.

Given the preponderance of microarray data available for most organisms of interest, it plays a key role in most function prediction systems. Methods for integrating it with other data types again include Bayesian networks (28, 80), kernel methods (81, 82), and a variety of network analyses (83). An illustrative example is provided by a method of data fusion developed by Aerts et al. (82) in which a variation on function prediction was used to prioritize candidate genes involved in human disease. A gold standard of known training genes was developed for each disease of interest, and for each dataset within each disease, one of two methods was used to rank the nontraining portion of the genome. For continuous data such as microarrays, standard Pearson correlation was used between the training set and each other gene. For discrete data (localization, domain presence/absence, binding motifs, etc.), Fisher's test was used. Thus, the genes within each dataset were ranked independently, and these ranks were combined to form a single list per disease using order statistics (84). The biological functions of genes with respect to a variety of human diseases were thus predicted by integrating microarray information with collections of other genomic data sources.

Many more methods have likewise been proposed for predicting functional relationships using diverse genomic data. Proposed techniques include kernel machines, Bayesian networks, and several types of graph analyses (85); as with function prediction, essentially any machine learner can be used to infer functional interaction networks (86). Popular implementations for various model organisms include GeneMANIA (87), STRING (88), bioPIXIE (89), HEFalMp (51), the "Net" series of tools (83), and FuncBase (90). Many of these share Bayesian methodologies similar to that described above for MEFIT, since the probability distribution $P_c(D_i|\mathrm{FR})$ can be computed easily for any type of dataset D_i and any gene set describing a context c. For example, consider integrating a microarray dataset D_1 with a protein–protein interaction dataset D_2. Each can be encoded as a set of data points representing experimental measurements between gene pairs. D_1 includes three values $d_{1,1}$ (anticorrelation), $d_{1,2}$ (no correlation), and $d_{1,3}$ (positive correlation); D_2 includes two values, $d_{2,1}$ (no interaction) and $d_{2,2}$ (interaction). Suppose our context of interest c includes three genes g_1 through g_3, and the entire genome contains ten genes through g_{10}. Thus, our gold standard contains three interacting gene pairs out of the 45 possible pairwise combinations of ten genes, making our prior $P_c(\mathrm{FR}) = 3/45 = 0.067$. Examining our microarray dataset D_1, we observe the following distribution of correlation values shown in Table 1.

Table 1
Known correlations in the gold standard of ten genes

	Unrelated	Related (g_1, g_2, g_3)
$d_{1,1}$ (Anticorrelated)	11	0
$d_{1,2}$ (Not correlated)	20	1
$d_{1,3}$ (Correlated)	11	2

Table 2
Known interactions in the gold standard of ten proteins

	Unrelated	Related (g_1, g_2, g_3)
$d_{2,1}$ (No interaction)	40	1
$d_{2,2}$ (Interaction)	2	2

Thus, $P_c(D_1 = d_{1,2}|\text{FR}) = 0.333$, $P_c(D_1 = d_{1,3}| \sim \text{FR}) = 0.262$, and so forth. Likewise, we observe interaction data D_2, shown in Table 2.

Suppose that g_4 is uncharacterized and that it is highly correlated with and physically interacts with g_3. Then the posterior is given by:

$$P_c(\text{FR}_{3,4}|D) = \frac{P_c(D|\text{FR})P_c(\text{FR})}{P_c(D)}, = \frac{P_c(D_1{=}d_{1,3}|\text{FR})P_c(D_2{=}d_{2,2}|\text{FR})P_c(\text{FR})}{P_c(D|\text{FR})P_c(\text{FR})+P_c(D|{\sim}\text{FR})P_c({\sim}\text{FR})},$$
$$= \frac{(2/3)(2/3)(3/45)}{(2/3)(2/3)(3/45) + (11/42)(2/42)(42/45)}, = 0.718. \quad (11.13)$$

Neither data source alone is a strong indicator that g_3 and g_4 are functionally related, but together they yield a relatively high probability of functional interaction. If g_4 is likewise correlated with g_1 and g_2 and physically interacts with g_2, this not only generates a set of high-confidence functional interactions using microarray data integration, it suggests that g_4 actually participates in biological process c based on guilt-by-association (91).

2.5. Summary

Microarrays, along with all other genomic data types, continue to accumulate at an exponential rate despite the ongoing reduction in the cost of high-throughput sequencing (86). RNA-seq results can, of course, be treated analogously in most cases to printed microarray data, and microarrays themselves continue to be used in settings ranging from clinical diagnostics (92) to metagenomics (93).

Integrative analyses of these data present a clear computational opportunity. Since experimental results are currently being generated at a rate that outpaces Moore's law, it is not enough to wait for faster computers—new bioinformatic tools must be developed with an eye to scalability and efficiency. However, the prospects for biological discovery are even more sweeping. Microarrays represent one of the best tools available for quickly and cheaply probing a biological system under many different conditions or for assaying many different members of a population. Since biology is, if anything, adaptive and ever-changing in response to a universe of environmental stimuli, each such measurement provides only a snapshot of the cell's underlying compendium of biomolecular activities. Considering microarrays integratively in tandem with other genomic data thus provides us with a more complete perspective on any target biological system.

3. Notes

1. We will consider primarily gene expression microarrays, but opportunities clearly exist to include information from tiling microarrays (e.g., copy number variation (94, 95) or ChIP results (96, 97)), from microarray-like uses of high-throughput sequencing (98), and from novel applications such as metagenomics (93); these will be referred to as other data types.
2. Broadly defined, a *meta-analysis* (32) is any process that combines the results of multiple studies, but the term has come to refer more specifically to a class of statistical procedures used to normalize and compare individual studies' results as effect sizes.
3. In any setting in which there are many more response variables p (i.e., genes) than there are samples n (i.e., microarray conditions), it can be difficult to distinguish reproducible biological activity from variations present in a study by chance. This has led to considerable contention regarding, for example, the reproducibility of genome-wide association studies (99, 100) and of gene expression biomarkers (37, 38) in which high-dimensional biomolecular variables (genetic polymorphisms or differentially regulated transcripts) are associated with a categorical (e.g., disease presence/absence) or continuous (e.g., survival) outcome of interest.
4. For example, one of the first major biomarker discovery publications in the field of microarray analysis was a comparison of acute myeloid leukemia (AML) patient samples with acute lymphoblastic leukemia (ALL) patients (4). This paper used 27 ALL and 11 AML samples to determine a 50-gene

biomarker distinguishing the two classes. The large number of genes relative to the small number of samples necessarily limits our confidence in any single component of the biomarker. A meta-analysis combining these with the dozens of additional subsequently published AML/ALL arrays (101) would effectively perform this experiment in replicate several times over. Any gene observed to be up- or downregulated in all of these many experiments is more likely to truly participate in the biology differentiating myeloid and lymphoblastic cancers, and the degree of confidence in such a reproducible result can be quantified statistically.

5. An *effect size* is a measure of the magnitude of the relationship between two variables – for example, between gene expression and phenotype or treatment, or between the coexpression of different genes.

6. The response variables of different studies may not be directly comparable for any of a number of reasons, for example, differences in array platform, patient cohorts, or experimental methodology.

7. Gene *function prediction* is the process of determining in which biochemical activities a gene product is involved, or to which environmental or intracellular stimuli it responds.

8. Microarray time courses are often used to better understand regulatory interactions, and these by definition involve integration of several time points. As one example, profiles of transcriptional activity as cells proceeded through the cell cycle were the subject of intense scrutiny (68, 69). These are often modeled using variations on continuous function fitting (sinusoids in the case of the cell cycle), allowing transcriptional activity to be understood in terms of a regulated response to a perturbation at time zero. Alternately, intergenic regulation can be inferred by determining which activity at time point $t + 1$ is likely to be a result of specific activities at time t (102, 103). Although these specific uses of microarrays are not discussed here (see ref. 54), the more general problem of coregulatory inference based on correlation analyses has also been deeply studied.

9. Using the rank products approach, a high rank in a single study can be enough to achieve a significant p-value, even if there is no apparent effect in one or more studies in the meta-analysis. If a more stringent test is desired, to identify only genes with an affect in all or most studies, the sum of ranks may be used instead; this is also implemented in the RankProd Bioconductor package. A gene with a moderate rank caused by a very small effect in several studies can also be significant.

10. Over the past decade of high-throughput biology, two main areas have developed in which microarray data is integrated in tandem with other genomic data sources: protein *function prediction* and *functional interaction* inference. Function prediction can include either the determination of the biochemical and enzymatic activities of a protein or the prediction of the cellular processes and biological roles in which it is used. For example, a protein may be predicted to function as a phosphatase, and it may also be predicted to perform that function as part of the mitotic cell cycle. Functional interactions (also referred to as functional linkages or functional relationships) occur between pairs of genes or gene products used in similar biological processes; for example, a phosphatase and a kinase both used to carry out the mitotic cell cycle would be functionally related.

Acknowledgments

The authors would like to thank the editors of this title for their gracious support, the laboratories of Olga Troyanskaya and Leonid Kruglyak for their valuable input, and all of the members of the Coller and Huttenhower laboratories. This research was supported by PhRMA Foundation grant 2007RSGl9572, NIH/NIGMS 1R01 GM081686, NSF DBI-1053486, NIH grant T32 HG003284, and NIGMS Center of Excellence grant P50 GM071508. H.A.C. was the Milton E. Cassel scholar of the Rita Allen Foundation.

References

1. Brazma A, Hingamp P, Quackenbush J et al (2001) Minimum information about a microarray experiment (MIAME)-toward standards for microarray data. Nat Genet 29: 365–371.
2. Rayner TF, Rocca-Serra P, Spellman PT et al (2006) A simple spreadsheet-based, MIAME-supportive format for microarray data: MAGE-TAB. BMC Bioinformatics 7:489.
3. Alon U, Barkai N, Notterman DA et al (1999) Broad patterns of gene expression revealed by clustering analysis of tumor and normal colon tissues probed by oligonucleotide arrays. Proc Natl Acad Sci U S A 96:6745–6750.
4. Golub TR, Slonim DK, Tamayo P et al (1999) Molecular classification of cancer: class discovery and class prediction by gene expression monitoring. Science 286:531–537.
5. Alizadeh AA, Eisen MB, Davis RE et al (2000) Distinct types of diffuse large B-cell lymphoma identified by gene expression profiling. Nature 403:503–511.
6. Gadbury GL, Garrett KA, Allison DB (2009) Challenges and approaches to statistical design and inference in high-dimensional investigations. Methods Mol Biol 553:181–206.
7. Leek JT, Scharpf RB, Bravo HC et al (2010) Tackling the widespread and critical impact

of batch effects in high-throughput data. Nat Rev Genet 11:733–739.

8. Hughes TR, Marton MJ, Jones AR et al (2000) Functional discovery via a compendium of expression profiles. Cell 102:109–126.
9. Beer MA, Tavazoie S (2004) Predicting gene expression from sequence. Cell 117:185–198.
10. Bonneau R, Reiss DJ, Shannon P et al (2006) The Inferelator: an algorithm for learning parsimonious regulatory networks from systems-biology data sets de novo. Genome Biol 7:R36.
11. Margolin AA, Wang K, Lim WK et al (2006) Reverse engineering cellular networks. Nat Protoc 1:662–671.
12. Faith JJ, Hayete B, Thaden JT et al (2007) Large-scale mapping and validation of *Escherichia coli* transcriptional regulation from a compendium of expression profiles. PLoS Biol 5:e8.
13. Barrett T, Troup DB, Wilhite SE et al (2009) NCBI GEO: archive for high-throughput functional genomic data. Nucleic Acids Res 37:D885–890.
14. Parkinson H, Kapushesky M, Kolesnikov N et al (2009) ArrayExpress update – from an archive of functional genomics experiments to the atlas of gene expression. Nucleic Acids Res 37:D868–872.
15. Kapushesky M, Emam I, Holloway E et al (2010) Gene expression atlas at the European bioinformatics institute. Nucleic Acids Res 38:D690–698.
16. Campain A, Yang YH (2010) Comparison study of microarray meta-analysis methods. BMC Bioinformatics 11:408.
17. Choi JK, Yu U, Kim S et al (2003) Combining multiple microarray studies and modeling interstudy variation. Bioinformatics 19: i84–90.
18. Rhodes DR, Yu, J, Shanker K et al (2004) Large-scale meta-analysis of cancer microarray data identifies common transcriptional profiles of neoplastic transformation and progression. Proc Natl Acad Sci U S A 101:9309–9314.
19. Cohen J (1988) Statistical Power Analysis for the Behavioral Sciences. Lawrence Erlbaum, New York, NY.
20. Marot G, Foulley J-L, Mayer C-D et al (2009) Moderated effect size and P-value combinations for microarray meta-analyses. Bioinformatics 25:2692–2699.
21. Smyth GK (2004) Linear models and empirical bayes methods for assessing differential expression in microarray experiments. Stat Appl Genet Mol Biol 3:Article3.
22. Irizarry RA, Hobbs B, Collin F et al (2003) Exploration, normalization, and summaries of high density oligonucleotide array probe level data. Biostatistics 4:249–264.
23. Wu Z, Irizarry RA (2004) Preprocessing of oligonucleotide array data. Nat Biotechnol 22: 656–658; author reply 658.
24. McCall MN, Bolstad BM, Irizarry RA (2009) Frozen robust multi-array analysis (fRMA), Johns Hopkins University, Baltimore, MD.
25. Aggarwal A, Guo DL, Hoshida Y et al (2006) Topological and functional discovery in a gene coexpression meta-network of gastric cancer. Cancer Res 66:232–241.
26. Hibbs MA, Hess DC, Myers CL et al (2007) Exploring the functional landscape of gene expression: directed search of large microarray compendia. Bioinformatics 23:2692–2699.
27. Wang K, Narayanan M, Zhong H et al (2009) Meta-analysis of inter-species liver co-expression networks elucidates traits associated with common human diseases. PLoS Comput Biol 5:e1000616.
28. Huttenhower C, Hibbs M, Myers C et al (2006) A scalable method for integration and functional analysis of multiple microarray datasets. Bioinformatics 22:2890–2897.
29. Choi JK, Yu U, Yoo OJ et al (2005) Differential coexpression analysis using microarray data and its application to human cancer. Bioinformatics 21:4348–4355.
30. Breitling R, Herzyk P (2005) Rank-based methods as a non-parametric alternative of the T-statistic for the analysis of biological microarray data. J Bioinform Comput Biol 3:1171–1189.
31. Hong F, Breitling R, McEntee CW et al (2006) RankProd: a bioconductor package for detecting differentially expressed genes in meta-analysis. Bioinformatics 22:2825–2827.
32. Rosner B (2005) Fundamentals of Biostatistics, Duxbury Press, Boston, USA.
33. DerSimonian R, Laird N (1986) Meta-analysis in clinical trials. Control Clin Trials 7:177–188.
34. Rhodes DR, Barrette TR, Rubin MA et al (2002) Meta-analysis of microarrays: interstudy validation of gene expression profiles

reveals pathway dysregulation in prostate cancer. Cancer Res 62:4427–4433.
35. Efron B (1994) An Introduction to the Bootstrap. Chapman and Hall/CRC, New York.
36. Benjamini Y, Hochberg Y (1995) Controlling the false discovery rate: a practical and powerful approach to multiple testing. J. Royal Statistical Society B 57:289–300.
37. Baggerly KA, Coombes KR (2009) Deriving chemosensitivity from cell lines: Forensic bioinformatics and reproducible research in high-throughput biology. Annals of Applied Statistics 3:1309–1334.
38. Ghosh D, Poisson LM (2009) "Omics" data and levels of evidence for biomarker discovery. Genomics 93:13–16.
39. Rosenthal R (1979) The file drawer problem and tolerance for null results. Psychological Bulletin 86:638–641.
40. Sutton AJ, Song F, Gilbody SM et al (2000) Modelling publication bias in meta-analysis: a review. Stat Methods Med Res 9:421–445.
41. Thornton A, Lee P (2000) Publication bias in meta-analysis: its causes and consequences. J Clin Epidemiol 53:207–216.
42. Simpson EH (1951) The Interpretation of Interaction in Contingency Tables. Journal of the Royal Statistical Society B 13:238–241.
43. Egger M, Smith GD, Sterne JA (2001) Uses and abuses of meta-analysis. Clin Med 1: 478–484.
44. Yuan Y, Hunt RH (2009) Systematic reviews: the good, the bad, and the ugly. Am J Gastroenterol 104:1086–1092.
45. Neapolitan RE (2004) Learning Bayesian Networks. Prentice Hall, Chicago, Illinois.
46. Ashburner M, Ball CA, Blake JA et al (2000) Gene ontology: tool for the unification of biology. The Gene Ontology Consortium. Nat Genet 25:25–29.
47. Kanehisa M, Goto S, Furumichi M et al (2010) KEGG for representation and analysis of molecular networks involving diseases and drugs. Nucleic Acids Res 38:D355–360.
48. Troyanskaya OG, Dolinski K, Owen AB et al (2003) A Bayesian framework for combining heterogeneous data sources for gene function prediction (in *Saccharomyces cerevisiae*). Proc Natl Acad Sci U S A 100:8348–8353.
49. Myers CL, Troyanskaya OG (2007) Context-sensitive data integration and prediction of biological networks. Bioinformatics 23:2322–2330.
50. Huttenhower C, Mutungu KT, Indik N et al (2009) Detailing regulatory networks through large scale data integration. Bioinformatics 25:3267–3274.
51. Huttenhower C, Haley EM, Hibbs MA et al (2009) Exploring the human genome with functional maps. Genome Res 19:1093–1106.
52. Huttenhower C, Hibbs MA, Myers CL et al (2009) The impact of incomplete knowledge on evaluation: an experimental benchmark for protein function prediction. Bioinformatics 25:2404–2410.
53. Huttenhower C, Hibbs M, Myers C et al (2010) Microarray Experiment Functional Integration Technology (MEFIT). Online. http://avis.princeton.edu/mefit/. Accessed 25 October, 2010.
54. Markowetz F, Spang R. (2007) Inferring cellular networks – a review. BMC Bioinformatics 8:S5.
55. Tompa M, Li N, Bailey TL et al (2005) Assessing computational tools for the discovery of transcription factor binding sites. Nat Biotechnol 23:137–144.
56. Griffiths-Jones S, Grocock RJ, van Dongen S et al (2006) miRBase: microRNA sequences, targets and gene nomenclature. Nucleic Acids Res 34:D140–144.
57. Lunde BM, Moore C, Varani G (2007) RNA-binding proteins: modular design for efficient function. Nat Rev Mol Cell Biol 8:479–490.
58. Segal E, Fondufe-Mittendorf Y, Chen L et al (2006) A genomic code for nucleosome positioning. Nature 442:772–778.
59. Margolin AA, Nemenman I, Basso K et al (2006) ARACNE: an algorithm for the reconstruction of gene regulatory networks in a mammalian cellular context. BMC Bioinformatics 7:S7.
60. van Steensel B (2005) Mapping of genetic and epigenetic regulatory networks using microarrays. Nat Genet 37:S18–24.
61. Farnham PJ (2009) Insights from genomic profiling of transcription factors. Nat Rev Genet 10:605–616.
62. Mathur D, Danford TW, Boyer LA et al (2008) Analysis of the mouse embryonic stem cell regulatory networks obtained by ChIP-chip and ChIP-PET. Genome Biol 9: R126.
63. Ouyang Z, Zhou Q, Wong WH (2009) ChIP-Seq of transcription factors predicts absolute and differential gene expression in embryonic stem cells. Proc Natl Acad Sci U S A 106:21521–21526.

64. Jiang C, Pugh BF (2009) Nucleosome positioning and gene regulation: advances through genomics. Nat Rev Genet 10:161–172.
65. Yeger-Lotem E, Sattath S, Kashtan N et al (2004) Network motifs in integrated cellular networks of transcription-regulation and protein-protein interaction. Proc Natl Acad Sci U S A 101:5934–5939.
66. Heintzman ND, Ren B (2009) Finding distal regulatory elements in the human genome. Curr Opin Genet Dev 19:541–549.
67. Visel A, Rubin EM, Pennacchio LA (2009) Genomic views of distant-acting enhancers. Nature 461:199–205.
68. Eisen MB, Spellman PT, Brown PO et al (1998) Cluster analysis and display of genome-wide expression patterns. Proc Natl Acad Sci U S A 95:14863–14868.
69. Spellman PT, Sherlock G, Zhang MQ et al (1998) Comprehensive identification of cell cycle-regulated genes of the yeast *Saccharomyces cerevisiae* by microarray hybridization. Mol Biol Cell 9:3273–3297.
70. Gollub J, Sherlock G (2006) Clustering microarray data. Methods Enzymol 411:194–213.
71. Bailey TL, Elkan C (1994) Fitting a mixture model by expectation maximization to discover motifs in biopolymers. Proc Int Conf Intell Syst Mol Biol 2:28–36.
72. Roth FP, Hughes JD, Estep PW et al (1998) Finding DNA regulatory motifs within unaligned noncoding sequences clustered by whole-genome mRNA quantitation. Nat Biotechnol 16:939–945.
73. Huttenhower C, Mutungu KT, Indik N et al (2009) Combinatorial Algorithm for Expression and Sequence-based Cluster Extraction (COALESCE). Online. http://imperio.princeton.edu/cm/coalesce/. Accessed 25 October, 2010.
74. Tanay A, Shamir R (2004) Multilevel modeling and inference of transcription regulation. J Comput Biol 11:357–375.
75. Kloster M, Tang C, Wingreen NS (2005) Finding regulatory modules through large-scale gene-expression data analysis. Bioinformatics 21:1172–1179.
76. Teixeira MC, Monteiro P, Jain P et al (2006) The YEASTRACT database: a tool for the analysis of transcription regulatory associations in *Saccharomyces cerevisiae*. Nucleic Acids Res 34:D446–451.
77. Reiss DJ, Baliga NS, Bonneau R (2006) Integrated biclustering of heterogeneous genome-wide datasets for the inference of global regulatory networks. BMC Bioinformatics 7:280.
78. Elemento O, Slonim N, Tavazoie S (2007) A universal framework for regulatory element discovery across all genomes and data types. Mol Cell 28:337–350.
79. Gama-Castro S, Jimenez-Jacinto V, Peralta-Gil M et al (2008) RegulonDB (version 6.0): gene regulation model of *Escherichia coli* K-12 beyond transcription, active (experimental) annotated promoters and Textpresso navigation. Nucleic Acids Res 36:D120–124.
80. Jansen R, Yu H, Greenbaum D et al (2003) A Bayesian networks approach for predicting protein–protein interactions from genomic data. Science 302:449–453.
81. Lanckriet GR, De Bie T, Cristianini N et al (2004) A statistical framework for genomic data fusion. Bioinformatics 20:2626–2635.
82. Aerts S, Lambrechts D, Maity S et al (2006) Gene prioritization through genomic data fusion. Nat Biotechnol 24:537–544.
83. Lee I, Date SV, Adai AT et al (2004) A probabilistic functional network of yeast genes. Science 306:1555–1558.
84. Stuart JM, Segal E, Koller D et al (2003) A gene-coexpression network for global discovery of conserved genetic modules. Science 302:249–255.
85. Troyanskaya OG (2005) Putting microarrays in a context: integrated analysis of diverse biological data. Brief Bioinform 6:34–43.
86. Huttenhower C, Hofmann O (2010) A quick guide to large-scale genomic data mining. *PLoS Comput Biol 6*:e1000779.
87. Warde-Farley D, Donaldson SL, Comes O et al (2010) The GeneMANIA prediction server: biological network integration for gene prioritization and predicting gene function. Nucleic Acids Res 38:W214–220.
88. Harrington ED, Jensen LJ, Bork P (2008) Predicting biological networks from genomic data. FEBS Lett 582:1251–1258.
89. Myers CL, Robson D, Wible A et al (2005) Discovery of biological networks from diverse functional genomic data. Genome Biol 6:R114.
90. Beaver JE, Tasan M, Gibbons FD et al (2010) FuncBase: a resource for quantitative gene function annotation. Bioinformatics 26:1806–1807.

91. Tian W, Zhang LV, Tasan M et al (2008) Combining guilt-by-association and guilt-by-profiling to predict *Saccharomyces cerevisiae* gene function. Genome Biol 9:S7.
92. Tillinghast GW (2010) Microarrays in the clinic. Nat Biotechnol 28:810–812.
93. Brodie EL, Desantis TZ, Joyner DC et al (2006) Application of a high-density oligonucleotide microarray approach to study bacterial population dynamics during uranium reduction and reoxidation. Appl Environ Microbiol 72:6288–6298.
94. Monni O, Barlund M, Mousses S et al (2001) Comprehensive copy number and gene expression profiling of the 17q23 amplicon in human breast cancer. Proc Natl Acad Sci U S A 98:5711–5716.
95. Muggerud AA, Edgren H, Wolf M et al (2009) Data integration from two microarray platforms identifies bi-allelic genetic inactivation of RIC8A in a breast cancer cell line. BMC Med Genomics 2:26.
96. Li H, Zhan M (2008) Unraveling transcriptional regulatory programs by integrative analysis of microarray and transcription factor binding data. Bioinformatics 24:1874–1880.
97. Youn A, Reiss DJ, Stuetzle W (2010) Learning transcriptional networks from the integration of ChIP-chip and expression data in a non-parametric model. Bioinformatics 26:1879–1886.
98. Wang Z, Gerstein M, Snyder M (2009) RNA-Seq: a revolutionary tool for transcriptomics. Nat Rev Genet 10:57–63.
99. Goldstein DB (2009) Common genetic variation and human traits. N Engl J Med 360:1696–1698.
100. McClellan J, King MC (2010) Genetic heterogeneity in human disease. Cell 141:210–217.
101. Bullinger L, Valk PJ (2005) Gene expression profiling in acute myeloid leukemia. J Clin Oncol 23:6296–6305.
102. Ong IM, Glasner JD, Page D (2002) Modelling regulatory pathways in *E. coli* from time series expression profiles. Bioinformatics 18: S241–248.
103. Zou M, Conzen SD (2005) A new dynamic Bayesian network (DBN) approach for identifying gene regulatory networks from time course microarray data. Bioinformatics 21:71–79.

Part III

Microarray Bioinformatics in Systems Biology (Bottom-Up Approach)

Chapter 12

Modeling Gene Regulation Networks Using Ordinary Differential Equations

Jiguo Cao, Xin Qi, and Hongyu Zhao

Abstract

Gene regulation networks are composed of transcription factors, their interactions, and targets. It is of great interest to reconstruct and study these regulatory networks from genomics data. Ordinary differential equations (ODEs) are popular tools to model the dynamic system of gene regulation networks. Although the form of ODEs is often provided based on expert knowledge, the values for ODE parameters are seldom known. It is a challenging problem to infer ODE parameters from gene expression data, because the ODEs do not have analytic solutions and the time-course gene expression data are usually sparse and associated with large noise. In this chapter, we review how the generalized profiling method can be applied to obtain estimates for ODE parameters from the time-course gene expression data. We also summarize the consistency and asymptotic normality results for the generalized profiling estimates.

Key words: Dynamic system, Gene regulation network, Generalized profiling method, Spline smoothing, Systems biology, Time-course gene expression

1. Introduction

Transcription is a fundamental biological process by which information in DNA is used to synthesize messenger RNA and proteins. Transcription is regulated by a set of transcription factors, which interact together to properly activate or inhibit gene expression. Transcription factors, their interactions, and targets compose a transcriptional regulatory network. Extensive research has been done to study transcriptional regulatory networks (1). Sun and Zhao provided a comprehensive review of various methods that have been developed to reconstruct regulatory networks from genomics data (2).

Transcriptional regulatory networks have been under extensive studies, and certain regulation patterns occur much more often than by chance. These patterns are called network motifs. One example of network motifs is the feed forward loop (FFL),

Junbai Wang et al. (eds.), *Next Generation Microarray Bioinformatics: Methods and Protocols*,
Methods in Molecular Biology, vol. 802, DOI 10.1007/978-1-61779-400-1_12, © Springer Science+Business Media, LLC 2012

which is composed of three genes X, Y, and Z, with gene X regulating the expressions of Y and Z, and gene Y regulating the expression of Z.

The dynamics of a regulation network can be modeled by a set of ordinary differential equations (ODEs). For example, Barkai and Leibler used ODEs to describe the cell-cycle regulation and signal transduction in simple biochemical networks (3). For an FFL with genes X, Y, and Z, let $X(t)$, $Y(t)$, and $Z(t)$ denote the expression levels of genes X, Y, and Z, respectively, at time t, the following ODEs were proposed in ref. *4* to model the FFL:

$$\frac{\mathrm{d}Y(t)}{\mathrm{d}t} = -\alpha_y Y(t) + \beta_y f(X(t), K_{xy}),$$

$$\frac{\mathrm{d}Z(t)}{\mathrm{d}t} = -\alpha_z Z(t) + \beta_z g(X(t), Y(t), K_{xz}, K_{yz}), \qquad (1)$$

where the regulation function is defined as $f(u, K) = (u/K)^H/(1 + (u/K)^H)$ when the regulation is activation, and this function is $f(u, K) = 1/(1 + (u/K)^H)$ when the regulation is repression. The parameter H controls the steepness of $f(u, K)$, and we choose $H = 2$ in our following analysis. The other parameter K defines the expression of gene X required to significantly regulate the expression of other genes. For example, when $u = K$, $f(u, K) = 0.5$. We assume genes X and Y regulate gene Z independently, and the regulation function from genes X and Y to gene Z is $g(t) = f(X(t), K_{xz})f(Y(t), K_{yz})$. The parameter α_y is the degradation and dilution rates of gene Y. If all regulations on gene Y stop at time $t = t^*$, then gene Y decays as $Y(t) = Y(t^*)\exp(-\alpha_y(t - t^*))$, and it reaches half of its peak expression at $t^* + \ln(2)/\alpha_y$. The parameter β_y, along with α_y, determins the maximal expression of gene Y, which is equal to β_y/α_y. Similar interpretations on parameters α_z and β_z apply to gene Z. The dynamic properties of this FFL were studied by Mangan and Alon (4).

With recent advances in genomics technologies, gene expression levels can be measured at multiple time points. These time-course gene expression data are often sparse, i.e., measured at a limited number of time points, and the measurements are also associated with substantial noises. Despite the noisy nature of the measured gene expression data, it is desirable to estimate the parameters in the FFL model from these data. Therefore, our objective is to make statistical inference about the parameters $\theta = (\beta_y, \beta_z, \alpha_y, \alpha_z, K_{xy}, K_{xz}, K_{yz})$ in the ODE model (1) from the noisy time-course gene expression data. In addition to many real data sets, the Dialogue for Reverse Engineering Assessments and Methods (DREAM) provides biologically plausible simulated gene expression data sets. These data sets allow researchers to evaluate various reverse engineering methods in an unbiased manner on their performance of deducing the structure of biological networks and predicting the outcomes of previously unseen experiments (5–7). These datasets can be found in the Web site (8).

It is a challenging problem to estimate ODE parameters from noisy data, since most ODEs do not have analytic solutions and solving ODE numerically is computationally intensive. Some methods have been proposed to address this problem. A two-step estimation procedure is proposed by Chen and Wu (9) to estimate time-varying parameters in ODE models, in which the derivative of the dynamic process is estimated by local polynomial regression in the first step, and the ODE parameters are estimated in the framework of nonlinear regression in the second step. Although this method is relatively easy to understand and implement, it is not easy to obtain accurate estimation for the derivative from noisy data. Ramsay et al. estimated ODE parameters with the generalized profiling method, and showed that this method can provide accurate estimates with low computation load (10). The asymptotic and finite-sample properties of the generalized profiling method were studied by Qi and Zhao (11).

Systems biology has also attracted much research on the identification of gene regulation dynamic process using ODE models. Transcriptional regulatory networks were inferred by Wang et al. (12) from gene expression data based on protein transcription complexes and mass action law. The Bayesian method was used by Rogers et al. (13), used to make the inference on ODE parameters. The transcription factor activity was estimated by Gao et al. (14) when the concentration of the activated protein cannot easily be measured. The Gaussian process was used by Aijo and Lahdesmaki (15) to estimate the nonparametric form of ODE models for the transcriptional-level regulation in the framework of Bayesian analysis. Gaussian process regression bootstrapping was applied by Kirk and Stumpf (16) to estimate an ODE model of a cell signaling pathway. Particularly, more than 40 benchmark problems were presented in (17) for ODE model identification of cellular systems.

In this chapter, we focus on the generalized profiling method, which is introduced in the next section. We also summarize the theoretical results in the next section. We then demonstrate the usefulness of this method through its application to estimate the parameters in the ODE model (1) from the noisy time-course gene expression data. We also provide a step-by-step description of using the Matlab function to estimate ODE parameters from the real gene expression data in the Web site (18). Some more details about the generalized profiling method can be found in (19).

2. Methods

Suppose the ODE model has I components and G ODEs:

$$\frac{dX_g(t)}{dt} = f_g(X_1(t), X_2(t), \cdots, X_I(t)|\theta), \quad g = 1, \cdots, G, \quad (2)$$

where the parametric form of the function $f_g(X_1(t), X_2(t), \cdots, X_I(t)|\theta)$ is known. Suppose we have noisy measurements for only $M \leq I$ components:

$$y_\ell(t_{\ell j}) = X_\ell(t_{\ell j}) + \epsilon_{\ell j},$$

where the measurement errors $\epsilon_{\ell j}$, $j = 1, 2, \cdots, n_\ell$ and $\ell = 1, 2, \cdots, M$, are assumed to be independent and identically distributed with the pdf $h(\cdot)$. The generalized profiling method estimates the ODE parameter θ in two nested levels of optimization. In the inner level, the ODE components are approximated with smoothing splines, conditional on the ODE parameter θ. So the fitted splines can be treated as an implicit function of θ. In the outer level, θ is estimated by maximizing the likelihood function.

2.1. Inner Level of Optimization

The ODE component $X_i(t)$, $i = 1, \cdots, I$, is approximated with a linear combination of K_i spline basis functions $\phi_k(t)$, $k = 1, \cdots, K_i$:

$$x_i(t) = \sum_{k=1}^{K_i} c_{ik}\phi_{ik}(t) = \phi_i(t)^T \mathbf{c}_i,$$

where $\phi_i = (\phi_{i1}, \cdots, \phi_{iK_i})^{\mathrm{T}}$ is a vector of spline basis functions and $\mathbf{c}_i = (c_{i1}, \cdots, c_{iK_i})^{\mathrm{T}}$ is a vector of spline coefficients. The nonparametric function $x_i(t)$ is required to be a tradeoff between fitting the noisy data and satisfying the ODE model (2).

Define the vector of spline coefficients $\mathbf{c} = (\mathbf{c}_1^{\mathrm{T}}, \cdots, \mathbf{c}_I^{\mathrm{T}})^{\mathrm{T}}$. The optimization criterion for estimating the spline coefficients $\mathbf{c}$ is chosen as the penalized likelihood function

$$J(\mathbf{c}|\theta) = -\sum_{\ell=1}^{M} \sum_{j=1}^{n_\ell} \omega_\ell \log(h(y_\ell(t_{\ell j}) - x_\ell(t_{\ell j}))) + \sum_{g=1}^{G} \lambda_g \omega_g \int \left[\frac{dx_g(t)}{dt} - f_g(x_1(t), x_2(t), \cdots, x_I(t)|\theta)\right]^2 dt, \quad (3)$$

where the first term measures the fit of $x_i(t)$ to the noisy data, and the second term measures the infidelity of $x_i(t)$ to the ODE model. The smoothing parameter $\lambda = (\lambda_1, \cdots, \lambda_g)$ controls the tradeoff between fitting the data and infidelity to the ODE model. The normalizing weight parameter ω_ℓ is used to keep different components having comparable scales. In this study, we set the values of ω_ℓ as the reciprocals of the variances taking over observations for the ℓth component.

In practice, the integration term in (3) as well as the integrations in the rest of this chapter are evaluated numerically. We use the composite Simpson's rule, which provides an adequate approximation to the exact integral (20). For an arbitrary function $u(t)$, the composite Simpson's rule is given by

$$\int_{t_1}^{t_n} u(t)\mathrm{d}t \approx \frac{a}{3}\left[u(s_0) + 2\sum_{q=1}^{Q/2-1} u(s_{2q} + 4\sum_{q=1}^{Q/2} u(s_{2q-1}) + u(s_Q)\right],$$

where the quadrature points $s_q = t_1 + qa$, $q = 0, \cdots, Q$, and $a = (t_n - t_1)/Q$.

The estimate $\hat{\mathbf{c}}$ can be treated as an implicit function of θ, which is denoted as $\hat{\mathbf{c}}(\theta)$. The derivative of $\hat{\mathbf{c}}$ with respect to θ is required to estimate θ in the next subsection. It can be obtained by using the implicit function theorem as follows. Taking the θ-derivative on both sides of the identity $\partial J/\partial \mathbf{c}|_{\hat{\mathbf{c}}} = 0$:

$$\frac{\mathrm{d}}{\mathrm{d}\theta}\left(\frac{\partial J}{\partial \mathbf{c}}\bigg|_{\hat{\mathbf{c}}}\right) = \frac{\partial^2 J}{\partial \mathbf{c}\partial\theta}\bigg|_{\hat{\mathbf{c}}} + \frac{\partial^2 J}{\partial \mathbf{c}^2}\bigg|_{\hat{\mathbf{c}}}\frac{\partial \hat{\mathbf{c}}}{\partial\theta} = 0.$$

Assuming that $\partial^2 J/\partial \mathbf{c}^2|_{\hat{\mathbf{c}}}$ is not singular, we get:

$$\frac{\partial \hat{\mathbf{c}}}{\partial\theta} = -\left[\frac{\partial^2 J}{\partial \mathbf{c}^2}\bigg|_{\hat{\mathbf{c}}}\right]^{-1}\left[\frac{\partial^2 J}{\partial \mathbf{c}\partial\theta}\bigg|_{\hat{\mathbf{c}}}\right]. \tag{4}$$

2.2. Outer Level of Optimization

The ODE parameter θ is estimated by maximizing the log likelihood function

$$H(\theta) = \sum_{\ell=1}^{M}\sum_{j=1}^{n_\ell} \omega_\ell \log(h(y_\ell(t_{\ell j}) - \hat{x}_\ell(t_{\ell j}))), \tag{5}$$

where the fitted curve $\hat{x}_\ell(t_{\ell j}) = \phi(t_{\ell j})^T\hat{\mathbf{c}}(\theta)$. The estimate $\hat{\theta}$ is obtained by optimizing $H(\theta)$ using the Newton–Raphson iteration method, which can run faster and is more stable if the gradient is given analytically. The analytic gradient is derived with the chain rule to accommodate $\hat{\mathbf{c}}$ being a function of θ:

$$\frac{\mathrm{d}H}{\mathrm{d}\theta} = \frac{\partial H}{\partial\theta} + \left(\frac{\partial\hat{\mathbf{c}}}{\partial\theta}\right)^T\frac{dH}{d\hat{\mathbf{c}}}.$$

2.3. Smoothing Parameter Selection

Our objective is to obtain the estimate $\hat{\theta}$ for the ODE parameters such that the solution of the ODEs with $\hat{\theta}$ fits the data. For each value of the smoothing parameter $\lambda = (\lambda_1, \cdots, \lambda_G)^T$, we obtain the ODE parameter estimate $\hat{\theta}$, so $\hat{\theta}$ may be treated as an implicit function of λ. The optimal value of λ is chosen by maximizing the likelihood function

$$F(\lambda) = \sum_{\ell=1}^{M} \sum_{j=1}^{n_\ell} \omega_\ell \log(h(y_\ell(t_{\ell j}) - s_\ell(t_{\ell j}|\hat{\theta}(\lambda)))), \qquad (6)$$

where $s_\ell(t_{\ell j}|\hat{\theta}(\lambda))$ is the ODE solution at the point $t_{\ell j}$ with the parameter $\hat{\theta}(\lambda)$ for the ℓth variable. The criterion (6) chooses the optimal value of λ such that the ODE solution with $\hat{\theta}(\lambda)$ is closest to the data.

2.4. Goodness-of-Fit of ODE Models

The goodness-of-fit of ODE models to noisy data can be assessed by solving ODEs numerically, and comparing the fit of ODE solutions to data. The initial values of the ODE variables are required to be specified for solving ODEs numerically. Because the ODE numerical solutions are sensitive to the initial values of the ODE variables, the estimates for the initial values have to be accurate. It is advisable to use the first observations for the ODE variables at the first time point as the initial values, which often have measurement errors. Moreover, some ODE variables may not be measurable, and no first observations are available.

A good byproduct of the parameter cascading method is that the initial values of the ODE variables can be estimated after obtaining the estimates for the ODE parameters. The parameter cascading method uses a nonparametric function to represent the dynamic process, hence the initial values of the ODE variables can be estimated by evaluating the nonparametric function at the first time point: $\hat{x}_g(t_0) = \hat{\mathbf{c}}_g^T \phi_g(t_0),\ g = 1, \cdots, G$. Our experience shows that the ODE solution with the estimated initial values tends to fit the data better than using the first observations directly.

2.5. Consistency and Asymptotic Normality

The asymptotic properties of the generalized profiling method were studied in ref. *11*. One novel feature of the generalized profiling method is that the true solutions to the ODEs are approximated by functions in a finite-dimensional space (e.g., the space spanned by the spline basis functions). Qi and Zhao defined a kind of distance, r, between the true solutions and the finite-dimensional space spanned by the basis functions (11). In the spline basis functions case, r depends on the number of knots. Hence, we can control the distance r by choosing an appropriate number of knots.

Qi and Zhao gave an upper bound on the uniform norm of the difference between the ODE solutions and their approximations in terms of the smoothing parameters λ and the distance r (11). Under some regularity conditions, if $\lambda \to \infty$ and $r \to 0$ as the sample size $n \to \infty$, the generalized profiling estimation is consistent. Furthermore, if we assume that

$$\frac{\lambda}{n^2} \to \infty \qquad \text{and} \qquad r = o_p\left(\frac{1}{n}\right), \qquad \text{as } n \to \infty,$$

we have asymptotic normality for the generalized profiling estimation and the asymptotic covariance matrix is the same as that of the maximum likelihood estimation. Therefore, the generalized profiling estimation is asymptotically efficient.

One innovative feature of the profiling procedure is that it incorporates a penalty term to estimate the coefficients in the first step. From the theory of differential equations, for such penalty, the bound on the difference between the approximations and the solutions will grow exponentially. As a result, if the time interval is large, the bound will be too large to be useful. However, for some ODEs (e.g., FitzHugh–Nagumo equations in ref. 10), the simulation studies indicate that when the smoothing parameter becomes large, the approximations to the solutions are very good. There is no trend of exponentially growing. To explain this phenomenon, Qi and Zhao fixed the sample and the approximation space, and studied the limiting situation as the smoothing parameter goes to infinity (11). Then they gave some conditions on the form of the ODEs under which they can give an upper bound without exponential increase.

2.6. Results

The time-course gene expression data in the yeast *Saccharomyces cerevisiae* are collected as described in ref. 21 under different conditions. Figure 1 displays the expression profiles of three genes (X: Gene GCN4; Y: Gene LEU3; Z: Gene ILV5) after the temperature is increased from 25 to 37°C. These three genes

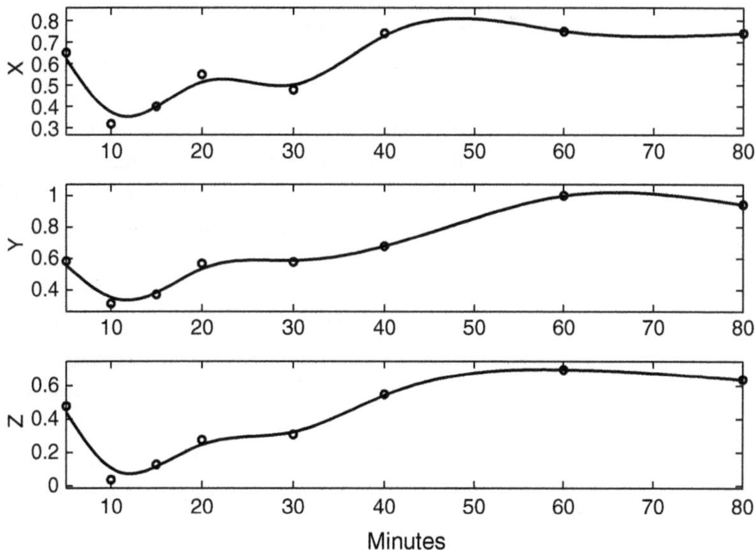

Fig. 1. The expression profiles of three genes (X: Gene GCN4; Y: Gene LEU3; Z: Gene ILV5) measured at 5, 10, 15, 20, 30, 40, 60, and 80 min. The data were collected by DNA microarrays from yeast after the temperature was increased from 25 to 37°C (21). The *solid lines* are the smooth curves estimated by penalized spline smoothing (The basis functions are cubic B-splines with 40 equally spaced knots, and the value of the smoothing parameter is 10).

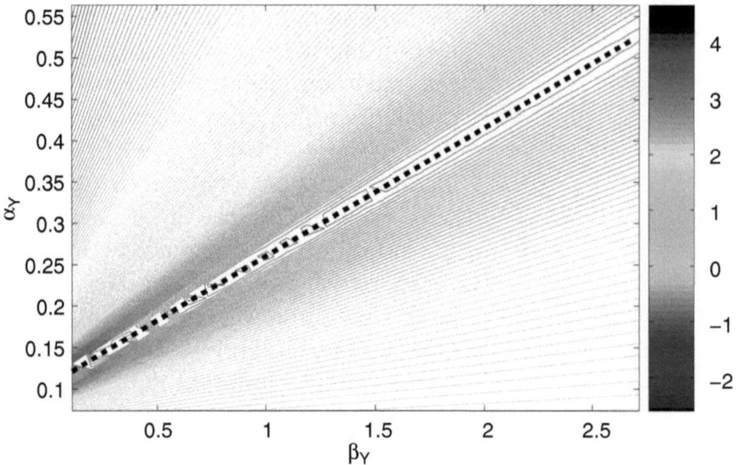

Fig. 2. The contour plot of the logarithm of the sums of squared differences between the measured expression of gene Y shown in Fig. 1 and the ODE (1) solution with different values of α_y and β_y. The value of K_{xy} is fixed as 0.93. The *dashed line* is $\alpha_y = 0.11 + 0.15 * \beta_y$.

compose a so-called Coherent Type 1 FFL, a type of FFL where X activates the expressions of Y and Z, and Y activates the expression of Z (4).

The ODE model (1) has seven parameters to estimate, but some preliminary analysis indicates that the estimates for α_y and β_y show strong collinearity, as well as the estimates for α_z and β_z. To demonstrate this, we fix the value of K_{xy}, vary values for α_y and β_y to solve the first ODE in (1), and compute the sum of squared differences between the ODE solution and the measured time-course expression of gene Y. Figure 2 is the contour plot of these logarithms of the sum squared differences. It shows that the values of α_y and β_y which lead to minimum sum squared differences are mostly located around the line $\alpha_y = 0.11 + 0.15\beta_y$. So in this application, the parameters β_y and β_z are fixed as 1, and we estimate the five parameters α_y, α_z, K_{xy}, K_{xz}, and K_{yz} from the time-course gene expression data.

The ODE model (1) is estimated for three different FFLs (FFL 1 is composed of X: Gene GCN4; Y: Gene LEU3; Z: Gene ILV5; FFL 2 is composed of X: Gene GCN4; Y: Gene LEU3; Z: Gene ILV1; FFL 3 is composed of X: Gene PDR1; Y: Gene PDR3; Z: Gene PDR5). The expression function for gene X, $X(t)$, is an input function in the ODE model and is estimated first by penalized spline smoothing. The parameters α_y, α_z, K_{xy}, K_{xz}, and K_{yz} are then estimated with the generalized profiling method from the time-course expression data of genes Y and Z. The expression functions for genes Y and Z, $Y(t)$ and $Z(t)$, are approximated by cubic B-splines with 40 equally spaced knots. The smoothing parameter is chosen as $\lambda = 1{,}000$.

Table 1
Parameter estimates and the standard errors for ODEs (1) and (2) from the measured expressions of genes Y and Z

FFL 1: X: Gene GCN4; Y: Gene LEU3; Z: Gene ILV5

Parameters	α_y	α_z	K_{xy}	K_{xz}	K_{yz}
Estimates	0.44	0.69	0.90	0.60	0.56
Standard errors	0.22	0.18	0.33	0.06	0.15

FFL 2: X: Gene GCN4; Y: Gene LEU3; Z: Gene ILV1

Parameters	α_y	α_z	K_{xy}	K_{xz}	K_{yz}
Estimates	0.44	0.90	0.90	0.75	1.21
Standard errors	0.22	0.01	0.33	0.44	0.74

FFL 3: X: Gene PDR1; Y: Gene PDR3; Z: Gene PDR5

Parameters	α_y	α_z	K_{xy}	K_{xz}	K_{yz}
Estimates	0.32	0.56	2.11	1.06	0.76
Standard errors	0.15	0.12	0.74	0.32	0.21

Each component is approximated by cubic B-splines with 40 equally spaced knots. The smoothing parameter $\lambda = 1{,}000$

The parameter estimates and their standard errors are displayed in Table 1. FFL 1 and FFL 2 have the same genes X and Y, and they are measured together in the same environmental changes (the temperature is increased from 25 to 37°C), so the parameters for gene Y to regulate gene X, α_y and K_{xy}, have the same values. The self-regulation parameter α_z for gene Z has different values, which means gene Z in FFL 2 is more self-repressed than gene Z in FFL 1. The parameter K_{yz} has a larger value in FFL 2 than FFL 1, so gene Y in FFL 2 has a higher level of threshold required to significantly activate the expression of gene Z. For FFL 3, K_{xy} and K_{xz} are relatively high, which indicates that gene X in FFL 3 has a high threshold to significantly activate the expression of genes Y and Z.

The goodness-of-fit of the ODE model (1) can be assessed by comparing time-course gene expression data with ODE solutions. Numerically solving ODEs requires the initial values for $Y(t)$ and $Z(t)$. These initial values are estimated by evaluating the spline curves at the start time point $t_0 = 5$, where the spline curves are estimated by minimizing penalized smoothing criterion (3).

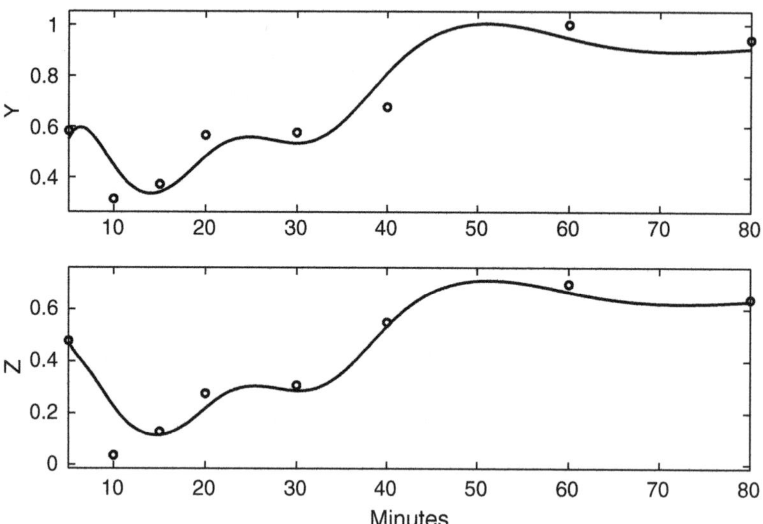

Fig. 3. The dynamic models for FFL 1 (X: Gene GCN4; Y: Gene LEU3; Z: Gene ILV5). The *circles* are the real expression profiles of three genes, and the *solid lines* are the numerical solutions to ODEs (1) and (2) with the ODE parameter estimates $\alpha_y = 0.44$, $\alpha_z = 0.69$, $K_{xy} = 0.90$, $K_{xz} = 0.60$, $K_{yz} = 0.56$ and the estimated initial values $Y(t_0) = 0.55$ and $Z(t_0) = 0.47$.

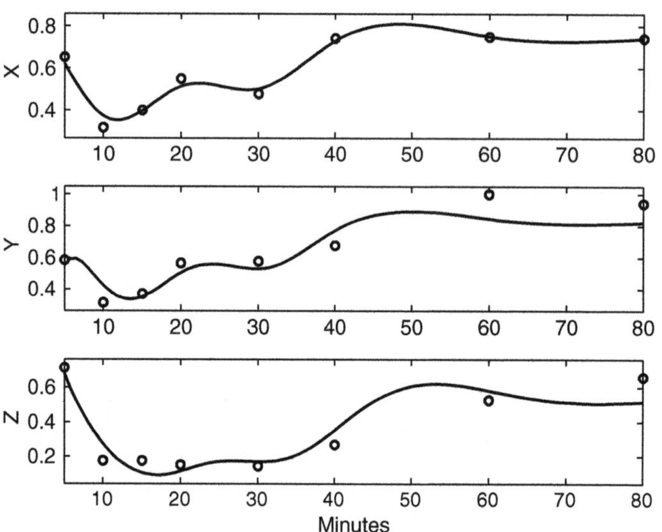

Fig. 4. The dynamic models for FFL 3 (X: Gene PDR1; Y: Gene PDR3; Z: Gene PDR5). The *circles* are the real gene expression profiles of three genes. The *solid lines* in the *top panel* is the estimated $\hat{X}(t)$, and the *solid lines* in the *bottom panels* are the ODE solutions to ODEs (1) and (2) with the ODE parameter estimates $\alpha_y = 0.32$, $\alpha_z = 0.56$, $K_{xy} = 2.11$, $K_{xz} = 1.06$, $K_{yz} = 0.76$ and the estimated initial values $Y(t_0) = 0.92$ and $Z(t_0) = 2.02$.

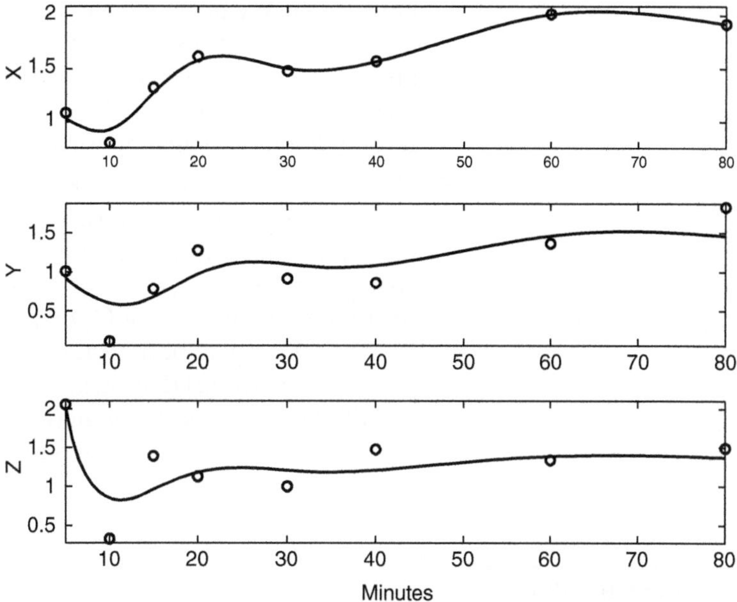

Fig. 5. The dynamic models for FFL 2 (X: Gene GCN4; Y: Gene LEU3; Z: Gene ILV1). The *circles* are the real gene expression profiles of three genes. The *solid lines* in the *top panel* is the estimated $\hat{X}(t)$, and the *solid lines* in the *bottom panels* are the ODE solutions to ODEs (1) and (2) with the ODE parameter estimates $\alpha_y = 0.44$, $\alpha_z = 0.90$, $K_{xy} = 0.90$, $K_{xz} = 0.75$, $K_{yz} = 1.21$ and the estimated initial values $Y(t_0) = 0.55$ and $Z(t_0) = 0.70$.

Figures 3–5 show the numerical solutions to the ODE model (1) with the ODE parameter estimates and the estimated initial values for the three FFLs. The ODE solutions are all close to the time-course expression data of genes Y and Z, which indicates that the ODE (1) is a good dynamic model for the FFL regulation network.

3. Notes

1. The regulation process of the FFL is modeled with two ODEs. The usefulness of the generalized profiling method is demonstrated by estimating parameters in the ODE model from time-course gene expression data. Although the ODE solution with the parameter estimates shows a satisfactory fit to the noisy data, we also find some limitations of the current data and method.
2. In our application, the expressions of three genes are only measured at eight time points. These data are too sparse to obtain precise estimates for ODE parameters. It will greatly

improve the accuracy of parameter estimates if more frequent data are collected, especially in the period when the dynamic process has sharp changes. In our application, more measurements are required in (0, 20), in which the gene expressions show a downward then upward trend.

3. The gene regulation networks usually contain hundreds of transcription factors and their targets. After figuring out the regulation connection among these genes, the dynamic system for the regulation of these genes can be modeled with the same number of ODEs, which may have the similar forms as (1). It will be a great challenge to infer thousands of parameters in the ODE model. Beyond this, it is even harder to identify the gene regulation networks directly using the ODE models from the sparse time-course gene expression data.

Acknowledgments

Qi and Zhao's research is supported by NIH grant GM 59507 and NSF grant DMS-0714817. Cao's research is supported by a discovery grant of the Natural Sciences and Engineering Research Council (NSERC) of Canada. The authors thank the invitation from the editors of this book.

References

1. Alon U (2007) An introduction to systems biology. Chapman & Hall/CRC, London.
2. Sun N, Zhao H (2009) Reconstructing transcriptional regulatory networks through genomics data. Statistical Methods in Medical Research 18:595–617.
3. Barkai N, Leibler S (1997) Robustness in simple biochemical networks. Nature 387:913–917.
4. Mangan S, Alon U (2003) Structure and function of the feed-forward loop network motif. Proceeding of the National Academy of Sciences 100:11980–11985.
5. Stolovitzky G, Monroe D, Califano A (2007) Dialogue on reverseengineering assessment and methods: The dream of high-throughput pathway inference. Annals of the New York Academy of Sciences 1115:11–22.
6. Stolovitzky G, Prill RJ, Califano A (2009) Lessons from the dream2 challenges. Annals of the New York Academy of Sciences 1158:159–195.
7. Prill RJ, Marbach D, Saez-Rodriguez J et al (2010) Towards a rigorous assessment of systems biology models: the dream3 challenges. PLoS One 5:e9202.
8. Dialogue for Reverse Engineering Assessments and Methods (DREAM), http://wiki.c2b2.columbia.edu/dream.
9. Chen J, Wu H (2008) Efficient local estimation for time-varying coefficients in deterministic dynamic models with applications to HIV-1 dynamics. Journal of the American Statistical Association 103(481):369–383.
10. Ramsay JO, Hooker G, Campbell D et al (2007) Parameter estimation for differential equations: a generalized smoothing approach (with discussion). Journal of the Royal Statistical Society, Series B 69:741–796.
11. Qi X, Zhao H (2010) Asymptotic efficiency and finite-sample properties of the generalized profiling estimation of parameters in ordinary differential equations. The Annals of Statistics 38:435–481.

12. Wang R, Wang Y, Zhang X et al (2007) Inferring transcriptional regulatory networks from high-throughput data. Bioinformatics 23:3056–3064.
13. Rogers S, Khanin R, Girolami M (2007) Bayesian model-based inference of transcription factor activity. BMC Bioinformatics 8:1–11.
14. Gao P, Honkela A, Rattray M et al (2008) Genomic expression programs in the response of yeast cells to environmental changes. Bioinformatics 24:i70–i75.
15. Aijo T, Lahdesmaki H (2009) Learning gene regulatory networks from gene expression measurements using non-parametric molecular kinetics. Bioinformatics 25:2937–2944.
16. Kirk PDW, Stumpf MPH (2009) Gaussian process regression bootstrapping: exploring the effects of uncertainty in time course data. Bioinformatics 25:1300–1306.
17. Gennemark P, Wedelin D (2009) Benchmarks for identification of ordinary differential equations from time series data. Bioinformatics 25:780–786.
18. Matlab codes for estimating parameters in the ODE models, http://www.stat.sfu.ca/~cao/Research.html.
19. Cao J, Zhao H (2008) Estimating dynamic models for gene regulation networks. Bioinformatics 24:1619–1624.
20. Burden RL, Douglas FJ (2000) Numerical Analysis. Brooks/Cole, Pacific Grove, California, seventh edition.
21. Gasch AP, Spellman PT, Kao CM et al (2000) Genomic expression programs in the response of yeast cells to environmental changes. Molecular Biology of the Cell 11:4241–4257.

Chapter 13

Nonhomogeneous Dynamic Bayesian Networks in Systems Biology

Sophie Lèbre, Frank Dondelinger, and Dirk Husmeier

Abstract

Dynamic Bayesian networks (DBNs) have received increasing attention from the computational biology community as models of gene regulatory networks. However, conventional DBNs are based on the homogeneous Markov assumption and cannot deal with inhomogeneity and nonstationarity in temporal processes. The present chapter provides a detailed discussion of how the homogeneity assumption can be relaxed. The improved method is evaluated on simulated data, where the network structure is allowed to change with time, and on gene expression time series during morphogenesis in *Drosophila melanogaster*.

Key words: Dynamic Bayesian networks (DBNs), Changepoint processes, Reversible jump Markov chain Monte Carlo (RJMCMC), Morphogenesis, *Drosophila melanogaster*

1. Introduction

There is currently considerable interest in structure learning of dynamic Bayesian networks (DBNs), with a variety of applications in signal processing and computational biology; see, e.g., refs. 1–3. The standard assumption underlying DBNs is that time series have been generated from a homogeneous Markov process. This assumption is too restrictive in many applications and can potentially lead to erroneous conclusions. While there have been various efforts to relax the homogeneity assumption for undirected graphical models (4, 5), relaxing this restriction in DBNs is a more recent research topic (1–3, 6–8). At present, none of the proposed methods is without its limitations, leaving room for further methodological innovation. The method proposed in (3, 8) for recovering changes in the network is non-Bayesian. This requires certain regularization parameters to be optimized "externally", by applying information criteria (such as AIC or BIC), cross-validation, or bootstrapping. The first approach is suboptimal, the latter approaches are computationally expensive. (See ref. 9 for a demonstration of the higher

Junbai Wang et al. (eds.), *Next Generation Microarray Bioinformatics: Methods and Protocols*,
Methods in Molecular Biology, vol. 802, DOI 10.1007/978-1-61779-400-1_13,

computational costs of bootstrapping over Bayesian approaches based on MCMC.) In the present chapter, we therefore follow the Bayesian paradigm like in refs. (1, 2, 6, 7). These approaches also have their limitations. The method proposed in (2) assumes a fixed network structure and only allows the interaction parameters to vary with time. This assumption is too rigid when looking at processes where changes in the overall structure of regulatory processes are expected, e.g., in morphogenesis or embryogenesis. The method proposed in (1) requires a discretization of the data, which incurs an inevitable information loss. The method also does not allow for individual nodes of the network to deviate from the homogeneity assumption in different ways. These limitations are addressed in (6, 7), which allows network structures associated with different nodes to change with time in different ways. However, this high flexibility causes potential problems when applied to time series with a low number of measurements, as typically available from systems biology, leading to overfitting or inflated inference uncertainty. The objective of the work described in this chapter is to propose a model that addresses the principled shortcomings of the three Bayesian methods mentioned above. Unlike ref. 1, our model is continuous and therefore avoids the information loss inherent in a discretization of the data. Unlike ref. 2, our model allows the network structure to change among segments, leading to greater model flexibility. As an improvement on (6, 7), our model introduces information sharing among time series segments, which provides an essential regularization effect.

2. Materials

2.1. Simulated Data

We generated synthetic time series, each consisting of K segments, as follows. A random network $\mathcal{M}^1$ is generated stochastically, with the number of incoming edges for each node drawn from a Poisson distribution with mean λ_1. To simulate a sequence of networks $\mathcal{M}^h$, $1 \leq h \leq K$, separated by changepoints, we sampled Δn_h from a Poisson distribution with mean λ_2, and then randomly changed Δn_h edges between $\mathcal{M}^h$ and $\mathcal{M}^{h+1}$, leaving the total number of existing edges unchanged.

Each directed edge from node j (the parent) to node i (the child) in segment h has a weight a_{ij}^h that determines the interaction strength, drawn from a Gaussian distribution. The signal associated with node i at time t, $y_i(t)$, evolves according to the nonhomogeneous first-order Markov process of equation (1). The matrix of all interaction strengths a_{ij}^h is denoted by $\mathbf{A}^h$. To ensure stationarity of the time series, we tested if all eigenvalues of $\mathbf{A}^h$ had a modulus ≤ 1 and removed edges randomly until this condition was met.

We randomly generated networks with 10 nodes each, with $\lambda_1 = 3$. We set $K = 4$ and $\lambda_2 \in \{0, 1\}$. For each segment, we generated a time series of length 15. The regression weights were drawn from a Gaussian $N(0, 1)$, and Gaussian observation noise $N(0, 1)$ was added. The process was repeated ten times to generate ten independent datasets.

2.2. Morphogenesis in Drosophila melanogaster

Drosophila and vertebrates share many common molecular pathways, e.g., embryonic segmentation and muscle development. As a simpler species than humans, *Drosophila* has fewer muscle types and each muscle type is composed of only one fibre type. We applied our method to the developmental gene expression time series for *Drosophila melanogaster* (fruit fly) obtained in (10). Expression values of 4,028 genes were measured with microarrays at 67 time points during the *Drosophila* life cycle, which contains the four distinct phases of embryo, larva, pupa, and adult. Initially, a homogeneous muscle development genetic network was proposed in (11) for a set of 20 genes reported to relate to muscle development (10, 12, 13). Following ref. 14 who inferred an undirected network specific to each of the four distinct phases of the *Drosophila* life cycle, and ref. 1, we concentrated on the subset of 11 genes corresponding to the largest connected component of this muscle development network in order to propose a nonhomogeneous network pointing out differences between the various *Drosophila* life phases.

3. Methods

This section summarizes briefly the nonhomogeneous DBN proposed in (6, 7), which combines the Bayesian regression model of (15) with multiple changepoint processes and pursues Bayesian inference with reversible jump Markov chain Monte Carlo (RJMCMC) (17). In what follows, we will refer to nodes as genes and to the network as a gene regulatory network. The method is not restricted to molecular systems biology, though. See Note 1 for a publicly available software implementation.

3.1. Model

Multiple changepoints. Let p be the number of observed genes, whose expression values $y = \{y_i(t)\}_{1 \leq i \leq p,\, 1 \leq t \leq N}$ are measured at N time points. $\mathcal{M}$ represents a directed graph, i.e., the network defined by a set of directed edges among the p genes. $\mathcal{M}_i$ is the subnetwork associated with target gene i, determined by the set of its parents (nodes with a directed edge feeding into gene i). The regulatory relationships among the genes, defined by $\mathcal{M}$, may vary across time, which we model with a multiple changepoint process. For each target gene i, an unknown number k_i of

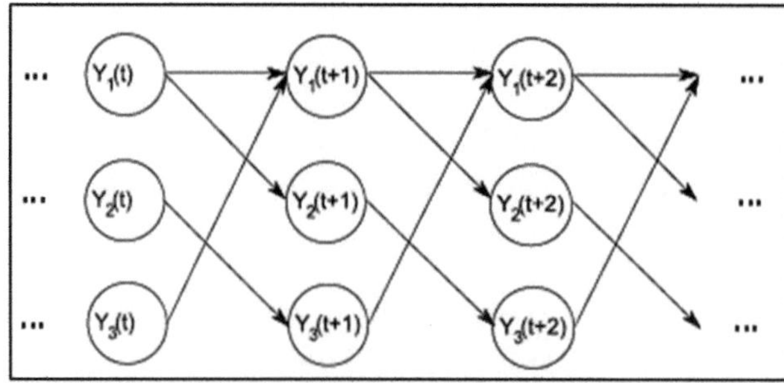

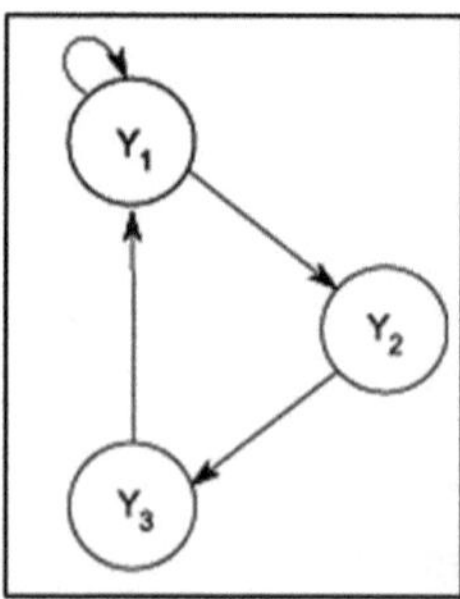

Fig. 1. *Left*: Structure of a dynamic Bayesian network. Three genes $\{Y_1, Y_2, Y_3\}$ are included in the network, and three time steps $\{t, t+1, t+2\}$ are shown. The *arrows* indicate interactions between the genes. *Right*: The corresponding state space graph, from which the structure on the left is obtained through the process of unfolding in time. Note that the state space graph is a recurrent structure, with two feedback loops: $Y_1 \rightarrow Y_2 \rightarrow Y_3 \rightarrow Y_1$, and a self-loop on Y_1.

changepoints define $k_i + 1$ nonoverlapping segments. Segment $h = 1, .., k_i + 1$ starts at changepoint ξ_i^{h-1} and stops before ξ_i^h, where $\xi_i = (\xi_i^0, \ldots, \xi_i^{h-1}, \xi_i^h, \ldots, \xi_i^{k_i+1})$ with $\xi_i^{h-1} < \xi_i^h$. To delimit the bounds, $\xi_i^0 = 2$ and $\xi_i^{k_i+1} = N + 1$. Thus, vector ξ_i has length $|\xi_i| = k_i + 2$. The set of changepoints is denoted by $\xi = \{\xi_i\}_{1 \leq i \leq p}$. This changepoint process induces a partition of the time series, $y_i^h = (y_i(t))_{\xi_i^{h-1} \leq t < \xi_i^h}$, with different structures $\mathcal{M}_i^h$ associated with the different segments $h \in \{1, \ldots, k_i + 1\}$. Identifiability is satisfied by ordering the changepoints based on their position in the time series.

Regression model. For all genes i, the random variable $\mathrm{Y_i(t)}$ refers to the expression of gene i at time t. Within any segment h, the expression of gene i depends on the p gene expression values measured at the previous time point through a regression model defined by (a) a set of s_i^h parents denoted by $\mathcal{M}_i^h = \{j_1, \ldots, j_{s_i^h}\} \subseteq \{1, \ldots, p\}$, $|\mathcal{M}_i^h| = s_i^h$, and (b) a set of parameters $((a_{ij}^h)_{j \in 0..p}, \sigma_i^h)$; $a_{ij}^h \in \mathbb{R}$, $\sigma_i^h > 0$. For all $j \neq 0$, $a_{ij}^h = 0$ if $j \notin \mathcal{M}_i^h$. For all genes i, for all time points t in segment h ($\xi_i^{h-1} \leq t < \xi_i^h$), random variable $\mathrm{Y_i(t)}$ depends on the p variables $\{\mathrm{Y_j(t-1)}\}_{1 \leq j \leq p}$ according to

$$\mathrm{Y_i(t)} = \mathrm{a_{i0}^h} + \sum\nolimits_{\mathrm{j} \in \mathcal{M}_i^h} \mathrm{a_{ij}^h} \, \mathrm{Y_j(t-1)} + \varepsilon_\mathrm{i}(\mathrm{t}), \quad (1)$$

where the noise $\varepsilon_i(t)$ is assumed to be Gaussian with mean 0 and variance $(\sigma_i^h)^2$, $\varepsilon_i(t) \sim N(0, (\sigma_i^h)^2)$. We define $a_i^h = (a_{ij}^h)_{j \in 0..p}$. Figure 1 illustrates the regression model and shows how the dynamic framework allows us to model feedback loops that would not otherwise be possible in a Bayesian network.

3.2. Prior

The $k_i + 1$ segments are delimited by k_i changepoints, where k_i is distributed a priori as a truncated Poisson random variable with mean λ and maximum $\bar{k} = N - 2 : P(k_i|\lambda) \propto \frac{\lambda^{k_i}}{k_i!} \mathbb{1}_{\{k_i \leq \bar{k}\}}$. Note that a restrictive Poisson prior encourages sparsity and is therefore comparable to a sparse exponential prior or to an approach based on the LASSO.

Conditional on k_i changepoints, the changepoint positions vector $\xi_i = (\xi_i^0, \xi_i^1, \ldots, \xi_i^{k_i+1})$ takes nonoverlapping integer values, which we take to be uniformly distributed a priori. There are $(N-2)$ possible positions for the k_i changepoints, thus vector ξ_i has prior density $P(\xi_i|k_i) = 1/\binom{N-2}{k_i}$. For all genes i and all segments h, the number s_i^h of parents for node i follows a truncated Poisson distribution with mean Λ and maximum $\bar{s} = 5 : P(s_i^h|\Lambda) \propto \frac{\Lambda^{s_i^h}}{s_i^h!} \mathbb{1}_{\{s_i^h \leq \bar{s}\}}$. Conditional on s_i^h, the prior for the parent set $\mathcal{M}_i^h$ is a uniform distribution over all parent sets with cardinality $s_i^h : P(\mathcal{M}_i^h \| \mathcal{M}_i^h| = s_i^h) = 1/\binom{p}{s_i^h}$. The overall prior on the network structures is given by marginalization:

$$P(\mathcal{M}_i^h|\Lambda) = \sum_{s_i^h=1}^{\bar{s}} P(\mathcal{M}_i^h|s_i^h) P(s_i^h|\Lambda). \tag{2}$$

Conditional on the parent set $\mathcal{M}_i^h$ of size s_i^h, the $s_i^h + 1$ regression coefficients, denoted by $a_{\mathcal{M}_i^h} = (a_{i0}^h, (a_{ij}^h)_{j \in \mathcal{M}_i^h})$, are assumed zero-mean multivariate Gaussian with covariance matrix $(\sigma_i^h)^2 \Sigma_{\mathcal{M}_i^h}$,

$$P(a_i^h|\mathcal{M}_i^h, \sigma_i^h) = \left|2\pi(\sigma_i^h)^2 \Sigma_{\mathcal{M}_i^h}\right|^{-\frac{1}{2}} \exp\left(-\frac{a_{\mathcal{M}_i^h}^{\dagger} \Sigma_{\mathcal{M}_i^h}^{-1} a_{\mathcal{M}_i^h}}{2(\sigma_i^h)^2}\right), \tag{3}$$

where the symbol † denotes matrix transposition, $\Sigma_{\mathcal{M}_i^h} = \delta^{-2} D_{\mathcal{M}_i^h}^{\dagger}(y) D_{\mathcal{M}_i^h}(y)$ and $D_{\mathcal{M}_i^h}(y)$ is the $(\xi_i^h - \xi_i^{h-1}) \times (s_i^h + 1)$ matrix whose first column is a vector of 1 (for the constant in model of equation (1)) and each $(j+1)th$ column contains the observed values $(y_j(t))_{\xi_i^{h-1}-1 \leq t < \xi_i^h - 1}$ for each factor gene j in $\mathcal{M}_i^h$ (15). Finally, the conjugate prior for the variance $(\sigma_i^h)^2$ is the inverse gamma distribution, $P((\sigma_i^h)^2) = \mathcal{IG}(\upsilon_0, \gamma_0)$. Following refs. 6, 7, we set the hyper-hyperparameters for shape, $\upsilon_0 = 0.5$, and scale, $\gamma_0 = 0.05$, to fixed values that give a vague distribution. The terms λ and Λ can be interpreted as the expected number of changepoints and parents, respectively, and δ^2 is the expected signal-to-noise ratio. These hyperparameters are drawn from vague conjugate hyperpriors, which are in the (inverse) gamma distribution family: $P(\Lambda) = P(\lambda) = \mathcal{G}a(0.5, 1)$ and $P(\delta^2) = \mathcal{IG}(2, 0.2)$.

3.3. Posterior

Equation (1) implies that

$$P\left(y_i^h|\xi_i^{h-1},\xi_i^h,\mathcal{M}_i^h,a_i^h,\sigma_i^h\right)=\left(\sqrt{2\pi}\sigma_i^h\right)^{-\left(\xi_i^h-\xi_i^{h-1}\right)}$$
$$\exp\left(-\frac{\left(y_i^h-D_{\mathcal{M}_i^h}(y)a_{\mathcal{M}_i^h}\right)^\dagger\left(y_i^h-D_{\mathcal{M}_i^h}(y)a_{\mathcal{M}_i^h}\right)}{2\left(\sigma_i^h\right)^2}\right). \quad (4)$$

From Bayes theorem, the posterior is given by the following equation, where all prior distributions have been defined above:

$$P(k,\xi,\mathcal{M},a,\sigma,\lambda,\Lambda,\delta^2|y)\propto P(\delta^2)P(\lambda)P(\Lambda)\prod_{i=1}^{p}P(k_i|\lambda)P(\xi_i|k_i)$$
$$\prod_{h=1}^{k_i}P(\mathcal{M}_i^h|\Lambda)P\left(\left[\sigma_i^h\right]^2\right)$$
$$P\left(a_i^h|\mathcal{M}_i^h,\left[\sigma_i^h\right]^2,\delta^2\right)P\left(y_i^h|\xi_i^{h-1},\xi_i^h,\mathcal{M}_i^h,a_i^h,\left[\sigma_i^h\right]^2\right). \quad (5)$$

3.4. Inference

An attractive feature of the chosen model is that the marginalization over the parameters a and σ in the posterior distribution of equation (5) is analytically tractable,

$$P(k,\xi,\mathcal{M},\lambda,\Lambda,\delta^2|y)=\int\int P(k,\xi,\mathcal{M},a,\sigma,\lambda,\Lambda,\delta^2|y)\,da\,d\sigma \quad (6)$$

$$=P(\delta^2)P(\lambda)P(\Lambda)\prod_{i=1}^{p}\int\int P(k_i,\xi_i,\mathcal{M}_i,a_i,\sigma_i|\lambda,\Lambda,\delta^2,y)\,da_i\,d\sigma_i \quad (7)$$

$$=P(\delta^2)P(\lambda)P(\Lambda)\prod_{i=1}^{p}P(k_i,\xi_i,\mathcal{M}_i|\lambda,\Lambda,\delta^2,y). \quad (8)$$

For each gene i, $P(k_i,\xi_i,\mathcal{M}_i,a_i,\sigma_i|\lambda,\Lambda,\delta^2,y)$ denotes the distribution of the quantities related to the changepoints (k_i, ξ_i), network structure ($\mathcal{M}_i$), interaction strengths (a_i), and noise levels (σ_i), conditional on the hyperparameters (λ, Λ, δ^2) and data y. The essence of the above equation is that the integral over the parameters a_i (normal distribution) and σ_i (inverse gamma distribution) can be solved in closed form to obtain an expression for the posterior distribution of the quantities related to the network structure and changepoints: $(k_i,\xi_i,\mathcal{M}_i)$ (see refs. 6, 7 for computational details).

The number of changepoints and their location, k, ξ, the network structure $\mathcal{M}$, and the hyperparameters λ, Λ, δ^2 can be sampled from the posterior distribution $P(k,\xi,\mathcal{M},\lambda,\Lambda,\delta^2|y)$ with RJMCMC (16). The RJMCMC scheme is outlined in Algorithm 1.

Algorithm 1: Outline of the RJMCMC procedure for nonhomogeneous DBN inference

1. Initialization:
   Define an initial network $\mathcal{M}$ with interaction parameters a, maximum number of regulators per node $\bar{s}$, noise level σ, and changepoint configurations (k,ξ).
2. Iteration l:
   Compute changepoint birth (b_k), death (d_k), and shift (v_k) probability.
   Sample $u \sim \mathcal{U}_{[0,1]}$.

   if $(u \leq b_k)$ **then**
   ⌊ carry out a changepoint birth move
   else if $(u \leq b_k + d_k)$ **then**
   ⌊ carry out a changepoint death move
   else if $(u \leq b_k + d_k + v_k)$ **then**
   ⌊ carry out a changepoint position shift
   else carry out a network structure change within segments.

   Accept or reject the move according to the Metropolis–Hastings criterion; see refs. 6, 7 for the specific expressions.
3. $l \leftarrow l + 1$ and go to 2.

The move for "network structure change within segments" is adapted from ref. 15. A complete description can be found in ref. 6, 7. The algorithm must be run until convergence is obtained to ensure that the sampled networks and changepoint locations correspond to a sample from the posterior distribution (see Note 2 for details about convergence criteria). Note that the generation of the regression model parameters (a_i, σ_i) is optional and only used when an estimation of their posterior distribution is wished for. Indeed, a changepoint birth or death acceptance is performed without generating the regression model parameters for the modified phase. Thus, the acceptance probability of the move does not depend on the regression model parameters (θ_i, σ_i) but only on the network topology in the phases delimited by the changepoint involved in the move.

3.5. Regularization via Information Coupling

Allowing the network structure to change between segments leads to a highly flexible model. However, this approach faces a conceptual and a practical problem. The *practical* problem is the Potential over flexibility of the model. If subsequent changepoints are close together, network structures have to be inferred from short time series segments. This will almost inevitably lead to overfitting (in a maximum likelihood context) or inflated inference uncertainty (in a Bayesian context). The *conceptual* problem is

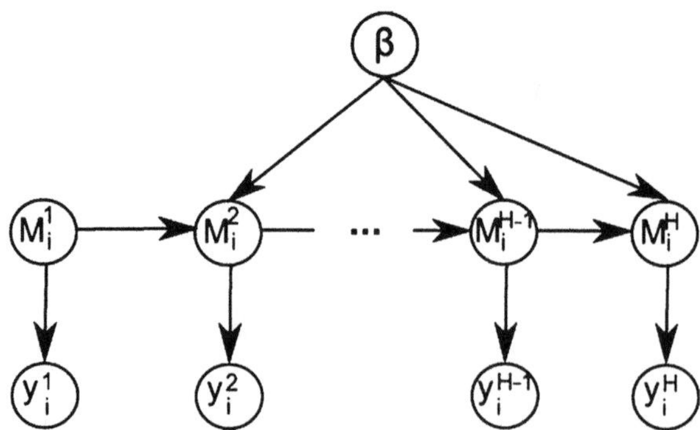

Fig. 2. Information sharing model with exponential prior. We couple each network segment $\mathcal{M}_i^h$ with $h > 1$ to the preceding segment $\mathcal{M}_i^{h-1}$ via an exponential prior on the number of structure differences between the two networks. The strength of the coupling is regulated by the inferred parameter β.

the underlying assumption that structures associated with different segments are a priori independent. This is not realistic. For instance, for the evolution of a gene regulatory network during embryogenesis, we would assume that the network evolves gradually and that networks associated with adjacent time intervals are a priori similar.

To address these problems, we propose a method of information sharing among time series segments, which is motivated by the work described in ref. 17 and is illustrated in Fig. 2.

Denote by $K_i := k_i + 1$ the total number of partitions in the time series, and recall that each time series segment y_i^h is associated with a separate subnetwork $\mathcal{M}_i^h$, $1 \leq h \leq K_i$. We impose a prior distribution $P(\mathcal{M}_i^h|\mathcal{M}_i^{h-1}, \beta)$ on the structures, and the joint probability distribution factorizes according to a Markovian dependence:

$$P(y_i^1, \ldots, y_i^{K_i}, \mathcal{M}_i^1, \ldots, \mathcal{M}_i^K, \beta) = \prod_{h=1}^{K_i} P(y_i^h|\mathcal{M}_i^h) P(\mathcal{M}_i^h|\mathcal{M}_i^{h-1}, \beta) P(\beta), \quad (9)$$

Similar to ref. 17 we define

$$P(\mathcal{M}_i^h|\mathcal{M}_i^{h-1}, \beta) = \frac{\exp(-\beta|\mathcal{M}_i^h - \mathcal{M}_i^{h-1}|)}{Z(\beta, \mathcal{M}_i^{h-1})}, \quad (10)$$

for $h \geq 2$, where β is a hyperparameter that defines the strength of the coupling between $\mathcal{M}_i^h$ and $\mathcal{M}_i^{h-1}$. In addition to coupling adjacent segments, sharing the same β parameter also provides a coupling over nodes by enforcing the same coupling strength for every node. For $h = 1$, $P(\mathcal{M}_i^h)$ is given by equation (2). The denominator $Z(\beta, \mathcal{M}_i^{h-1})$ in equation (10) is a normalizing constant, also known as the partition function: $Z(\beta) = \sum_{\mathcal{M}_i^h \in \mathbb{M}} e^{-\beta|\mathcal{M}_i^h - \mathcal{M}_i^{h-1}|}$

where $\mathbb{M}$ is the set of all valid subnetwork structures. If we ignore any fan-in restriction that might have been imposed a priori (via $\bar{s}$), then the expression for the partition function can be simplified: $Z(\beta) \approx \prod_{j=1}^{p} Z_j(\beta)$, where $Z_j(\beta) = \sum_{e_j^h=0}^{1} e^{-\beta|e_j^h - e_j^{h-1}|} = 1 + e^{-\beta}$ and hence $Z(\beta) = \left(1 + e^{-\beta}\right)^p$. Inserting this expression into equation (10) gives:

$$P(\mathcal{M}_i^h|\mathcal{M}_i^{h-1}, \beta) = \frac{\exp(-\beta|\mathcal{M}_i^h - \mathcal{M}_i^{h-1}|)}{(1 + e^{-\beta})^p}. \tag{11}$$

It is straightforward to integrate the proposed model into the RJMCMC scheme of refs. 6, 7 as described in Subheading 3.4. When proposing a new network structure $\mathcal{M}_i^h \rightarrow \tilde{\mathcal{M}}_i^h$ for segment h, the prior probability ratio has to be replaced by: $\frac{P(\mathcal{M}_i^{h+1}|\tilde{\mathcal{M}}_i^h,\beta)P(\tilde{\mathcal{M}}_i^h|\mathcal{M}_i^{h-1},\beta)}{P(\mathcal{M}_i^{h+1}|\mathcal{M}_i^h,\beta)P(\mathcal{M}_i^h|\mathcal{M}_i^{h-1},\beta)}$. An additional MCMC step is introduced for sampling the hyperparameters β from the posterior distribution. For a proposal move $\beta \rightarrow \tilde{\beta}$ with symmetric proposal probability $Q(\tilde{\beta}|\beta) = Q(\beta|\tilde{\beta})$, we get the following acceptance probability:

$$A(\tilde{\beta}|\beta) = \min\left\{\frac{P(\tilde{\beta})}{P(\beta)}\prod_{i=1}^{p}\prod_{h=2}^{K_i}\frac{\exp(-\tilde{\beta}|\mathcal{M}_i^h - \mathcal{M}_i^{h-1}|)}{\exp(-\beta|\mathcal{M}_i^h - \mathcal{M}_i^{h-1}|)}\frac{(1+e^{-\beta})^p}{\left(1+e^{-\tilde{\beta}}\right)^p}, 1\right\}, \tag{12}$$

where in our study the hyperprior $P(\beta)$ was chosen as the uniform distribution on the interval [0, 10].

3.6. Results

3.6.1. Comparative Evaluation on Simulated Data

We compared the network reconstruction accuracy on the simulated data described in Subheading 2.1. Figure 3 shows the network reconstruction performance in terms of AUROC and AUPRC scores. (See Notes 3 and 4 for a definition and interpretation.) Information sharing with exponential prior (HetDBN-Exp) shows a clear improvement in network reconstruction over no information sharing (HetDBN-0), as confirmed by paired t-tests ($p < 0.01$). We chose to draw the number of changes from a Poisson distribution with mean 1 for each node. We investigated two different situations, the case where all segment structures are the same (although edge weights are allowed to vary) and the case where changes are applied sequentially to the segments. Information sharing is most beneficial for the first case, but even when we introduce changes we still see an increase in the network reconstruction scores compared to HetDBN-0. When the segments are different, HetDBN-Exp still outperforms HetDBN-0 ($p < 0.05$).

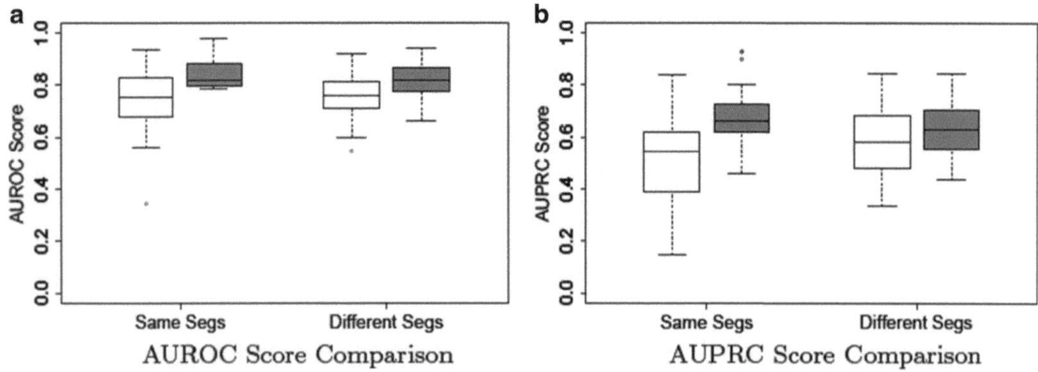

Fig. 3. Network reconstruction performance comparison of AUROC and AUPRC reconstruction scores without information sharing (*white*), and with sequential information sharing via an exponential prior (*light grey*). The *boxplots* show the distributions of the scores for ten datasets with four network segments each, where the *horizontal bar* shows the median, the *box margins* show the 25th and 75th percentiles, the *whiskers* indicate data within two times the interquartile range, and *circles* are outliers. "Same Segs" means that all segments in a dataset have the same structure, whereas "Different Segs" indicates that structure changes are applied to the segments sequentially.

3.6.2. Morphogenesis in Drosophila melanogaster

We applied our methods to a gene expression time series for 11 genes involved in the muscle development of *Drosophila melanogaster*, described in Subheading 2.2. The microarray data measured gene expression levels during all four major stages of morphogenesis: embryo, larva, pupa, and adult. First, we investigated whether our methods were able to infer the correct changepoints corresponding to the known transitions between stages.

The left panel in Fig. 4 shows the marginal posterior probability of the inferred changepoints during the life cycle of *Drosophila melanogaster*. We present the changepoints found without information sharing (HetDBN-0) and using sequential information sharing with an exponential prior as described in Subheading 3.5 (HetDBN-Exp). For a comparison, we applied the method proposed in ref. 3, using the authors' software package TESLA. Note that this model depends on various regularization parameters, which were optimized by maximizing the BIC score, as in ref. 3. The results are shown in the right panel of Fig. 4, where the graph shows the L1-norm of the difference of the regression parameter vectors associated with adjacent time points. Robinson and Hartemink (1) applied their discrete nonhomogeneous DBN to the same data set, and a plot corresponding to the left panel of Fig. 4 can be found in their paper. A comparison of these plots suggests that our method is the only one that clearly detects all three morphogenic transitions: embryo → larva, larva → pupa, and pupa → adult. The right panel of Fig. 4 indicates that the last transition, pupa → adult, is less clearly detected with TESLA, and it is completely missing in ref. 1. Both TESLA and our methods, HetDBN-0 and HetDBN-Exp, detect additional transitions during the embryo stage, which are missing in ref. 1. We would argue

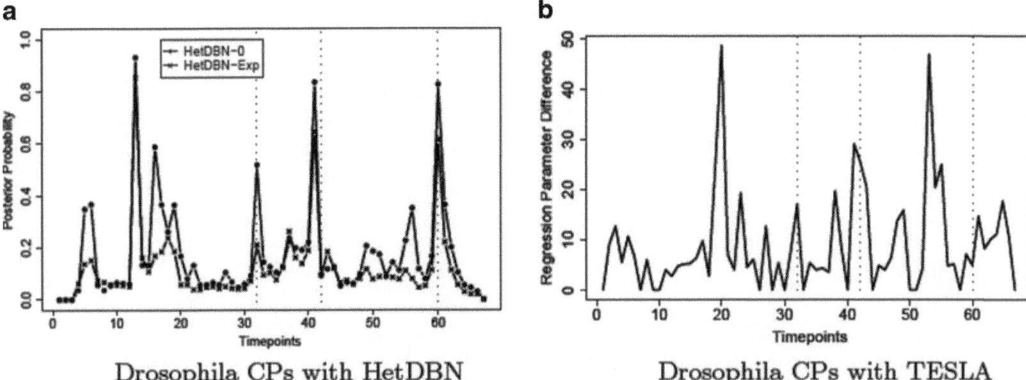

Fig. 4. Changepoints inferred on gene expression data related to morphogenesis in *Drosophila melanogaster*. (**a**): Changepoints for Drosophila using HetDBN-0 (no information sharing) and HetDBN-Exp (sequential information sharing via exponential prior). We show the posterior probability of a changepoint occurring for any node, plotted against time. (**b**): TESLA, L1-norm of the difference of the regression parameter vectors associated with two adjacent time points, plotted against time.

that a complex gene regulatory network is unlikely to transit into a new morphogenic phase all at once, and some pathways might have to undergo activational changes earlier in preparation for the morphogenic transition. As such, it is not implausible that additional transitions at the gene regulatory network level occur. However, a failure to detect known morphogenic transitions can clearly be seen as a shortcoming of a method, and on these grounds our model appears to outperform the two alternative ones.

In addition to the changepoints, we have inferred network structures for the morphogenic stages of embryo, larva, pupa, and adult (Fig. 5). An objective assessment of the reconstruction accuracy is not feasible due to the limited existing biological knowledge and the absence of a gold standard. However, our reconstructed networks show many similarities with the networks discovered by Robinson and Hartemink (1), Guo et al. (14), and Zhao et al. (11). For instance, we recover the interaction between two genes, *eve* and *twi*. This interaction is also reported in refs. 14 and refs. 11, while ref. 1 seem to have missed it. We also recover a cluster of interactions among the genes *myo61f*, *msp300*, *mhc*, *prm*, *mlc1*, and *up* during all morphogenic phases. This result makes sense, as all genes (except *up*) belong to the myosin family. However, unlike ref. 1, we find that *actn* also participates as a regulator in this cluster. There is some indication of this in ref. 11, where *actn* is found to regulate *prm*. As far as changes between the different stages are concerned, we found an important change in the role of *twi*. This gene does not have an important role as a regulator during the early phases, but functions as a regulator of five other genes during the adult phase: *mlc1, gfl, actn, msp300,* and *sls*. The absence of a regulatory role for *twi* during the earlier

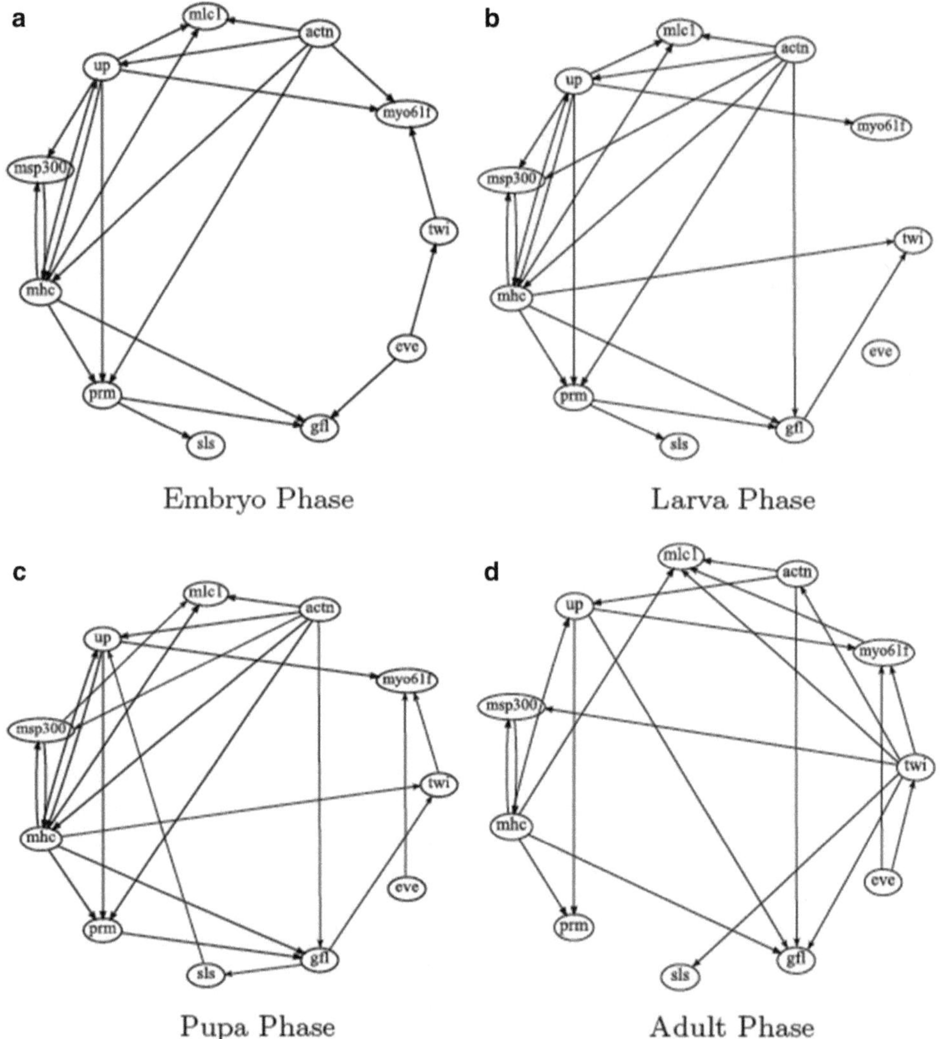

Fig. 5. Network structures inferred by our method for a set of muscle development genes during the four major phases in morphogenesis of *Drosophila melanogaster*. The structures were inferred using the sequential information sharing prior from Subheading 3.5 in order to conserve similarities among different phases.

phases is consistent with ref. 18, who found that another regulator, *mef2* (not included in the dataset), controls the expression of *mlc1*, *actn*, and *msp300* during early development.

3.7. Conclusions

We have proposed a novel nonhomogeneous DBN, which has various advantages over existing schemes: it does not require the data to be discretized (as opposed to ref. 1); it allows the network structure to change with time (as opposed to ref. 2); it includes a regularization scheme based on inter-time segment information sharing (as opposed to refs. 6, 7); and it allows all hyperparameters to be inferred from the data via a consistent Bayesian inference scheme (as opposed to ref. 3). An evaluation on synthetic data has demonstrated an improved performance over refs. 6, 7. The application of our method to gene expression time series taken during the life cycle of *Drosophila melanogaster* has revealed better agreement with known morphogenic transitions than the methods of refs. 1 and refs. 3, and we have detected changes in gene regulatory interactions that are consistent with independent biological findings.

4. Notes

1. *Software implementation.* The methods described in this chapter have been implemented in R, based on the program ARTIVA (Auto Regressive TIme VArying network inference) from ref. 6, 7. Our program sets up an RJMCMC simulation to sample the network structure, the changepoints, and the hyperparameters from the posterior distribution. The software will be made available from the Comprehensive R Archive Network Web site (19). The package will include a reference manual and worked examples of how to use each function. To use the package, proceed as follows:
   (a) Set the hyperparameters and the initial network (or use default settings).
   (b) Run the RJMCMC algorithm until convergence (see Note 2 for more details about the convergence criteria).
   (c) Get an approximation of the posterior distribution for the quantity of interest; e.g., an approximation of the probability $P(k = l \mid D)$ for having l changepoints (i.e., $l + 1$ segments) is obtained as follows: $\hat{P}(k = l|D) = \frac{Number\ of\ samples\ with\ l\ changepoints}{Total\ number\ of\ samples}$, where the number of samples refers to the number of configurations obtained from the MCMC sampling phase, that is after convergence has been reached (see Note 2).
2. *Convergence criterion.* As a convergence diagnostic, we monitor the potential scale reduction factor (PSRF) (20), computed from the within-chain and between-chain variances of marginal edge posterior probabilities. Values of PSRF ≤ 1.1 are usually taken as indication of sufficient convergence. In our simulations, we extended the burn-in phase until a value of PSRF ≤ 1.05 was reached, and then sampled 1,000 network and changepoint configurations in intervals of 200 RJMCMC steps. From these samples, we compute the marginal posterior probabilities of all potential interactions, which define a ranking of the edges. From this ranking, we can compute receiver operating characteristic (ROC) and precision–recall (PR) curves as described in Note 3.

3. *Results evaluation.* If we select a threshold, then all edges with a posterior probability above the threshold correspond to predicted interactions, and all edges with posterior probability below the threshold correspond to non-edges. When the true network is known, this allows us to compute, for each choice of the threshold, the number of true positive (*TP*), false positive (*FP*), true negative (*TN*), and false negative (*FN*) interactions. From these counts, various quantities can be computed. The *sensitivity* or *recall* is defined by $TP/(TP+FN)$ and describes the proportion of true non-interactions that have been correctly identified. The *specificity*, defined by $TN/(TN+FP)$, describes the proportion of non-interactions that have been correctly identified. Its complement, 1-specificity, is called the *complementary specificity.* It is given by $FP/(TN+FP)$ and describes the false prediction rate, i.e., the proportion of non-interactions that are erroneously predicted to be true interactions. Finally, the *precision* is defined by $TP/(TP+FP)$ and describes the proportion of predicted interactions that are true interactions. If we plot, for all threshold values, the sensitivity on the vertical axis against the complementary specificity on the horizontal axis, we obtain what is called a ROC curve. A diagonal line from (0,0) to (1,1) corresponds to random expectation; the area under this curve is 0.5. The perfect prediction is given by a graph along the coordinate axes: $(0,0) \rightarrow (0,1) \rightarrow (1,1)$. This curve, which covers an area of 1, indicates a perfect prediction, where a threshold is found that allows the recovery of all true interactions without incurring any spurious ones. In general, ROC curves are between these two extremes, with a larger *area under the curve* (AUC) indicating a better performance. It is recommended to also plot PR curves (see Note 4).
4. *ROC curve versus PR curve.* While ROC curves have a sound statistical interpretation, they are not without problems (21). The total number of non-interactions (TN) usually increases proportionally to the square of the number of nodes. Hence, for a large number of nodes, ROC curves are often dominated by the TN count, and the differences in network reconstruction performance between two alternative methods are not sufficiently clearly indicated. For that reason, precision–recall (PR) curves have become more popular lately (22). Here, the precision is plotted against the recall for all values of the threshold; note that both quantities are independent of TN. Like for ROC curves, larger AUC scores indicate a better performance. A more detailed comparison between ROC and PR curves is discussed in ref. (22).

References

1. Robinson JW, Hartemink AJ (2009) Non-stationary dynamic Bayesian networks. In Koller D, Schuurmans D, Bengio Y et al editors, Advances in Neural Information Processing Systems (NIPS), volume 21, 1369–1376. Morgan Kaufmann Publishers.
2. Grzegorczyk M, Husmeier D (2009) Non-stationary continuous dynamic Bayesian networks. In Bengio Y, Schuurmans D, Lafferty J et al editors, Advances in Neural Information Processing Systems (NIPS), volume 22, 682–690.
3. Ahmed A, Xing EP (2009) Recovering time-varying networks of dependencies in social and biological studies. Proceedings of the National Academy of Sciences 106:11878–11883.
4. Talih M, Hengartner N (2005) Structural learning with time-varying components: Tracking the cross-section of financial time series. Journal of the Royal Statistical Society B 67(3):321–341.
5. Xuan X, Murphy K (2007) Modeling changing dependency structure in multivariate time series. In Ghahramani Z editor, Proceedings of the 24th Annual International Conference on Machine Learning (ICML 2007), 1055–1062. Omnipress.
6. Lèbre S (2007) Stochastic process analysis for Genomics and Dynamic Bayesian Networks inference. Ph.D. thesis, Université d'Evry-Val-d'Essonne, France.
7. Lèbre S, Becq J, Devaux F et al. (2010) Statistical inference of the time-varying structure of gene-regulation networks. BMC Systems Biology 4(130).
8. Kolar M, Song L, Xing E (2009) Sparsistent learning of varying-coefficient models with structural changes. In Bengio Y, Schuurmans D, Lafferty J et al editors, Advances in Neural Information Processing Systems (NIPS), volume 22, 1006–1014.
9. Larget B, Simon DL (1999) Markov chain Monte Carlo algorithms for the Bayesian analysis of phylogenetic trees. Molecular Biology and Evolution 16(6):750–759.
10. Arbeitman M, Furlong E, Imam F et al. (2002) Gene expression during the life cycle of *Drosophila melanogaster*. Science 297 (5590):2270–2275.
11. Zhao W, Serpedin E, Dougherty E (2006) Inferring gene regulatory networks from time series data using the minimum description length principle. Bioinformatics 22(17):2129.
12. Giot L, Bader JS, Brouwer C et al (2003) A protein interaction map of *drosophila melanogaster*. Science 302:1727–1736.
13. Yu J, Pacifico S, Liu G et al. (2008) DroID: the Drosophila Interactions Database, a comprehensive resource for annotated gene and protein interactions. BMC Genomics 9(461).
14. Guo F, Hanneke S, Fu W et al. (2007) Recovering temporally rewiring networks: A model-based approach. In Proceedings of the 24th international conference on Machine learning page 328. ACM.
15. Andrieu C, Doucet A (1999) Joint Bayesian model selection and estimation of noisy sinusoids via reversible jump MCMC. IEEE Transactions on Signal Processing 47(10):2667–2676.
16. Green P (1995) Reversible jump Markov chain Monte Carlo computation and Bayesian model determination. Biometrika 82:711–732.
17. Werhli AV, Husmeier D (2008) Gene regulatory network reconstruction by Bayesian integration of prior knowledge and/or different experimental conditions. Journal of Bioinformatics and Computational Biology 6 (3):543–572.
18. Elgar S, Han J, Taylor M (2008) mef2 activity levels differentially affect gene expression during Drosophila muscle development. Proceedings of the National Academy of Sciences 105 (3):918.
19. http://cran.r-project.org.
20. Gelman A, Rubin D (1992) Inference from iterative simulation using multiple sequences. Statistical science 7(4):457–472.
21. Hand DJ (2009) Measuring classifier performance: a coherent alternative to the area under the roc curve. Machine Learning 77:103–123.
22. Davis J, Goadrich M (2006) The relationship between precision-recall and ROC curves. In ICML '06: Proceedings of the 23rd international conference on Machine Learning 233–240. ACM, New York, NY, USA. ISBN 1-59593-383-2. doi: http://doi.acm.org/10.1145/1143844.1143874.

Chapter 14

Inference of Regulatory Networks from Microarray Data with R and the Bioconductor Package qpgraph

Robert Castelo and Alberto Roverato

Abstract

Regulatory networks inferred from microarray data sets provide an estimated blueprint of the functional interactions taking place under the assayed experimental conditions. In each of these experiments, the gene expression pathway exerts a finely tuned control simultaneously over all genes relevant to the cellular state. This renders most pairs of those genes significantly correlated, and therefore, the challenge faced by every method that aims at inferring a molecular regulatory network from microarray data, lies in distinguishing direct from indirect interactions. A straightforward solution to this problem would be to move directly from bivariate to multivariate statistical approaches. However, the daunting dimension of typical microarray data sets, with a number of genes p several orders of magnitude larger than the number of samples n, precludes the application of standard multivariate techniques and confronts the biologist with sophisticated procedures that address this situation. We have introduced a new way to approach this problem in an intuitive manner, based on limited-order partial correlations, and in this chapter we illustrate this method through the R package `qpgraph`, which forms part of the Bioconductor project and is available at its Web site (1).

Key words: Molecular regulatory network, Microarray data, Reverse engineering, Network inference, Non-rejection rate, qpgraph

1. Introduction

The genome-wide assay of gene expression by microarray instruments provides a high-throughput readout of the relative RNA concentration for a very large number of genes p across a typically much smaller number of experimental conditions n. This enables a fast systematic comparison of all expression profiles on a gene-by-gene basis by analysis techniques such as differential expression. However, the simultaneous assay of all genes embeds in the microarray data a pattern of correlations projected from the regulatory interactions forming part of the cellular

Junbai Wang et al. (eds.), *Next Generation Microarray Bioinformatics: Methods and Protocols*,
Methods in Molecular Biology, vol. 802, DOI 10.1007/978-1-61779-400-1_14, © Springer Science+Business Media, LLC 2012

state of the samples, and therefore, estimating this pattern from the data can aid in building a network model of the transcriptional regulatory interactions.

Many published solutions to this problem rely on pairwise measures of association based on bivariate statistics, such as Pearson correlation or mutual information (2). However, marginal pairwise associations cannot distinguish direct from indirect (that is, spurious) relationships and specific enhancements to this pairwise approach have been made to address this problem (*see*, for instance, (3) and (4)).

A sensible approach is to try to apply multivariate statistical methods such as undirected Gaussian graphical modeling (5) and compute partial correlations which are a measure of association between two variables while controlling for the remaining ones. However, these methods require inverting the sample covariance matrix of the gene expression profiles and this is only possible when $n > p$ (6). This has led to the development of specific inferential procedures, which try to overcome the ***small n and large p problem*** by exploiting specific biological background knowledge on the structure of the network to be inferred. From this viewpoint, the most relevant feature of regulatory networks is that they are sparse, that is the direct regulatory interactions between genes represent a small proportion of the edges present in a fully connected network (see, for instance, (7)). Statistical procedures for inference on sparse networks include, among others, a Bayesian approach with sparsity inducing prior (8), the lasso estimate of the inverse covariance matrix (see, among others, (9) and (10)), the shrinkage estimate of the covariance matrix (11) and procedures based on limited-order partial correlations (see, for instance, (12) and (13)).

In (14) a procedure is proposed for the statistical learning of sparse networks based on a quantity called the non-rejection rate. The computation of the non-rejection rate requires carrying out a large number of hypothesis tests involving limited-order partial correlations, nonetheless that procedure is not affected by the multiple testing problem. Furthermore, in (15) it is shown that averaging non-rejection rates obtained through different orders of the partial correlations is an effective strategy to release the user from making an educated guess on the most suitable order. In the same article, a method based on the concept of functional coherence is introduced, for the comparison of the functional relevance of different inferred networks and their regulatory modules. In the rest of this chapter we show how to apply this entire methodology by using the statistical software R and the Bioconductor package `qpgraph`.

2. Materials

2.1. The Non-rejection Rate

We represent the molecular regulatory network we want to infer by means of a mathematical object called a *graph*. A graph is a pair $G = (V, E)$, where $V = \{1,2,\ldots,p\}$ is a finite set of *vertices* and E is a subset of pairs of vertices, called the *edges* of G. In this context, vertices are genes and edges are direct regulatory interactions (see Note 1). Nevertheless, the graphs we consider here have no multiple edges and no loops; furthermore, they are undirected so that both $(i,j) \in E$ and $(j, i) \in E$ are an equivalent way to write that the vertices i and j are linked by an edge. A basic feature of graphs is that they are visual objects. In the graphical representation, vertices may be depicted with circles while undirected edges are lines joining pairs of vertices. For example, the graph $G = (V, E)$ with $V = \{1, 2, 3\}$ and $E = \{(1, 2), (2, 3)\}$ can be represented as ①—②—③. A *path* in G from i to j is a sequence of vertices such that i and j are the first and last vertex of the sequence, respectively, and every vertex in the sequence is linked to the next vertex by an edge. The subset $Q \subseteq V$ is said to *separate* i from j if all paths from i to j have at least one vertex in Q. For instance, in the graph of the example above the sequence (1, 2, 3) is a path between 1 and 3, whereas the sequence (1, 3, 2) is not a path. Furthermore, the set $Q = \{2\}$ separates 1 from 3.

The random vector of gene expression profiles is indexed by the set V and denoted by $X_V = (X_1, X_2, \ldots, X_p)^T$ and, furthermore, we denote by $\rho_{ij.V\backslash\{i,j\}}$ the *full-order partial correlation* between the genes i and j, that is the correlation coefficient between the two genes adjusted for all the remaining genes $V/\{i, j\}$. We assume that X_V belongs to a Gaussian graphical model with graph $G = (V, E)$ and refer to (5) for a full account on these models. Here, we recall that in a Gaussian graphical model X_V is assumed to be multivariate normal and that the vertices i and j are not linked by an edge if and only if $\rho_{ij.V\backslash\{i,j\}} = 0$. It follows that the sample version of full-order partial correlations plays a key role in statistical procedures for inferring the network structure from data. However, these quantities can be computed only if n is larger than p and this has precluded the application of standard techniques in the context of regulatory network inference from microarray data. On the other hand, if the edge between the genes i and j is missing from the graph then possibly a large number of limited-order partial correlations are equal to zero. More specifically, for a subset $Q \subset V\backslash\{i,j\}$ we denote by $\rho_{ij.Q}$ the *limited-order partial correlation*, that is the correlation coefficient between i and j adjusted for the genes in Q. It can be shown that if Q separates i and j in G, then $\rho_{ij.Q}$ is equal to zero. This is a useful result because the sample version of $\rho_{ij.Q}$ can be computed whenever $n > q + 2$

and, if the distribution of X_V is *faithful* to G (see (14) and references therein), then $\rho_{ij.Q} = 0$ also implies that the vertices i and j are not linked by an edge in G.

In sparse graphs, one should expect a high degree of separation between vertices, and therefore, limited-order partial correlations are useful tools for inferring sparse molecular regulatory networks from data. There are, however, several difficulties related to the use of limited-order partial correlations because for every pair of genes i and j there are a huge number of potential subsets Q, and this leads to computational problems as well as to multiple testing problems. In (14) the authors propose to use a quantity based on partial correlations of order q that they call the non-rejection rate. The *non-rejection rate* for vertices i and j is denoted by $NRR(i,j|q)$ and it is the probability of not rejecting, on the basis of a suitable statistical test, the hypothesis that $\rho_{ij.Q} = 0$ where Q is a subset of q genes randomly selected from $V\backslash\{i,j\}$. Hence, the non-rejection rate is a probability associated to every pair of vertices, genes in the context of this chapter, and takes values between zero and one, with larger values providing stronger evidence that an edge is not present in G. The procedure introduced in (15) amounts to estimating the non-rejection rate for every pair of vertices, ranking all the possible edges of the graph according to these values and then removing those edges whose non-rejection rate values are above a given threshold. Different methods for the choice of the threshold are discussed in the forthcoming sections where the graph inferred with this method will be called the *qp-graph*; we refer to (14) and (15) for technical details. Here we recall that the computation of the non-rejection rate requires the specification of a value q corresponding to the dimension of the potential separator, with q ranging from the value 1 to the value $n - 3$. Obviously, a key question when using the non-rejection rate with microarray data is what value of q should be employed. We know that a larger value of q increases the probability that a randomly chosen subset Q separates i and j, but this could compromise the statistical power of the tests which depends on $n - q$. In (15) a simple and effective solution to this question was introduced and consists of averaging (taking the arithmetic mean), for each pair of genes, the estimates of the non-rejection rates for different values of q spanning its entire range from 1 to somewhere close to $n - 3$. These authors also showed that the average non-rejection rate is more stable than the non-rejection rate, avoids having to specify a particular value of q and it behaves similarly to the non-rejection rate for connected pairs of vertices in the true underlying graph G (i.e., for directly interacting genes in the underlying molecular regulatory network). They also pointed out that the drawback of averaging is that a disconnected pair of vertices (i,j) in a graph G whose indirect relationship is mediated by a large number of other vertices, will be easier to identify with the non-rejection rate using a sufficiently large value of q than with the average non-rejection rate.

However, in networks showing high degrees of modularity and sparseness the number of genes mediating indirect interactions should not be very large, and therefore, the average non-rejection rate should be working well, just as they observed in the empirical results reported in (15).

2.2. Functional Coherence

A critical question when estimating a molecular regulatory network from data is to know the extent to which the inferred regulatory relationships reflect the functional organization of the system under the experimental conditions employed to generate the microarray data. The authors in (15) addressed this question using the Gene Ontology (GO) database (16), which provides structured functional annotations on genes for a large number of organisms including *Escherichia coli* (*E. coli*). The approach followed consists of assessing the functional coherence of every regulatory module within a given network. Assume a *regulatory module* is defined as a transcription factor and its set of regulated genes. The functional coherence of a regulatory module is estimated by relying on the observation that, for many transcription factor genes, their biological function, beyond regulating transcription, is related to the genes they regulate. Note that different regulatory modules can form part of a common pathway and thus share some more general functional annotations, which can lead to some degree of functional coherence between target genes and transcription factors of different modules. However, in (15) it is shown that for the case of *E. coli* data, the degree of functional coherence within a regulatory module is higher than between highly correlated but distinct modules. This observation allowed them to conclude that functional coherence constitutes an appealing measure for assessing the discriminative power between direct and indirect interactions and therefore can be employed as an independent measure of accuracy.

The way in which the authors in (15) estimated functional coherence is as follows. Using GO annotations, concretely those that refer to the biological process (BP) ontology, two GO graphs are built such that vertices are GO terms and (directed) links are GO relationships. One GO graph is induced (i.e., grown toward vertices representing more generic GO terms) from GO terms annotated on the transcription factor gene discarding those terms related to transcriptional regulation. The other GO graph is induced from GO terms overrepresented among the regulated genes in the estimated regulatory module which, to try to avoid spuriously enriched GO terms, we take it only into consideration if it contains at least five genes. These overrepresented GO terms can be found, for instance, by using the conditional hypergeometric test implemented in the Bioconductor package `GOstats` (17) on the *E. coli* GO annotations from the `org.EcK12.eg.db` Bioconductor package. Finally, the level of functional coherence of the regulatory module is estimated as the degree of similarity between the two GO graphs, which in this case amounts to a comparison of the two corresponding subsets of vertices. The level of functional coherence of the entire network is determined by the distribution of the functional coherence values of all the regulatory modules for which this measure was calculated (see Note 2).

2.3. *Escherichia coli* Microarray Data

In this chapter, we describe our procedure through the analysis of an *E. coli* microarray data set from (18) and deposited at the NCBI Gene Expression Omnibus (GEO) with accession GDS680. It contains 43 microarray hybridizations that monitor the response from *E. coli* during an oxygen shift targeting the a priori most relevant part of the network by using six strains with knockouts of key transcriptional regulators in the oxygen response (*ΔarcA*, *ΔappY*, *Δfnr*, *ΔoxyR*, *ΔsoxS*, and the double knockout *ΔarcAΔfnr*). We will infer a network starting from the full gene set of *E. coli* with $p = 4{,}205$ genes (see the following subsection for details on filtering steps).

2.4. *Escherichia coli* Functional and Microarray Data Processing

We downloaded the Release 6.1 from RegulonDB (19) formed by an initial set of 3,472 transcriptional regulatory relationships. We translated the Blattner IDs into Entrez IDs, discarded those interactions for which an Entrez ID was missing in any of the two genes and did the rest of the filtering using Entrez IDs. We filtered out those interactions corresponding to self-regulation and among those conforming to feedback-loop interactions we discarded arbitrarily one of the two interactions. Some interactions were duplicated due to a multiple mapping of some Blattner IDs to Entrez IDs, in that case we removed the duplicated interactions arbitrarily. We finally discarded interactions that did not map to genes in the array and were left with 3,283 interactions involving a total of 1,428 genes.

We have obtained RMA expression values for the data in (18) using the `rma()` function from the `affy` package in Bioconductor. We filtered out those genes, for which there was no Entrez ID and when two or more probesets were annotated under the same Entrez ID we kept the probeset with highest median expression level. These filtering steps left a total number of $p = 4{,}205$ probesets mapped one-to-one with *E. coli* Entrez genes.

3. Methods

3.1. Running the Bioconductor Package qpgraph

The methodology briefly described in this chapter is implemented in the software called `qpgraph`, which is an add-on package for the statistical software R (20). However, unlike most other available software packages for R, which are deposited at the Comprehensive R Archive Network – CRAN – (21), the package qpgraph forms part of the Bioconductor project (see (22) and (1)) and it is deposited in the Bioconductor Web site instead. The version of the software employed to illustrate this chapter runs over R 2.12 and thus forms part of Bioconductor package bundle version 2.7 (see Note 3). Among the packages that get installed by default with R and Bioconductor, qpgraph will automatically load some of them when calling certain functions but one of these, Biobase, should be explicitly loaded to manipulate microarray expression data through the ExpressionSet class of objects. Therefore, the initial sequence of commands to successfully start working with qpgraph through the example illustrated in this chapter is as follows:

```
> library(Biobase)
> library(qpgraph)
```

Additionally, we may consider the fact that most modern desktop computers come with four or more core processors and that it is relatively common to have access to a cluster facility with dozens, hundreds, or perhaps thousands of processors scattered through an interconnected network of computer nodes. The qpgraph package can take advantage of such a multiprocessor hardware by performing some of the calculations in parallel. In order to enable this feature, it is necessary to install the R packages snow and rlecuyer from the CRAN repository and load them prior to using the qpgraph package. The specific type of cluster configuration that will be employed will depend on whether additional packages providing such a specific support are installed. For example, if the package Rmpi is installed, then the cluster configuration will be that of an MPI cluster (see (23) and Note 4 for details on this subject). Thus, if we want to take advantage of an available multiprocessor infrastructure we should additionally write the following commands:

```
> library(snow)
> library(rlecuyer)
```

Once these packages have been successfully loaded, to perform calculations in parallel it is necessary to provide an argument, called clusterSize, to the corresponding function indicating the number of processors that we wish to use. In this chapter we assume we can use eight processors, which should allow the longest calculation illustrated in this chapter to finish in less than 15 min. During long calculations it is convenient to monitor their progress and this is possible in most of the functions from the qpgraph package if we set the argument verbose = TRUE, which by default is set to FALSE.

3.2. A Quick Tour Through the qpgraph Package

In this section we illustrate the minimal function calls in the qpgraph package that allow one to infer a molecular regulatory network from microarray data. We need first to load the data described in the previous section and which is included as an example data set in the qpgraph package.

```
> data(EcoliOxygen)
```

The previous command will load on our current R default environment two objects, one of them called gds680.eset, which is an object of the class ExpressionSet and contains the *E. coli* microarray data described in the previous section. We can see these objects in the workspace with the function ls() and figure out the dimension of this particular microarray data set with dim(), as follows:

```
> ls()
[1] "filtered.regulon6.1" "gds680.eset"
> dim(gds680.eset)
Features  Samples
    4205       43
```

When we have a microarray data set, either as an ExpressionSet object or simply as a matrix of numeric values, we can immediately proceed to estimate non-rejection rates with a q-order of, for instance, $q = 3$ with the function qpNrr():

```
> nrr <- qpNrr(gds680.eset, q=3, clusterSize=8)
```

This function returns a symmetric matrix of non-rejection rate values with its diagonal entries set to NA. Using this matrix as input to the function qpGraph() we can directly infer a molecular regulatory network by setting a non-rejection rate cutoff value above which edges are removed from an initial fully connected graph. The selection of this cutoff could be done, for instance, on the basis of targeting a graph of specific density which can be examined by calling first the function qpGraphDensity(), whose result is displayed in Fig. 1a and from which we consider retrieving a graph of 7% density by using a 0.1 cutoff value:

```
> qpGraphDensity(nrr, title="", breaks=10)
> g <- qpGraph(nrr, threshold=0.1, return.type="graphNEL")
> g
A graphNEL graph with undirected edges
Number of Nodes = 4205
Number of Edges = 644036
```

By default, the qpGraph() function returns an adjacency matrix but, by setting return.type = “graphNEL“ we obtain a graphNEL-class object as a result, which, as we shall see later, is amenable for processing with functions from the

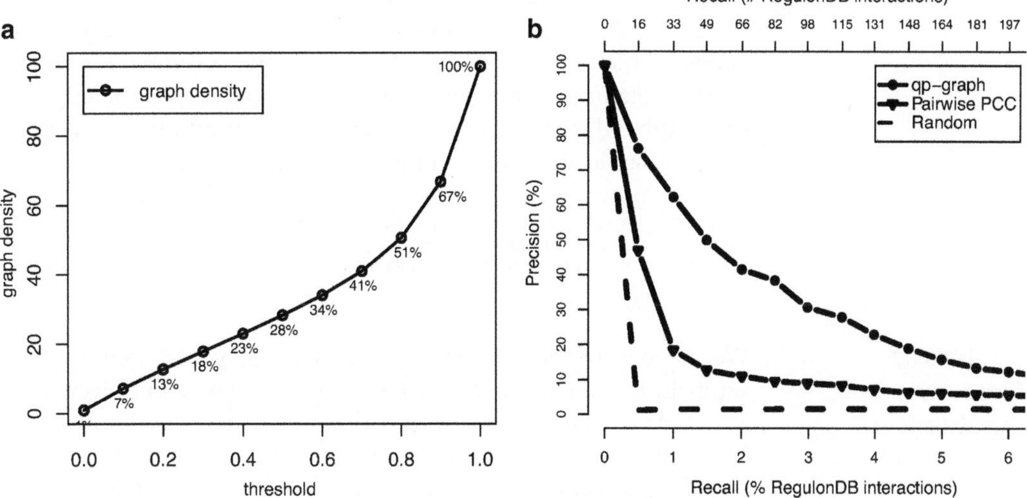

Fig. 1. Performance comparison on the oxygen deprivation *E. coli* data with respect to RegulonDB. (**a**) Graph density as function of the non-rejection rate estimated with $q = 3$. (**b**) Precision–recall curves comparing a random ranking of the putative interactions, a ranking made by absolute Pearson correlation (Pairwise PCC) and a ranking derived from the average non-rejection rate (qp-graph).

Bioconductor packages `graph` and `Rgraphviz`. We can conclude this quick tour through the main cycle of the task of inferring a network from microarray data by showing how we can extract a ranking of the strongest edges in the network with the function `qpTopPairs()`:

```
> qpTopPairs(nrr)
       i      j x
1 947758 947761 0
2 948517 948512 0
3 944834 944794 0
4 948517 944797 0
5 948512 944797 0
6 945112 945108 0
```

where the first two columns, called `i` and `j`, correspond to the identifiers of the pair of variables and the third column `x` corresponds, in this case, to non-rejection rate values. An immediate question is whether the value of $q = 3$ was appropriate for this data set and while we may try to find an answer by exploring the estimated non-rejection rate values in a number of ways described in ref. 14, an easy solution introduced in ref. 15 consists of estimating the so-called average non-rejection rates whose corresponding function, `qpAvgNrr()`, is called in an analogous way to `qpNrr()` but without the need to specify a value for q.

In (15) a comparison of this procedure with other widely used techniques is carried out. Here, we restrict the comparison to a simple procedure based on sample Pearson correlation coefficients and, furthermore, to the worst performing strategy which consists of setting association values uniformly at random to every pair of genes (which we shall informally call the *random association method*) leading to a completely random ranking of the edges of the graph. All these quantities can be computed using two functions available also through the `qpgraph` package:

```
> allGenes <-  featureNames(gds680.eset)
> pcc <- qpPCC(gds680.eset)
> rndcor <- qpUnifRndAssociation(length(allGenes), allGenes)
```

3.3. Avoiding Unnecessary Calculations

We saw before that as part of the `EcoliOxygen` example data set included in the `qpgraph` package, there was an object called `filtered.regulon6.1`. This object is a `data.frame` and contains pairs of genes corresponding to curated transcriptional regulatory relationships from *E. coli* retrieved from the 6.1 version of the RegulonDB database. Each of these relationships indicates that one transcription factor gene activates or represses the transcription of the other target gene. If we are interested in just this kind of transcriptional regulatory interactions, i.e., associations involving at least one transcription factor gene, we can substantially speed up calculations by restricting them to those pairs of genes suitable to form such an association. In order to illustrate this feature, we start here by extracting from the RegulonDB data what genes form the subset of transcription factors:

```
> regulonTFgenes <- unique(filtered.regulon6.1[, "EgID_TF"])
```

In general, this kind of functional information about genes is available for many organisms through different on-line databases (24). Once we have a list of transcription factor genes, restricting the pairs that include at least one of them can be done through the arguments `pairup.i` and `pairup.j` in both functions, `qpNrr()` and `qpAvgNrr()`. We use here the latter to estimate average non-rejection rates that will help us to infer a transcriptional regulatory network without having to specify a particular q-order value. Since the estimation of non-rejection rates is carried out by means of a Monte Carlo sampling procedure, to allow the reader to reproduce the exact numbers shown here we will set a specific seed to the random number generator before estimating average non-rejection rates.

```
> set.seed(123)
> avgnrr <- qpAvgNrr(gds680.eset, pairup.i=regulonTFgenes,
                     pairup.j=allGenes, clusterSize=8)
```

The default settings for the function `qpAvgNrr()` employ 4 q-values uniformly distributed along the available range of q values. In this example, these correspond to $q = \{1, 11, 21, 31\}$. However, we can change this default setting by using the argument `qOrders`.

3.4. Network Accuracy with Respect to a Gold-Standard

E. coli is the free-living organism with the largest fraction of its transcriptional regulatory network supported by some sort of experimental evidence. As a result of an effort in combining all this evidence the database RegulonDB (19) provides a curated set of transcription factor and target gene relationships that we can use as a gold-standard to, as we shall see later, calibrate a nominal precision or recall at which we want to infer the network or compare the performance of different parameters and network inference methods. This performance is assessed in terms of *precision–recall curves.*

Every network inference method that we consider here provides a ranking of the edges of the fully connected graph, that is, of all possible interactions. Then a threshold is chosen and this leads to a partition of the set of all edges into a set of *predicted edges* and a set of *missing edges.* On the other hand, the set of RegulonDB interactions are a subset of the set of all possible interactions and a predicted edge that belongs to the set of RegulonDB interactions is called a *true positive.* Following the conventions from (25), when using RegulonDB interactions for comparison the *recall* (also known as sensitivity) is defined as the fraction of true positives in the set of RegulonDB interactions and the *precision* (also known as positive predictive value) is defined as the number of true positives over the number of predicted edges whose genes belong to at least one transcription factor and target gene relationship in RegulonDB. For a given network inference method, the precision–recall curve is constructed by plotting the precision against the recall for a wide range of different threshold values. In the *E. coli* dataset we analyze, precision–recall curves should be calculated on the subset of 1,428 genes forming the 3,283 RegulonDB interactions and this can be achieved with the `qpgraph` package through the function `qpPrecisionRecall()` as follows:

```
> regulonGenes <- unique(c(filtered.regulon6.1$EgID_TF,
                           filtered.regulon6.1$EgID_TG))
> avgnrr.pr <- qpPrecisionRecall(avgnrr,
                   refGraph=filtered.regulon6.1[, c("EgID_TF",
"EgID_TG")],
                   decreasing=FALSE,
                   pairup.i=regulonTFgenes,
                   pairup.j=regulonGenes,
                   recallSteps=c(seq(0,0.1,0.005),
                                 seq(0.1,1.0,0.1)))
```

The previous lines calculate the precision–recall curve for the ranking derived from the average non-rejection rate values. The calculation of these curves for the other two rankings derived from Pearson coefficients and uniformly random association values would require replacing the first argument by the corresponding matrix of measurements in absolute value since these two methods provide values ranging from -1 to $+1$. We can plot the resulting precision–recall curve for the average non-rejection rate stored in `avgnrr.pr` as follows:

```
> plot(100*avgnrr.pl[, 1:2], type="b", pch=19, xlim=c(0, 6),
      xlab="Recall (% RegulonDB interactions)", ylab="Precision
(%)")
```

In Fig. 1b this plot is shown jointly with the other calculated curves, where the comparison of the average non-rejection rate (labeled qp-graph) with the other methods yields up to 40% improvement in precision with respect to using absolute Pearson correlation coefficients and observe that for precision levels between 50% and 80% the qp-graph method doubles the recall. We shall see later that this has an important impact when targeting a network of a reasonable nominal precision in such a data set with $p = 4{,}205$ and $n = 43$.

3.5. Inference of Molecular Regulatory Networks of Specific Size

Given a measure of association for every pair of genes of interest, the most straightforward way to infer a network is to select a number of top-scoring interactions that conform a resulting network of a specific size that we choose. We showed before such a strategy by looking at the graph density as a function of threshold, however, we can also extract a network of specific size by using the argument `topPairs` in the call to the `qpGraph()` and `qpAnyGraph()` functions where the call for the random association values would be analogous to the one of Pearson correlations.

```
> qpg1000sze <- qpGraph(avgnrr, threshold=NULL, topPairs=1000)
> pcc1000sze <- qpAnyGraph(abs(pcc$R), threshold=NULL,
                  topPairs=1000, decreasing=TRUE,
                  pairup.i=regulonTFgenes,
                  pairup.j=allGenes)
```

In the example above we are extracting networks formed by the top-scoring 1,000 interactions.

3.6. Inference of Molecular Regulatory Networks at Nominal Precision and Recall Levels

When a gold-standard network is available we can infer a specific molecular regulatory network using a nominal precision and/or using a nominal recall. This is implemented in the `qpgraph` package by calling first the function `qpPRscoreThreshold()` which, given a precision–recall curve calculated with `qpPrecisionRecall()`, will calculate for us the score that attains the desired nominal level. In this particular example, and considering the precision–recall curve of Fig. 1b, we will employ nominal values of 50% precision and 3% recall:

```
> avgnrr.50p.thr <- qpPRscoreThreshold(avgnrr.pr, level=0.50,
                      recall.level=FALSE, max.score=0)
> avgnrr.3r.thr <- qpPRscoreThreshold(avgnrr.pr, level=0.03,
                     recall.level=TRUE, max.score=0)
```

where the thresholds for the other methods would be analogously calculated replacing the first argument by the object storing the corresponding curve returned by `qpPrecisionRecall()`.

Next, we apply these nominal precision and recall thresholds to obtain the networks by using the functions `qpGraph()` for the average non-rejection rate and `qpAnyGraph()` for any other type of association measure, here illustrated only with Pearson correlation coefficients:

```
> qpg50pre <- qpGraph(avgnrr, avgnrr.50p.thr)
> pcc50pre <- qpAnyGraph(abs(pcc$R), abspcc.50p.thr,
                remove="below", pairup.i=regulonTFgenes,
                pairup.j=allGenes)
> qpg3rec <- qpGraph(avgnrr, avgnrr.3r.thr)
> pcc3rec <- qpAnyGraph(abs(pcc$R), abspcc.3r.thr,
               remove="below", pairup.i=regulonTFgenes,
               pairup.j=allGenes)
```

3.7. Estimation of Functional Coherence

In order to estimate functional coherence we need to install a Bioconductor package with GO functional annotations associated to the feature names (genes, probes, etc.) of the microarray data. For this example, we require the *E. coli* GO annotations stored in the package `org.EcK12.eg.db`. It will be also necessary to have installed the `GOstats` package to enable the GO enrichment analysis. The function `qpFunctionalCoherence()` will allow us to estimate functional coherence values as we illustrate here below for the case of the nominal 50%-precision network obtained with the qp-graph method. The estimation for the other networks would require replacing only the first argument by the object storing the corresponding network:

```
> library(GOstats)
> library(org.EcK12.eg.db)
> qpg50preFC <- qpFunctionalCoherence(qpg50pre,
                   TFgenes=regulonTFgenes,
                   chip="org.EcK12.eg.db",
                   minRMsize=5, clusterSize=8)
```

This function returns a `list` object storing the transcriptional network and the values of functional coherence for each regulatory module. These values can be examined by means of a boxplot as follows:

```
> boxplot(qpg50preFC$functionalCoherenceValues,
          col=grey(0.50), ylim=c(0,1), ylab="Functional coherence")
```

In Fig. 2 we see the boxplots for the functional coherence values of all networks obtained from each method and selection strategy. Through the three different strategies, the networks obtained with the qp-graph method provide distributions of functional coherence with mean and median values larger than those obtained from networks built with Pearson correlation or simply at random.

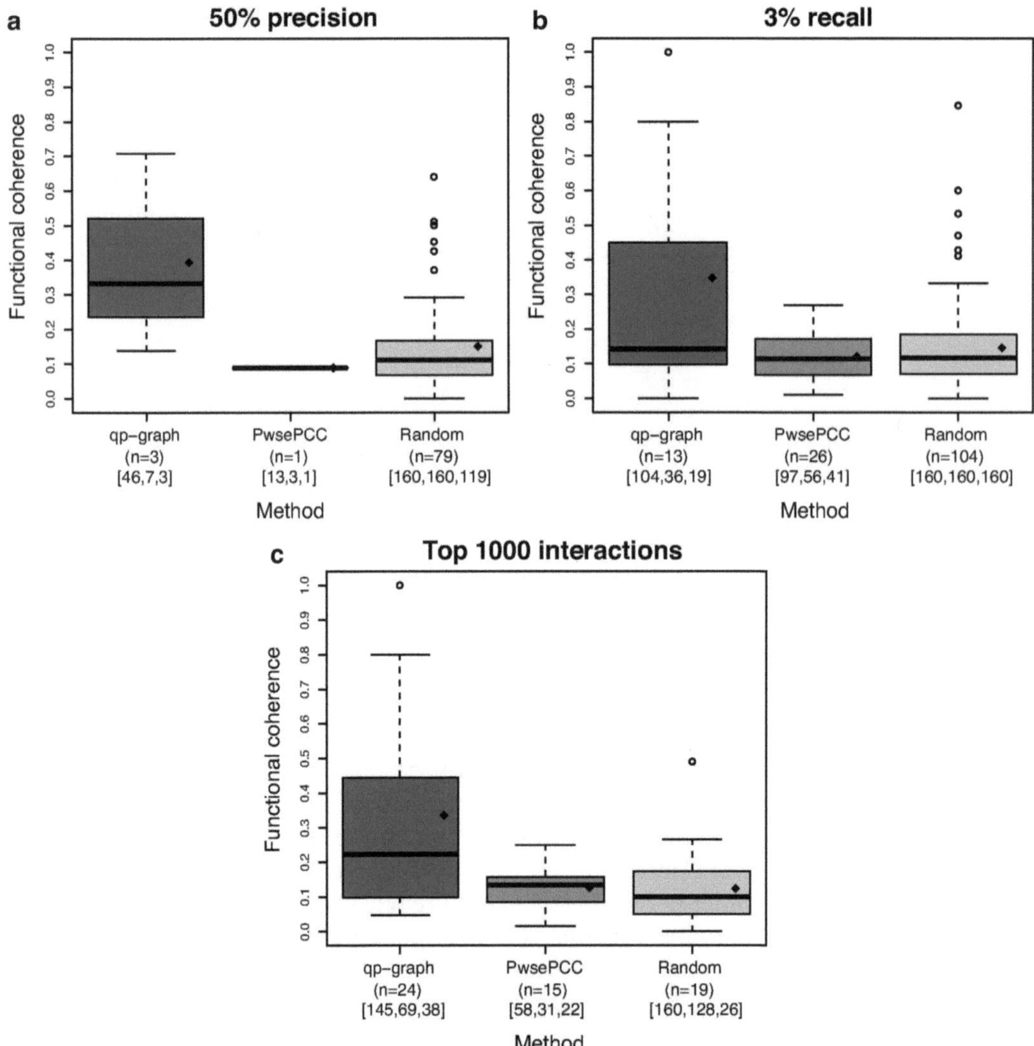

Fig. 2. Functional coherence estimated from networks derived with different strategies and methods. (**a**) A nominal RegulonDB-precision of 50%, (**b**) a nominal RegulonDB-recall of 3%, and (**c**) using the top ranked 1,000 interactions. On the *x*-axis and between *square brackets*, under each method, are indicated, respectively, the total number of regulatory modules of the network, the number of them with at least five genes and the number of them with at least five genes with GO-BP annotations. Among this latter number of modules, the number of them where the transcription factor had GO annotations beyond transcription regulation is noted above between parentheses by *n* and corresponds to the number of modules on which functional coherence could be calculated.

3.8. The 50%-Precision qp-graph Regulatory Network

We are going to examine in detail the 50%-precision qp-graph transcriptional regulatory network. A quick glance at the pairs with strongest average non-rejection rates including the functional coherence values of their regulatory modules within this 50%-precision network can be obtained with the function `qpTopPairs()` as follows:

```
> qpTopPairs(measurementsMatrix=avgnrr, refGraph=qpg50pre,
             pairup.i=regulonTFgenes, pairup.j=allGenes,
             annotation="org.EcK12.eg.db", fcOutput=qpg50preFC,
             fcOutput.na.rm=TRUE)
       i      j iSymbol jSymbol    x funCoherence
1 948797 945585    appY    appC 0.01         0.33
2 947466 945938    glcC    mhpR 0.01         0.14
3 945938 946255    mhpR    astC 0.01         0.71
4 947466 948336    glcC    fadB 0.02         0.14
5 948797 947547    appY    appB 0.02         0.33
6 948797 946206    appY    appA 0.02         0.33
```

The previous function call admits also a `file` argument that would allow us to store these information as a tab-separated column text file, thus more amenable for automatic processing when combined with the argument $n =$ `Inf` since by default this is set to a limited number ($n =$ 6) of pairs being reported.

For many other types of analysis, it is useful to store the network as an object of the `graphNEL` class, which is defined in the `graph` package. This is obtained by calling the `qpGraph()` function setting properly the argument `return.type` as follows:

```
> g <- qpGraph(avgnrr, threshold=avgnrr.50p.thr,
return.type="graphNEL")
> g
A graphNEL graph with undirected edges
Number of Nodes = 147
Number of Edges = 120
```

As we see from the object's description, the 50%-precision qp-graph network consists of 120 transcriptional regulatory relationships involving 147 different genes. A GO enrichment analysis on this subset of genes can give us insights into the main molecular processes related to the assayed conditions. Such an analysis can be performed by means of a conditional hypergeometric test using the Bioconductor package `GOstats` as follows:

```
> goHypGparams <- new("GOHyperGParams",
                      geneIds=nodes(g),
                      universeGeneIds=allGenes,
                      annotation="org.EcK12.eg.db", ontology="BP",
                      pvalueCutoff=0.05, conditional=TRUE,
                      testDirection="over")
> goHypGcond <- hyperGTest(goHypGparams)
```

where the object `goHypGcond` stores the result of the analysis which can be examined in R through the `summary()` function whose output is displayed in Table 1. The GO terms enriched by the subset of 147 genes reflect three broad functional categories one being transcription, which is the most enriched but it is also probably a byproduct of the network models themselves that are anchored on transcription factor genes. The other two are metabolism and response to an external stimulus, which are central among the biological processes that are triggered by an oxygen shift. Particularly related to this, is the fatty acid oxidation process

Table 1
Gene Ontology biological process terms enriched (*P*-value $\leq$ 0.05) among the 147 genes forming the 50%-precision qp-graph network inferred from the oxygen deprivation data in (18)

GO term identifier	*P*-value	Odds ratio	Exp. count	Count	Size	Term
GO:0006350	<0.0001	4.76	13.79	39	292	Transcription
GO:0009059	0.0004	2.14	27.81	43	589	Macromolecule biosynthetic process
GO:0019395	0.0022	5.34	1.42	6	30	Fatty acid oxidation
GO:0030258	0.0022	5.34	1.42	6	30	Lipid modification
GO:0044260	0.0035	1.84	38.15	51	808	Cellular macromolecule metabolic process
GO:0044238	0.0073	2.08	66.10	76	1,400	Primary metabolic process
GO:0006542	0.0096	8.92	0.47	3	10	Glutamine biosynthetic process
GO:0006578	0.0124	20.62	0.19	2	4	Betaine biosynthetic process
GO:0009268	0.0124	20.62	0.19	2	4	Response to pH
GO:0006807	0.0398	1.50	43.44	52	920	Nitrogen compound metabolic process
GO:0042594	0.0428	4.44	0.80	3	17	Response to starvation

since fatty acid metabolism is crucial to allow the cell to adapt quickly to environmental changes and allows *E. coli* to grow under anaerobic conditions (26).

Finally, using the `graphNEL` representation of our network stored in the variable *g* and the function `connComp()` from the `graph` package we can easily look up the distribution of sizes of the connected components:

```
> table(sapply(connComp(g), length))

 2  3  4  6  8  9 17 19
24  6  4  2  1  1  1  1
```

and observe that two of them, formed by 17 and 19 genes, are distinctively larger than the rest, thus corresponding to the more complex part of the network. In order to examine in more detail these two subnetworks, we can plot them using the Bioconductor package `Rgraphviz` (see Note 5) and calling the function `qpPlotNetwork()` which will output Fig. 3a:

```
> library(Rgraphviz)
> qpPlotNetwork(g, minimumSizeConnComp=17, pairup.i=regulonTFgenes,
                pairup.j=allGenes, annotation="org.EcK12.eg.db")
```

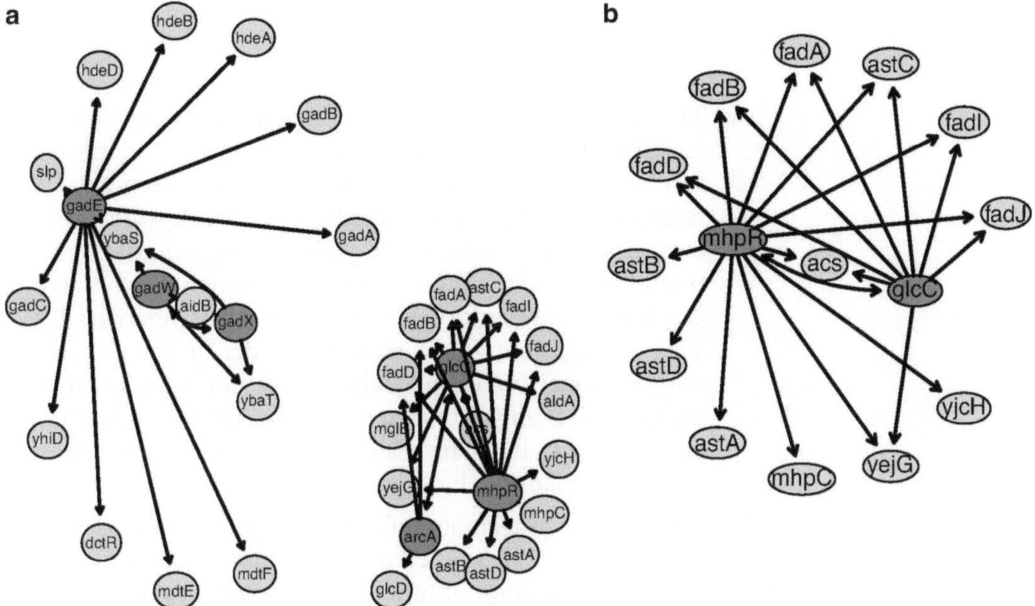

Fig. 3. The 50%-precision qp-graph transcriptional network. (**a**) The two largest connected components. (**b**) The *mhpR* regulatory module in detail.

Often the visualization of many interacting genes is difficult to interpret as, for instance in this case, the module regulated by *mhpR*. We can also visualize the part of the network connected to *mhpR* by using the arguments `vertexSubset` and `boundary` as follows and obtain the result shown in Fig. 3b:

```
> g_mhpR <- qpPlotNetwork(g, vertexSubset="mhpR", boundary=TRUE,
                          pairup.i=regulonTFgenes, pairup.j=allGenes,
                          annotation="org.EcK12.eg.db")
```

Note that we have assigned the result of this function to a variable named `g_mhpR`. This will store the graph we have just visualized into this variable as a `graphNEL` object and can be useful to extract the list of edges forming this subnetwork again through the function `qpTopPairs()`:

```
> qpTopPairs(measurementsMatrix=avgnrr[nodes(g_mhpR), nodes(g_mhpR)],
             refGraph=g_mhpR, pairup.i=regulonTFgenes,
pairup.j=allGenes,
             annotation="org.EcK12.eg.db", fcOutput=qpg50preFC)
       i      j iSymbol jSymbol    x funCoherence
1 945938 947466    mhpR    glcC 0.01         0.71
2 945938 946255    mhpR    astC 0.01         0.71
3 947466 948336    glcC    fadB 0.02         0.14
4 945938 944954    mhpR    mhpC 0.02         0.71
5 947466 946255    glcC    astC 0.02         0.14
6 945938 948823    mhpR    fadI 0.02         0.71
```

This last step allows us to see that the two strongest associations occur within the *mhpR* regulatory module, which also has a very high value of functional coherence, thus constituting two promising candidates for a follow up study.

4. Notes

1. The underlying method assumes that it is estimating an undirected Gaussian graphical model, which is a well-defined statistical model. However, our biological interpretation of this model as a transcriptional regulatory network will lead us to discard interactions between genes where none of them is a transcription factor, and to put directions in the resulting graph from transcription factor genes to their putative targets. This provides us with a network model of the underlying transcriptional regulation, which does not have a statistical interpretation anymore in terms of, for instance, conditional independence, but which allows one to formulate educated guesses on plausible biological hypotheses.
2. The limited availability of GO functional annotations for genes outside well-studied model organisms can compromise a reliable estimation of functional coherence values.
3. Bioconductor release versions are synchronized with R software release versions and thus updated twice a year. It is always recommended to work with the latest versions. For a detailed explanation on how to install and update the R and Bioconductor software please visit the Web site (27).
4. The installation of the package `Rmpi` requires a prior installation and configuration of an MPI library. For further details on this issue please visit the Web site (28).
5. The installation of the package `Rgraphviz` requires a prior installation of the software `graphviz` available at the Web site (29).

Acknowledgments

This work is supported by the Spanish Ministerio de Ciencia e Innovación (MICINN) [TIN2008-00556/TIN] and the ISCIII COMBIOMED Network [RD07/0067/0001]. R.C. is a research fellow of the “Ramon y Cajal” program from the Spanish MICINN [RYC-2006-000932]. A.R. acknowledges support from the Ministero dell’Università e della Ricerca [PRIN-2007AYHZWC].

References

1. http://www.bioconductor.org
2. Butte AJ, Tamayo P, Slonim D et al (2000) Discovering functional relationships between RNA expression and chemotherapeutic susceptibility using relevance networks. Proc Natl Acad Sci U S A 97:12182–12186.
3. Basso K, Margolin AA, Stolovitzky G et al (2005) Reverse engineering of regulatory networks in human B cells. Nat Genet 37:382–390.
4. Faith JJ, Hayete B, Thaden JT et al (2007) Large-scale mapping and validation of Escherichia coli transcriptional regulation from a compendium of expression profiles. PLoS Biol 5:e8.
5. Edwards D (2000) Introduction to graphical modelling. Springer, New York.
6. Dykstra RL (1970) Establishing Positive Definiteness of Sample Covariance Matrix. Ann Math Statist 41:2153–2154.
7. Barabasi A-L, Oltvai ZN (2004) Network biology: understanding the cell's functional organization. Nat Rev Genet 5:101–113.
8. Dobra A, Hans C, Jones B et al (2004) Sparse graphical models for exploring gene expression data. J. Multivariate. Anal. 90:196–212.
9. Friedman J, Hastie T, Tibshirani R (2008) Sparse inverse covariance estimation with the graphical lasso. Biostatistics 9:432–441.
10. Yuan M, Lin Y (2007) Model selection and estimation in the Gaussian graphical model. Biometrika 94:19–35.
11. Schäfer J, Strimmer K (2005) A shrinkage approach to large-scale covariance matrix estimation and implications for functional genomics. Stat. Appl. Genet. Mol. Biol. 4:1–32.
12. de la Fuente A, Bing N, Hoeschele I et al (2004) Discovery of meaningful associations in genomic data using partial correlation coefficients. Bioinformatics 20:3565–3574.
13. Wille A, Bühlmann P (2006) Low-order conditional independence graphs for inferring genetic networks. Stat. Appl. Genet. Mol. Biol. 5:1.
14. Castelo R, Roverato A (2006) A robust procedure for Gaussian graphical model search from microarray data with p larger than n. J Mach Learn Res 7: 2621–2650.
15. Castelo R, Roverato A (2009) Reverse engineering molecular regulatory networks from microarray data with qp-graphs. J Comput Biol 16:213–227.
16. http://www.geneontology.org
17. Falcon S, Gentleman R (2007) Using GOstats to test gene lists for GO term association. Bioinformatics 23:257–258.
18. Covert MW, Knight EM, Reed JL et al (2004) Integrating high-throughput and computational data elucidates bacterial networks. Nature 429:92–96.
19. Gama-Castro S, Jimenez-Jacinto V, Peralta-Gil M et al (2008) RegulonDB (version 6.0): gene regulation model of Escherichia coli K-12 beyond transcription, active (experimental) annotated promoters and Textpresso navigation. Nucleic Acids Res 36:D120–124.
20. http://www.r-project.org
21. http://cran.r-project.org
22. Gentleman RC, Carey VJ, Bates DM et al (2004) Bioconductor: open software development for computational biology and bioinformatics. Genome Biol 5: R80.
23. Schmidberger M, Morgan M, Eddelbuettel D et al (2009) State-of-the-art in Parallel Computing with R, Journal of Statistical Software 31:i01.
24. Wasserman WW, Sandelin A (2004) Applied bioinformatics for the identification of regulatory elements. Nat Rev Genet 5:276–287.
25. Fawcett T (2006) An introduction to ROC analysis. Pattern Recogn Lett 27: 861–874.
26. Cho, B.-K., Knight, E. M., and Palsson, B. O. (2006) Transcriptional regulation of the fad regulon genes of Escherichia coli by arcA., *Microbiology 152*, 2207–2219.
27. http://www.bioconductor.org/install
28. http://www.stats.uwo.ca/faculty/yu/Rmpi
29. http://www.graphviz.org

Chapter 15

Effective Non-linear Methods for Inferring Genetic Regulation from Time-Series Microarray Gene Expression Data

Junbai Wang and Tianhai Tian

Abstract

Owing to the quick development of high-throughput techniques and the generation of various "omics" datasets, it creates a prospect of performing the study of genome-wide genetic regulatory networks. Here, we present a sophisticated modelling framework together with the corresponding inference methods for accurately estimating genetic regulation from time-series microarray data. By applying our non-linear model on human p53 microarray expression data, we successfully estimated the activities of transcription factor (TF) p53 as well as identified the activation/inhibition status of p53 to its target genes. The predicted top 317 putative p53 target genes were supported by DNA sequence analysis. Our quantitative model can not only be used to infer the regulatory relationship between TF and its downstream genes but also be applied to estimate the protein activities of TF from the expression levels of its target genes.

Key words: Microarray, Genetic regulation, Non-linear model, Genetic algorithm, Inference

1. Introduction

Current advance in high-throughput technologies such as DNA microarrays, together with the availability of whole genome sequence for several species, enable us to study the genome-wide genetic regulatory networks in a cost-effect way. The heterogeneous functional genomic datasets have been used to acquire, catalogue and infer genetic regulatory networks in a "top-down" fashion (1–3). On the contrary, another principal research method, namely, the "bottom-up" approach, builds detailed mathematical models for small-scaled genetic regulatory networks based on extensive experimental observations. There are various models to accomplish the "bottom-up" approach: for example, differential equation models with continuous-time and continuous-variables, Bayesian network models with discrete-time and continuous-variables, and

Junbai Wang et al. (eds.), *Next Generation Microarray Bioinformatics: Methods and Protocols*,
Methods in Molecular Biology, vol. 802, DOI 10.1007/978-1-61779-400-1_15, © Springer Science+Business Media, LLC 2012

Boolean network models with discrete-time and discrete-variables (4). One of the major challenges of using a "bottom-up" approach to infer genetic regulation from microarray datasets is the lack of information for both protein concentrations and activities. Most of previous works were based on the assumption that the expression levels of a gene are consistent with its protein activities, though we know that is not always the case. An earlier practice to rectify above assumption is a hidden variable dynamic modelling (HVDM) method, which is a linear dynamic model designed to estimate the activities of a TF by using the expression activities of its target genes (5). Later, the HVDM method was extended to a non-linear one by using the Michaelis–Menten function (6). In addition, mathematical models with time delay were used to elucidate the time difference between the activities of TFs and the expression profiles of target genes (7). Recently, a more sophisticated inference method, which considers both the time delay and protein-DNA binding structure, has been developed for inferring the regulatory relationship between TF and its downstream genes (8).

Several earlier "bottom-up" researches used the "master" gene such as p53 networks to validate their proposed inference methodologies, as well as to investigate the regulatory function of the "master" gene (5). Although many experimental methods have been employed to identify the transcriptional target genes of p53 (e.g. the clustering analysis of microarray data, protein expression profiles, and Chip-PET identification of transcriptional-factor binding sites (9, 10)), it is imperative to use more sophisticated mathematical models to precisely describe the p53 regulation. Here, we devote to apply the proposed non-linear differential equation model on inferring genetic regulation of p53 from time-series microarray experiments.

2. Materials

2.1. Microarray Data

The work is based on a published microarray dataset which was generated from the Human All Origin, MOLT4 cells carrying wild-type p53. Cell was γ-irradiated and harvested every 2 h over a 12-h period (5). We obtained the ionizing radiation Affymetrix dataset (5) from ArrayExpress (E-MEXP-549).

3. Methods

3.1. Microarray Data Analysis

First, the raw microarray dataset was pre-processed by an R BioConduct package (11), in which probes with bad signal quality and less variation across all the time points were removed.

Table 1
A comparison of significantly differential gene selections between the pair-wise Fisher's linear discriminant method and maSigPro method

(*Q*, *R*)	Genes selected by maSigPro	Genes overlap with our selection	Percentage of overlapping
(0.05, 0.3)	1,165	646	55
(0.05, 0.4)	1,084	616	57
(0.05, 0.5)	661	455	69
(0.05, 0.6)	306	263	86
(0.05, 0.7)	139	131	94
(0.05, 0.8)	43	40	93
(0.05, 0.9)	14	12	86

This resulted in ~8,737 probes from a total of 22,284 probes. The pre-processed probes were then further median centred within each array and transformed to *Z*-scores before using the pair-wise Fisher's linear discriminant method (12) to screen probes with the most relevant response to ionizing radiation. The top 15% of the most relevant response probes (~1,312 probes) were selected as the input data to our non-linear model. All gene symbols were obtained from the NETAFFX (13).

To assess the robustness of our selected 1,312 probes, we compared the gene selections between the pair-wise Fisher's linear discriminant method and maSigPro method (14). The maSigPro method is an R package especially designed for analyzing time-course microarray experiments, which was applied to the same preprocessed microarray dataset. The parameter settings of the maSigPro method are a false discovery value (Q) that equals to 0.05 and an R-squared threshold (R) whose value ranges from 0.3 to 0.9. Table 1 suggests that both methods converged when a higher R-squared threshold (e.g. $R > 0.5$ represents a good model fitting in the original paper of the maSigPro method (14)) is used. Particularly, with a higher R-squared threshold, genes provided by the maSigPro method overlap more (>85%) with that selected by the Fisher's method.

Consequently, the selected 1,312 probes were assigned to 40 co-expressed gene modules by using a published computational approach (3, 12). Each gene module represents a set of co-expressed genes that are stimulated by either a specific experimental condition or a common trans-regulatory input. From a functional analysis of the 40 gene modules, we found that the co-expressed gene modules might contain genes with either heterogeneous or homogeneous biological functions, which are irrelevant to the number of genes in each module. Rather, it may reflect the complex mechanisms that control the transcription regulation. Therefore, in order to infer putative target genes of p53, we applied our non-linear model on the profile of each individual gene instead of the mean centre of each gene module. Detailed information of 1,312 probes and the corresponding 40 co-expressed gene modules are available in our earlier publication (8).

3.2. Non-linear Model

We have proposed a general type of the *cis*-regulatory functions that includes both positive and negative regulation, time delay, number of DNA-binding sites, and the cooperative binding of TFs (8). The dynamics of gene transcription is represented as

$$\frac{\mathrm{d}x_i}{\mathrm{d}t} = c_i + k_i f_i(x_j(t-\tau_{ij}), \ldots, x_k(t-\tau_{ik})) - d_i x_i, \qquad (1)$$

where c_i is the basal transcriptional rate, k_i is the maximal expression rate and d_i is the degradation rate. Here we use one value τ_{ij} to represent regulatory delays of gene j related to the expression of gene i. The *cis*-regulatory function $f_i(x_j, \ldots, x_k)$ includes both positive and negative regulations, given by

$$f_i(X) = \left[1 - \prod_{j \in R_i^+} g(x_j, n_j, m_j, k_j)\right] \prod_{j \in R_i^-} g(x_j, n_j, m_j, k_j),$$

and R_i^+ and R_i^- are subsets of positive and negative regulations of the total regulation set R, respectively. For each TF, the regulation is realized by

$$g(x, n, m, k) = \frac{1}{(1 + kx^n)^m},$$

where m is the number of DNA-binding site and n represents the cooperative binding of the TF. The present model is a more general approach which includes the proposed *cis*-regulatory function model when $n = 1$ (7), the Michaelis–Menten function model when $m = n = 1$ (6), and the Hill function model when $n>1$. Based on the structure of TF p53, the transcription of a p53 target gene is represented by

$$\frac{\mathrm{d}x_i(t)}{\mathrm{d}t} = c_i + k_i \frac{[p(t-\tau_i)]^{4\delta_i}}{K_i^4 + [p(t-\tau_i)]^4} - d_i x_i(t), \qquad (2)$$

where $x_i(t)$ is the expression level of gene i and $p(t)$ is the p53 activity at time t. Here δ_i is an indicator of the feedback regulation, namely, $\delta_i = 0$ if p53 inhibit the transcription of gene i or $\delta_i = 1$ if the transcription is induced by p53. The Hill coefficient was chosen to be 4 since p53 is in the form of tetramer as a transcriptional factor (15).

The model assumed that a TF regulates the expression of N target genes, which can be used to infer the activities of the TF from the expression levels of these N target genes. A system thus has N differential equations; and each equation represents the expression process of a specific gene. This system contains unknown parameters including the kinetic rates $(c_i, k_i, K_i, d_i, \tau_i, \delta_i)$ $(i = 1, \ldots, N)$ together with the TF activities $(p_j = p(t_j))$ at M measurement time points $(t_1, \ldots, t_M)$. Using an optimization method such as the genetic algorithm (16), we can search the optimal model parameters to match the expression levels $\{x_{ij}, i = 1, \ldots, N, \quad j = 1, \ldots, M\}$ of these N target genes at M measurement time points of the microarray experiments. The estimated values $\{p_j\}$ from the optimization method are our predicted TF activities.

3.3. Estimation of p53 Activities

Here we provide an example of using the non-linear model to predict the p53 activities from a set of five training target genes $(N = 5)$. A system of five differential equations was used to represent the expression of five training genes. The unknown parameters of the system are rate constants $(c_i, k_i, K_i, d_i, \tau_i, \delta_i)$ $(i = 1, \ldots, 5)$ and p53 activities $(p_j = p(t_j), \quad t_j = 2, 4, \ldots, 12)$ at 6 time points. The activities of p53 at other time points will be obtained by the natural spline interpolation. In total, there are 26 unknown parameters in the system, the p53 activities at 6 time points is our inference result.

We used a MATLAB toolbox of the genetic algorithm (16) to search the optimal values of these 26 parameters. The search space of each parameter is $[0, W_{\max}]$ and the values of $W_{\max}$ are $[5, 5, 5, 2]$ for $[c_i, k_i, K_i, d_i]$. For p53 activity p_i, the values of $W_{\max}$ are unit one. After a set of unknown parameters is created by the genetic algorithm, a program developed in MATLAB was used to simulate the non-linear system of five equations and calculate the objective value. The program is described below.

1. Create an individual of p53 activity $(p_i, \quad i = 1, \ldots, 6)$ and regulatory parameters (c_i, k_i, K_i, d_i) $(i = 1, \ldots, 5)$ from the genetic algorithm.
2. Use the natural spline interpolation to calculate p53 activity of time points [0, 12].
3. Solve the system of five equations 15.2 by using the fourth order classic Runge–Kutta method from the initial expression level $u_{i0}(= x_{i0})$, and find the simulated levels u_{ij} $(j = 1, \ldots, 6)$.
4. Calculate the estimation error of gene i as $e_i = \sum_{j=1}^{6} |u_{ij} - x_{ij}|/|x_{ij}|$ (see Note 1), where x_{ij} is the microarray expression level. Finally, the objective value is $e = \sum_{i=1}^{5} e_i$.

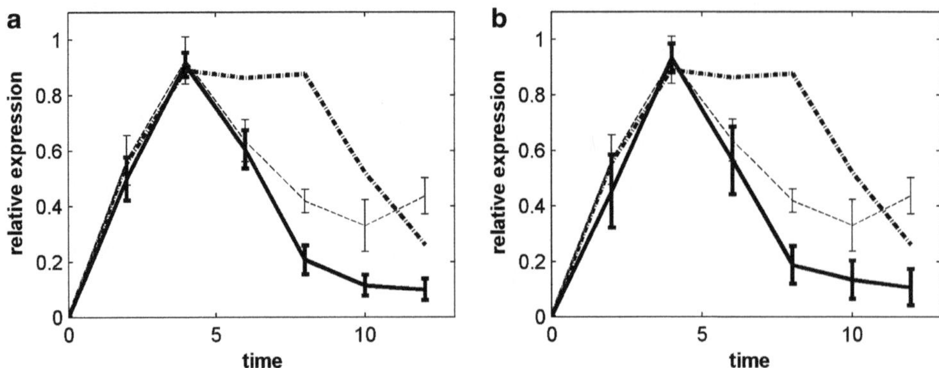

Fig. 1. Estimated p53 activity and the 95% confidence intervals based on five training genes (DDB2, PA26, TNFRSF10b, p21, and Bik) that are positively regulated by P53. (**a**) Estimates from the three replicates of microarray expression data. (**b**) Estimates from the mean of the three-replicate expression data. (*Dash-dot line*: p53 activities measured by Western blot [5]. The protein level p53 activation come a time-course immunoblot examination of p53 phosphorylated on S15; *dash line*: estimate of the HVDM method; *solid line*: prediction of the non-linear model).

In an earlier work, a linear model provided good estimation of p53 activities by using five known p53 target genes (5). To evaluate the performance of our non-linear model, we used the same p53 targets (i.e. DDB2, PA26, TNFRSF10b, p21, and Bik which are all positively regulated by p53) to predict the activities of p53. Here the time delay was assumed to be zero due to performing a consistent comparison study between the two models. Ten sets of the p53 activities at 6 time points were estimated from each replicate of the three microarray experiments and also from the average of these three microarray time courses. Figure 1a presents the mean and 95% confidence interval of the 30 sets of the predicted p53 activities from three microarray experiments, and Fig. 1b shows the results of the ten predictions from the averaged time courses of three microarray experiments. The relative error of the estimate in Fig. 1b is 2.70, which is slightly larger than both that in Fig. 1a (2.70) and that obtained by the linear model (1.89). Figure 1 indicates that the new non-linear model achieves the same goal as the linear model for predicting p53 activities.

To determine the influence of training genes on the estimation of p53 activities, we selected various sets of five training genes to infer the p53 activities. Estimation results indicated that there is slight difference between the estimated p53 activities by using different sets of training genes. One of the tests is shown in Fig. 2, where the estimated p53 activities were based on five training genes (RAD21, CDKN3, PTTG1, MKI67, and IFITM1) that are negatively regulated by p53 (17, 18). Similar to the study presented in Fig. 1, ten sets of the p53 activities were estimated from each replicate of the three microarray experiments and also from the average of these three microarray time courses. The mean and 95% confidence interval of both estimates are

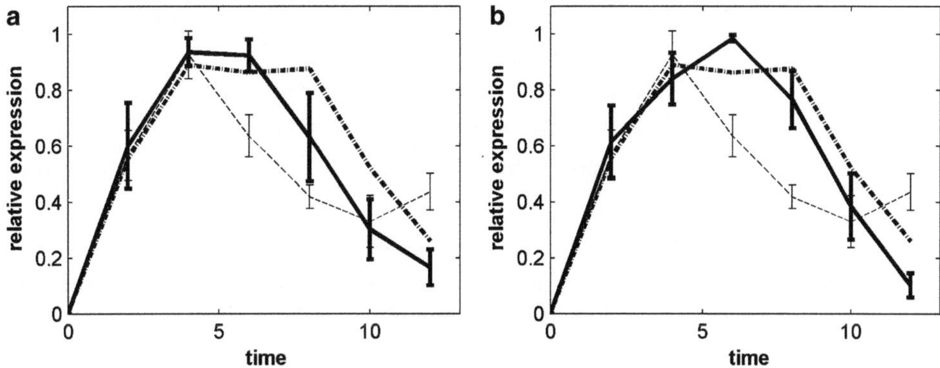

Fig. 2. Estimated p53 activity and the 95% confidence intervals based on five training genes (RAD21, CDKN3, PTTG1, MKI67, and IFITM1) that are negatively regulated by P53. (**a**) Estimates from the three replicates of microarray expression data. (**b**) Estimates from the mean of the three-replicate expression data. (*Dash-dot line*: p53 activities measured by Western blot [5]; *dash line*: estimate of the HVDM method; *solid line*: prediction of the non-linear model).

presented in Fig. 2a, b, respectively. The relative error of the estimate in Fig. 2b is 1.28, which is very close to that in Fig. 2a (1.30) but smaller than that obtained by the linear model (1.89) in Fig. 1. In this case, the estimated p53 activities are very close to the measured ones. It suggests that our proposed non-linear model is capable of making reliable predictions for the TF activities from the training genes that are all either positively or negatively regulated by the TF p53.

3.4. Prediction of Putative Target Genes by Using the Non-linear Model

Here we used the newly inferred p53 activity in Fig. 2b and the non-linear model 15.2 to infer the genetic regulation of p53 target genes. There are six unknown parameters for each gene's regulation, namely, $(c_i, k_i, K_i, d_i, \tau_i, \delta_i)$. The genetic algorithm was used to search for the optimal values of these six parameters. The value of δ_i is determined by another parameter η_i whose search area is $[-1, 1]$; and parameter η_i indicates either positive ($\eta_i>0$, $\delta_i = 1$) or negative ($\eta_i<0$, $\delta_i = 0$) regulation from p53. The time delay τ_i is treated as one of the unknown parameter and its value was searched by the genetic algorithm. The maximal possible time delay was set to 2.5 h because the experimentally determined time delay for p53 target genes is up to 2 h (9) (see Notes 2 and 3). Ten estimates $(c_{ij}, k_{ij}, K_{ij}, d_{ij}, \tau_{ij}, \delta_{ij})(j = 1, \ldots, 10)$ were obtained from different implementations of the genetic algorithm; and we selected the set of parameters that has the smallest estimation error as the final estimate. The following algorithm was developed to estimate the model parameters.

1. Create an individual of the regulatory parameter $(c_i, k_i, K_i, d_i, \tau_i, \delta_i)$ from the genetic algorithm.
2. Determine the value of δ_i in Eq. 2. If $\eta_i>0$, $\delta_i = 1$. Otherwise $\delta_i = 0$.

3. Determine the p53 activity based on activity in Fig. 2b and the time delay τ_i. $p(t - \tau_i) = 0\ (t \leqslant \tau_i)$.
4. Simulate model 15.2 by using the initial level $u_{i0} (= x_{i0})$ and find the simulated expression levels u_{ij} $(j = 1, \ldots, m)$.
5. Calculate the objective value $e_i = \sum_{j=1}^{m} |u_{ij} - x_{ij}|/|x_{ij}|$ (see Note 1).

All genes considered here are ranked by the model error e_i. Genes with smaller model error will be selected as the putative target genes for further study (see Note 1).

3.5. Selection of p53 Target Genes

To reduce variations in estimated parameters, we used a natural spline interpolation to expand the measurements from the original 7 time points to 25 time points, by adding three equidistant measurement points between each pair of measured time points. In addition, we used the genetic algorithm to infer the p53 mediated genetic regulation twice for each gene (e.g. either with or without time delay), and selected a final regulation result which has the smallest model estimation error. Then both the event method (19) and correlation approach (20) were used to infer the activation/inhibition of the p53 regulation. By comparing the consistency of inferred regulation relationships among above three mentioned methods, we only focused on the top 656 (~50%) predicted genes. Among these putative p53 target genes, ~64% are positively regulated by p53, while the rest are negatively regulated. A GO functional study of these 656 putative p53 target genes indicates that ~16% of them have unknown functions and these genes are excluded from our further study.

To provide more criteria for selecting putative p53 target genes, we searched for the p53 binding motif on the upstream non-coding region of the top 656 genes. This is because a physical interaction between p53 and its targets is essential for its role as a controller of the genetic regulation (20). Thus, for each putative target, we extracted the corresponding 10 kb DNA sequences located directly upstream of the transcription start site from ref. 21. Among the 656 putative p53 target genes, we found the upstream DNA sequences for 511 of them. Then, a motif discovery program MatrixREDUCE (22) was applied to search for the p53 consensus binding site. The results indicate that ~72.0% (366 out of 511 genes) of putative p53 targets have at least two copies of the p53 binding motif (perfect match counts of p53 binding site), while only ~10% (47 out of 511 genes) and ~20% (98 out of 511 genes) of them have zero and one p53 monomer, respectively. Based on the model estimation error and upstream TF-binding information of the 656 putative p53 target genes, we further narrowed down the number of possible p53 targets. In addition, for any gene that has more than one probe, we chose only the probe that has the smallest estimation error. We also excluded

Table 2
Comparison of the p53 consensus motif distributions in the four sets of putative p53 target genes obtained by the MVDM method (5), gene expression analysis (GEA) (9), Chip-PET analysis (10) and the non-linear model (8)

# of perfect match	MVDM	GEA	Chip-PET	Non-linear
0_p53_motif (5k)	0.41	0.24	0.28	0.22
1_p53_motif (5k)	0.22	0.38	0.32	0.33
2_p53_motif (5k)	0.24	0.20	0.25	0.23
>2 p53 motif (5k)	0.14	0.18	0.16	0.20
0_p53 motif (10k)	0.25	0.08	0.06	0.05
1_p53_motif (10k)	0.14	0.19	0.24	0.15
2_p53 motif (10k)	0.20	0.27	0.23	0.22
>2 p53 motif (10k)	0.41	0.46	0.47	0.58

genes with very small parameter k_i in model 15.2 because p53 may not have much influence on them (5). A final list containing ~317 putative p53 targets covers around ~24% of the total studied probes (~1,312) (see Notes 4 and 5; see also ref. 8).

3.6. Protein Binding Motif Analysis for Putative p53 Target Genes

The lack of common p53 targets among the four different predictions (5, 8–10) leads us to investigate whether the four lists of putative p53 targets share the same p53 binding motif distribution on the upstream non-coding region (see Note 6). By collecting the p53 binding motif counts on the gene upstream regions for the four predictions, Table 2 indicates that putative targets predicted by the gene expression analysis, the Chip-PET analysis, and our non-linear model, share a similar p53 binding preference. For example, there is an even distribution (~20%) of zero, one, two, and more than two p53 binding sites on the 5 kb region. However, there are more p53 binding motifs on the 10 kb upstream region than those on the 5-kb region. In addition, ~46–58% of putative p53 targets have more than two p53 binding sites on the 10 kb upstream region but only ~16–20% of targets have multiple binding sites on the 5 kb region. Furthermore, less than 10% of targets do not have p53 binding sites on the 10 kb region. The similar binding preference among various predictions suggests that the majority of putative p53 targets (~70%) may be directly controlled by remote p53 transcription factors but less than 30% of them may be the second effect targets.

A functional analysis of above four lists of putative p53 targets tells us that all works identified the same core biological functions of p53 (e.g. cell cycle, cell death, cell proliferation, and response to

DNA damage stimulus). However, there are a few gene functional categories that were only predicted by individual studies. For example, the lists from the gene expression analysis and Chip-PET analysis contain blood coagulation, body fluids, response to wound, muscle and signal transduction genes. However, only the list from the Chip-PET analysis is enriched by cell motility, cell localization and enzyme activity genes. In addition, high enrichment of metabolism, biosynthetic process and immune system process exclusively appear in our prediction. Although our results indicate that most of the p53 targets share the same p53 binding preference, their functional roles are conditionally specific, and their biological functions span to various functional categories with the dependence of intrinsic and extrinsic conditions. The functional differences among the four lists of putative p53 targets may partially explain the reason for the poor overlapping among them.

4. Conclusions

This chapter presented a non-linear model for inferring genetic regulation from time-series microarray data. This "bottom-up" method was designed not only to infer the regulation relationship between TF and its downstream genes but also to estimate the upstream protein activities based on the expression levels of the target genes. The major feature of the method is the inclusion of the cooperative binding of TFs, time delay and non-linearity by which we can study the non-linear properties of gene expression in a sophisticated way. The proposed method has been validated by comparing the estimated TF p53 activities with experimental data. In addition, the predicted putative p53 target genes from the non-linear model were supported by DNA sequence analysis.

5. Notes

1. The relative error was used in this work to compare the errors of different genes but the model estimation error may be large if the gene expression is weak. For that reason, a number of discovered p53 target genes were not included in our prediction, even though their simulations matched well the gene expression profiles. Therefore, it is worthy to further evaluate the influence of the error measurement on both the predictions of the TF activities and genetic regulation to the putative target genes (23).

2. Since the activities of all the promoters in the transcriptional machinery are modelled as those of TF, the estimated TF activities may be slightly different from one another if various sets of training target genes were used and consequently alter the prediction of putative target genes.

3. This is a practical approach to study the time delay effect of each individual p53 target gene by simplifying all kinds of time delay effects into a single factor. Therefore, the estimated time delay of each gene may differ.

4. Currently the Michaelis–Menten function has been widely used to model genetic regulation; but more precise estimates may be obtained by using a more sophisticated synthesis function which requires TFs' cooperative binding and/or binding sites information.

5. It is also important to develop stochastic models and the corresponding stochastic inference methods (24) to investigate the impact of gene expression noise on the accuracy of the modelling inference because there are noisy in microarray experiments.

6. A comparison study of different predictions obtained from different methods indicated the overlapping among the different predictions is quite poor (8). The discrepancy of p53 target gene predictions among various studies may be mainly caused by either pre-processing of microarray data or condition-specific gene regulation.

References

1. Sun N, Carroll RJ, Zhao H (2006) Bayesian error analysis model for reconstructing transcriptional regulatory networks. Proc Natl Acad Sci USA 103:7988–7993.
2. Wang J, Cheung LW, Delabie J (2005) New probabilistic graphical models for genetic regulatory networks studies. J Biomed Inform. 38:443–455.
3. Wang J (2007) A new framework for identifying combinatorial regulation of transcription factors: A case study of the yeast cell cycle. J Biomed Inform. 40:707–725.
4. de Jong H (2002) Modelling and simulation of genetic regulatory systems: A literature review. J. Comput. Biol. 9:67–103.
5. Barenco M, Tomescu D, Brewer D et al (2006) Ranked prediction of p53 targets using hidden variable dynamic modeling. Genome Biol. 7:R25.
6. Rogers S, Khanin R, Girolami M (2007) Bayesian model-based inference of transcription factor activity. BMC Bioinformatics 8:S2.
7. Goutsias J, Kim S (2006) Stochastic transcriptional regulatory systems with time delay: a mean-field approximation. J. Comput. Biol. 13:1049–1076.
8. Wang J, Tian T (2010) Quantitative model for inferring dynamic regulation of the tumour suppressor gene p53. BMC Bioinform. 11:36.
9. Zhao RB, Gish K, Murphy M et al (2000) Analysis of p53-regulated gene expression patterns using oligonucleotide arrays. Genes Deve. 14:981–993.
10. Wei CL, Wu Q, Vega VB et al (2006) A global map of p53 transcription-factor binding sites in the human genome. Cell 124:207–219.
11. Gentleman RC, Carey VJ, Bates DM et al (2004) Bioconductor: open software development for computational biology and bioinformatics. Genome Biol. 5:R80.
12. Wang J, Bo TH, Jonassen I et al (2003) Tumor classification and marker gene prediction by feature selection and fuzzy c-means

clustering using microarray data. BMC Bioinformatics 4:60.

13. Liu G, Loraine AE, Shigeta R et al (2003) NetAffx: Affymetrix probesets and annotations. Nucleic Acids Res. 31:82–86.
14. Conesa A, Nueda MJ, Ferrer A et al (2006) maSigPro: a method to identify significantly differential expression profiles in time-course microarray experiments. Bioinformatics 22:1096–1102.
15. Ma L, Wagner J, Rice JJ et al (2005) A plausible model for the digital response of p53 to DNA damage. Proc Natl Acad Sci USA 102:14266–14271.
16. Chipperfield A, Fleming PJ, Pohlheim H (1994) A Genetic Algorithm Toolbox for MATLAB. Proc. Int. Conf. Sys. Engineering: p.200-207.
17. Kho PS, Wang Z, Zhuang L et al (2004) p53-regulated Transcriptional Program Associated with Genotoxic Stress-induced Apoptosis. J. Biol. Chem. 279:21183–21192.
18. Wu Q, Kirschmeier P, Hockenberry T et al (2002) Transcriptional regulation during p21WAF1/CIP1-induced apoptosis in human ovarian cancer cells. J. Biol. Chem. 277:36329–36337.
19. Kwon AT, Hoos HH, Ng R (2003) Inference of transcriptional regulation relationships from gene expression data. Bioinformatics 19:905–912.
20. El-Deiry WS, Kern SE, Pietenpol JA et al (1992) Definition of a consensus binding site for p53. Nat Genet. 1:45–49.
21. Aach J, Bulyk ML, Church GM et al (2001) Computational comparison of two draft sequences of the human genome. Nature 409:856–859.
22. Moorman C, Sun LV, Wang J et al (2006) Hotspots of transcription factor colocalization in the genome of Drosophila melanogaster. Proc Natl Acad Sci USA 103:12027–12032.
23. Moles CG, Mendes P, Banga JR (2003) Parameter estimation in biochemical pathways: A comparison of global optimization methods. Genome Res. 13:2467–2474.
24. Tian T, Xu S, Gao J et al (2007) Simulated maximum likelihood method for estimating kinetic rates in genetic regulation. Bioinformatics 23:84–91.

Part IV

Next Generation Sequencing Data Analysis

Chapter 16

An Overview of the Analysis of Next Generation Sequencing Data

Andreas Gogol-Döring and Wei Chen

Abstract

Next generation sequencing is a common and versatile tool for biological and medical research. We describe the basic steps for analyzing next generation sequencing data, including quality checking and mapping to a reference genome. We also explain the further data analysis for three common applications of next generation sequencing: variant detection, RNA-seq, and ChIP-seq.

Key words: Next generation sequencing, Read mapping, Variant detection, RNA-seq, ChIP-seq

1. Introduction

In the last decade, a new generation of sequencing technologies revolutionized DNA sequencing (1). Compared to conventional Sanger sequencing using capillary electrophoresis, the massively parallel sequencing platforms provide orders of magnitude more data at much lower recurring cost. To date, several so-called next generation sequencing platforms are available, such as the 454-FLX (Roche), the Genome Analyzer (Illumina/Solexa), and SOLiD (Applied Biosystems); each having its own specifics. Based on these novel technologies, a broad range of applications has been developed (see Fig. 1).

Next generation sequencing generates huge amounts of data, which poses a challenge both for data storage and analysis, and consequently often necessitates the use of powerful computing facilities and efficient algorithms. In this chapter, we describe the general procedures of next generation sequencing data analysis with a focus on sequencing applications that use a reference sequence to which the reads can be aligned. After describing how to check the sequencing quality, preprocess the sequenced reads, and map the sequenced reads to a reference, we briefly

Junbai Wang et al. (eds.), *Next Generation Microarray Bioinformatics: Methods and Protocols*,
Methods in Molecular Biology, vol. 802, DOI 10.1007/978-1-61779-400-1_16, © Springer Science+Business Media, LLC 2012

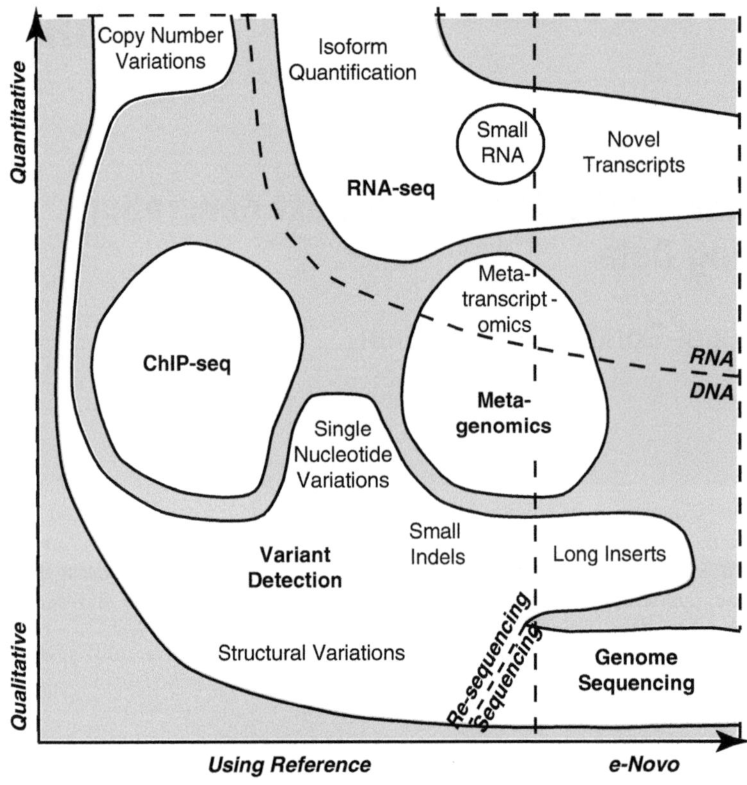

Fig. 1. Illustration of some common applications based on next generation sequencing. The decoding of new genomes is only one of various possibilities to use sequencing. Variant detection, ChIP-seq, and RNA-seq are discussed in this book. Metagenomics (16) is a method to study communities of microbial organisms by sequencing the whole genetic material gathered from environmental samples.

discuss three of the most common applications for next generation sequencing.

1. *Variant detection* (2) means to find genetic differences between the studied sample and the reference. These differences range from single nucleotide variants (SNVs) to large genomic deletions, insertions, or rearrangements.
2. *RNA-seq* (3) can be used to determine the expression level of annotated genes as well as to discover novel transcripts.
3. *ChIP-seq* (4) is a method for genome-wide screening protein–DNA interactions.

2. Methods

2.1. General Read Processing

Current next generation sequencing technologies based on photochemical reactions recorded on digital images, which are further processed to get sequences (reads) of nucleotides or, for SOLiD, dinucleotide "colors" (5) (base/color calling). The sequencing data analysis starts from files containing DNA sequences and quality values for each base/color.

1. Check the overall success of the sequencing process by counting the raw reads, i.e., spots (clusters/beads) on the images, and the fraction of reads accepted after base calling (filtered reads). These counts could be looked up in a results file generated by the base calling software. A low number of filtered reads could be caused by various problems during the library preparation or sequencing procedure (see Note 1). Only the filtered reads should be used for further processing. For more ways to test the quality of the sequencing process see Notes 2 and 3.
2. Sequencing data are usually stored in proprietary file formats. Since some mapping software tools do not accept these formats as input, a script often has to be employed to convert the data into common file formats such as FASTA or FASTQ.
3. The sequenced DNA fragments are sometimes called "inserts" because they are wrapped by sequencing adapters. The adapters are partially sequenced if the inserts are shorter than the read length, for example, in small RNAs sequencing (see Subheading 2.4, step 5). In these occasions, it is necessary to remove the sequenced parts of the adapter from the reads, which could be achieved by removing all read suffixes that are adapter prefixes (see Note 4).

2.2. Mapping to a Reference

Many applications of next generation sequencing require a reference sequence to which the sequenced reads could be aligned. Read mapping means to find the position in the reference where the read matches with a minimum number of differences. This position is hence most likely the origin of the sequenced DNA fragment (see Note 5).

1. There are numerous tools available for read mapping (6). Select a tool that is appropriate for mapping reads of the given kind (see Note 6). Some applications may require special read mapping procedures that, for example, allow small insertions and deletions (indels) or account for splicing in RNA-seq.
2. Select an appropriate maximum number of allowed errors (see Note 7).
3. For most applications you only need uniquely mapped reads, i.e., reads matching to a single "best" genomic position. If nonuniquely mapped reads could also be useful, then consider to specify an upper bound for the number of reported mapping positions, because otherwise the result list is blown up by reads mapping to highly repetitive regions.

4. Most mapping tools create output files in proprietary formats, so we advice to convert the mapping output into a common file format such as BED, GFF, or SAM (7, 8).
5. Count the percentage of all reads which could be mapped to at least one position in the reference. A low amount of mappable reads could indicate a low sequencing quality (see Note 3) or a failed adapter removal (see Note 4).
6. Some pieces of DNA could be overamplified during library preparation (PCR artifacts) resulting in a stack of redundant reads that are mapped to the same genomic position and same strand. If it is necessary to get rid of such redundancy, discard all but one read mapped to the same position and on the same strand.
7. Transform SOLiD reads into nucleotide space after mapping.

2.3. Application 1: Variant Detection

The detection of different variation types requires different sequencing formats and analysis strategies. Tools are available for the detection of most variant types (2) (see Note 8).

1. For detecting SNVs, search the mapped reads for bases that are different from the reference sequence. Since there will probably be more sequencing errors than true SNVs, each SNV candidate must be supported by several independent reads. A sufficient coverage is therefore required (see Note 9). Note that some SNVs might be heterozygous, which means that they occur only in some of the reads spanning them.
2. Structural variants can be detected by sequencing both ends of DNA fragments (paired-end sequencing) (see Fig. 2) (9). After mapping the individual reads independently to the reference, estimate the distribution of fragment lengths. Then search for read pairs which were mapped to different chromosomes or have abnormal distance, ordering, or strand orientation. Search for a most parsimonious set of structural variants explaining all discordant read pairs. The more read pairs can be explained by the same variant, the more reliable this variant is and the more precise the break point(s) could be determined. If only one end of a DNA fragment could be mapped to the reference, the other end is possibly part of a (long) insertion. Given a suitable coverage, the sequence of the insertion can possibly be determined by assembling the unmapped reads.

2.4. Application 2: RNA-seq

The experimental sequencing protocols and hence the data analysis procedures are usually different for longer RNA molecules such as mRNA (Subheading 2.4 steps 2 and 3) and small RNA such as miRNA (Subheading 2.4 steps 5 and 6).

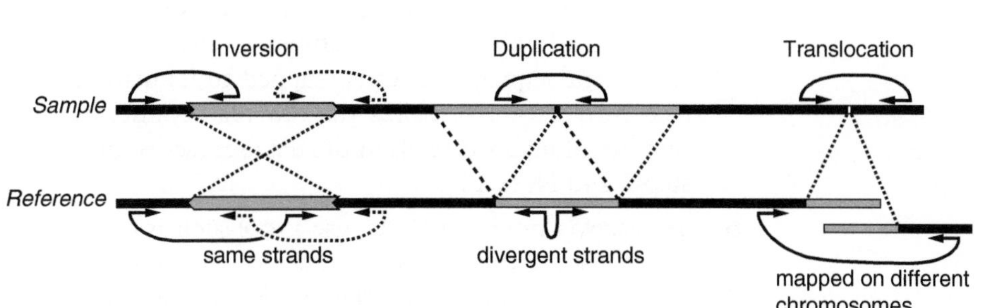

Fig. 2. Different variant types detected by paired-end sequencing (9). (1) Deletion: The reference contains a sequence that is not present in the sample. (2–3) Insertion and Long Insertion: The sample contains a sequence that does not exist in the reference. (4) Inversion: A part of the sample is reverse compared to the reference. (5) Duplication: A part of the reference occurs twice in the sample (tandem repeat). (6) Translocation: The sample is a combination of sequences coming from different chromosomes in the reference. Note that the pattern for concordant reads varies depending on the sequencing technologies and the library preparation protocol.

1. Check the data quality. Classify the mapped reads on the basis of available genome annotation into different functional groups such as exons, introns, rRNA, intergenic, etc. For example, in the case of sequencing polyA-RNA, only a small fraction of reads should be mapped to rRNA.
2. Determine the expression level of annotated genes by counting the reads mapped to the corresponding exons, and then divide these counts by the cumulated exon lengths (in kb) and the total number of mapped reads (in millions). The resulting RPKM ("reads per kilobase of transcript per million mapped reads") can be used for comparing expression levels of genes in different data sets (10).
3. To quantify different splicing isoforms, select reads belonging exclusively to certain isoforms, for example, reads mapping to exons or crossing splicing junctions present only in a single isoform. From the amounts of these reads infer a maximum likelihood estimation of the isoform expression levels.
4. To discover novel transcripts or splicing junctions, use a spliced alignment procedure to map the RNA-seq reads to a reference genome. Then find a most parsimonious set of transcripts that explains the data. Alternatively, you could first assemble the sequencing reads and then align the assembled

contigs to the genome (11). In both cases, it is advisable to sequence long paired-end reads.

5. Small RNA-seq reads are first preprocessed to remove adapter sequences (see Subheading 2.1, step 3). To profile known miRNA, the reads could then be mapped either to the genome or to the known miRNA precursor sequences (12). Do not remove redundant reads (see Subheading 2.2, step 6) when analyzing this kind of data. The expression level of a specific miRNA could be estimated given the number of redundant sequencing reads mapped to its mature sequence (see Note 10). Normalize the raw read counts by the total number of mapped reads in the data set (see Subheading 2.4, step 2 and Note 11).
6. To discover novel miRNAs, use a tool such as miRDeep (13), which uses a probabilistic model of miRNA biogenesis to score compatibility of the position and frequency of sequenced RNA with the secondary structure of the miRNA precursor.

2.5. Application 3: ChIP-seq

In ChIP-seq, chromatin immunoprecipitation uses antibodies to specifically select the proteins of interest together with any piece of randomly fragmented DNA bound to them. Then the precipitated DNA fragments are sequenced. Genomic regions binding to the proteins consequently feature an increased number of mapped sequencing reads.

1. Use a "peak calling" tool to search for enriched regions in the ChIP-seq data (10) (see Note 12). ChIP-seq data should be evaluated relative to a control data set obtained either by sequencing the input DNA without ChIP or by using an antibody with unspecific binding such as IgG (see Note 9).
2. An alternative way to analyze the data that is especially suited for profiling histone modifications is to determine the normalized read density (RPKM) of certain genomic areas such as genes or promoter regions. This method is similar to the analysis of RNA-seq data (see Subheading 2.4, step 2).

3. Notes

1. In some cases, the sequencing results could be improved by manually restarting the base calling using nondefault parameters. For example, choosing a better control lane when starting the Illumina offline base caller could boost up the number of successfully called sequencing reads. Candidates for good control lanes feature a nearly uniform base

distribution (see Note 2). Note that for this reason a flow cell should never be filled completely with, e.g., small RNA libraries, since these are not expected to produce uniform base distributions.

2. Check the base/color distribution over the whole read length. If the sequenced DNA fragments are randomly sampled from the genome – for example, sequencing genomic DNA, ChIP-seq, or (long) RNA-seq libraries – then the bases should be nearly uniformly distributed for all sequencing cycles. The software suite provided by the instrument vendors usually creates all relevant plots.

3. The base caller annotates each base with a value reflecting its putative quality. These values could be used to determine the number of high/low quality bases for each cycle. The overall quality of sequenced bases normally declines slowly toward the end of the read. A drop of quality for a single cycle could be a hint for a temporary problem during the sequencing.

4. Since the sequenced adapter could contain errors, it is reasonable to allow some mismatches during the adapter search. Note that there is a trade-off between the sensitivity and the specificity of this search.

5. In order to avoid wrongly mapped reads, it is important to use a reference as accurate and complete as possible. All possible sources of reads should be present in the reference.

6. Not all tools can handle SOLiD reads in dinucleotide color space; Roche 454 reads may contain typical indels in homopolymer runs. When mapping the relatively short reads created by or Illumina Genome Analyzer or SOLiD, it is usually sufficient to consider only mismatches, unless it is planned to detect small indels.

7. We recommend to choose a mapping strategy that guarantees accurate mappings rather than to maximize the mere number of mapped reads. Next generation sequencing usually generates huge quantities of reads, so a negligible loss of reads is certainly affordable. Consequently, most mapping tools are optimized to allow only a small number of mismatches. Higher error numbers are only necessary if the reads are long or if we are especially interested in variations between reads and reference.

8. Check the success of your experiment by comparing your results to already known variants deposited in public data bases such as dbSNP (14) and the Database of Genomic Variants (15).

9. Sequencing reads are *never* uniformly distributed throughout the genome, and any statistical analysis assuming this is inaccurate. Some parts of the genome usually are covered by much more reads than expected, whereas some other parts are not sequenced at all. The experimenter should be aware of this fact, for example, when planning the required read coverage for variant detection. Moreover, this effect certainly impacts quantitative measurements such as ChIP-seq or RNA-seq. ChIP-seq assays, for example, should always include a control library (see Subheading 2.5, step 1), and in a RNA-seq experiment, it is easier to compare expression levels of the same gene in different circumstances rather than the expression level of different genes in the same sample.
10. Note that the actual sequenced mature miRNA could be shifted by some nucleotides compared to the annotation in the miRNA databases.
11. One problem of this normalization method is that sometimes few miRNAs get very high read counts, which means that any change of their expression level could affect the read counts of all other miRNAs. In some cases, a more elaborated normalization method could therefore be necessary.
12. Most tools for analyzing ChIP-seq data focus on finding punctuate binding sites (peaks) typical for transcription factors. For ChIP-seq experiments targeting broader binding proteins, like polymerases or histone marks such as H3K36me3, use a tool that can also find larger enriched regions. In order to precisely identify protein binding sites, it is often necessary to determine the average length of the sequenced fragments. Some ChIP-seq data analysis tools estimate the fragment length from the sequencing data. Keep in mind that this is not trivial, because ChIP-seq data usually consist of single-end sequencing reads. Therefore, always check whether the estimated length is plausible according to the experimental design.

References

1. Shendure J, Ji H (2008) Next-generation DNA sequencing. Nature Biotechnology 26:1135–1145
2. Medvedev P, Stanciu M, Brudno M (2009) Computational methods for discovering structural variation with next-generation sequencing. Nature Methods 6:S13-S20
3. Mortazavi A, Williams BA, McCue K et al (2008) Mapping and quantifying mammalian transcriptomes by RNA-Seq. Nature Methods 5:621–628
4. Johnson DS, Mortazavi A, Myers RM et al (2007) Genome-Wide Mapping of in Vivo Protein-DNA Interactions. Science 316 (5830):1497–1502
5. Fu Y, Peckham HE, McLaughlin SF et al. SOLiD Sequencing and 2-Base Encoding. http://appliedbiosystems.com
6. Flicek P, Birney E (2009) Sense from sequence reads: methods for alignment and assembly. Nature Methods 6:S6-S12
7. UCSC Genome Bioinformatics. Frequently Asked Questions: Data File Formats. http://genome.ucsc.edu/FAQ/FAQformat.html
8. Sequence Alignment/Map (SAM) Format. http://samtools.sourceforge.net/SAM1.pdf
9. Korbel JO, Urban AE, Affourtit JP et al. (2007) Paired-End Mapping Reveals Extensive Structural Variation in the Human Genome. Science 318 (5849):420–426
10. Pepke S, Wold B, Mortazavi A (2009) Computation for ChIP-seq and RNA-seq studies. Nature Methods 6:S22-S32
11. Haas BJ, Zody MC (2010) Advancing RNA-seq analysis. Nature Biotechnology 28:421–423
12. Griffiths-Jones S, Grocock RJ, van Dongen S et al (2006) miRBase: microRNA sequences, targets and gene nomenclature. Nucleic Acids Research 34:D140-D144. http://microrna.sanger.ac.uk
13. Friedländer MR, Chen W, Adamidi C et al (2008) Discovering microRNAs from deep sequencing data using miRDeep. Nature Biotechnology 26:407–415
14. dbSNP. http://www.ncbi.nlm.nih.gov/projects/SNP
15. Database of Genomic Variants. http://projects.tcag.ca/variation
16. Handelsman J, Rondon MR, Brady SF et al (1998) Molecular biological access to the chemistry of unknown soil microbes: a new frontier for natural products. Chemistry & Biology 5:245–249

References

Chapter 17

How to Analyze Gene Expression Using RNA-Sequencing Data

Daniel Ramsköld, Ersen Kavak, and Rickard Sandberg*

Abstract

RNA-Seq is arising as a powerful method for transcriptome analyses that will eventually make microarrays obsolete for gene expression analyses. Improvements in high-throughput sequencing and efficient sample barcoding are now enabling tens of samples to be run in a cost-effective manner, competing with microarrays in price, excelling in performance. Still, most studies use microarrays, partly due to the ease of data analyses using programs and modules that quickly turn raw microarray data into spreadsheets of gene expression values and significant differentially expressed genes. Instead RNA-Seq data analyses are still in its infancy and the researchers are facing new challenges and have to combine different tools to carry out an analysis. In this chapter, we provide a tutorial on RNA-Seq data analysis to enable researchers to quantify gene expression, identify splice junctions, and find novel transcripts using publicly available software. We focus on the analyses performed in organisms where a reference genome is available and discuss issues with current methodology that have to be solved before RNA-Seq data can utilize its full potential.

Key words: RNA-Seq, Genomics, Tutorial

1. Introduction

Recent advances in high-throughput DNA sequencing have enabled new approaches for transcriptome analyses, collectively named RNA-Seq (RNA-Sequencing) (1). Variations in library preparation protocols allow for the enrichment or exclusion of specific types of RNAs, e.g. an initial polyA$^+$ enrichment step will efficiently remove nonpolyadenylated transcripts (2, 3). Alternative protocols retain both polyA$^+$ and polyA$^-$ RNAs while excluding ribosomal RNAs (4, 5). Protocols have also been developed for direct targeting of actively transcribed (6) or translated (7) RNA.

* Daniel Ramsköld and Ersen Kavak contributed equally to this work.

Junbai Wang et al. (eds.), *Next Generation Microarray Bioinformatics: Methods and Protocols*,
Methods in Molecular Biology, vol. 802, DOI 10.1007/978-1-61779-400-1_17, © Springer Science+Business Media, LLC 2012

These sequence libraries are then sequenced at great depth (often tens of millions) on Illumina, SOLiD, or Helicos platforms (8).

Compared to microarrays, RNA-Seq data is richer in several ways. RNA-Seq at low depth is similar to gene microarrays, but without cross-hybridization and with a larger dynamic range (1, 2, 9). This makes RNA-Seq considerably more sensitive, making present/absent calls more meaningful. At higher sequence depths, RNA-Seq resembles exon junction arrays, but analyses of differential RNA processing, such as alternative splicing (2, 3), are simplified and more powerful due to the larger number of independent observations and the nucleotide-level resolution over exon–exon junctions. In addition, the RNA-Seq data can be used to find novel exons and junctions, since it does not require probe selection. Indeed, paired-end sequencing at great depth enabled the first cell-type specific transcript maps to be reconstructed *de novo* (10, 11). For these reasons, as sequencing capacity is set up at core facilities or external companies and the RNA-Seq data analyses become easier for the end users, we expect sequencing to gradually replace hybridization-based gene expression analyses. With more RNA-Seq data being generated using a variety of experimental protocols, we will soon have an unprecedented detailed picture of what parts of the genome is transcribed, at what expression level, and the full extent of RNA diversity stemming from alternative RNA processing pathways.

This chapter is written for researchers starting with RNA-Seq data analyses. We provide a tutorial for the analyses of raw sequence data, expression level estimates, differentially expressed genes, novel gene predictions, and visual coverage of genomic regions. We discuss different analyses approaches and highlight current challenges and caveats. Although this tutorial focuses mainly on RNA-Seq data generated on the Illumina and SOLiD platforms, many of the steps will be directly analogous for other types of RNA-Seq data. Many of the tools we discuss are run from the command line. In Windows, the default command line interpreter is the "Command prompt." In other operating systems, it is typically named "Terminal."

2. Methods

2.1. Sequence Reads and Their Formats

We first discuss the file formats used for RNA-Seq data and where publicly available data can be found. High-throughput sequence data is stored mainly in NCBI's Sequence Read Archive (SRA) (12). More processed versions of the data (such as sequence reads mapped to the genome or calculated expression levels) are often found in gene expression omnibus (GEO) (13). For the Illumina Genome Analyzer, data downloaded from SRA or obtained from

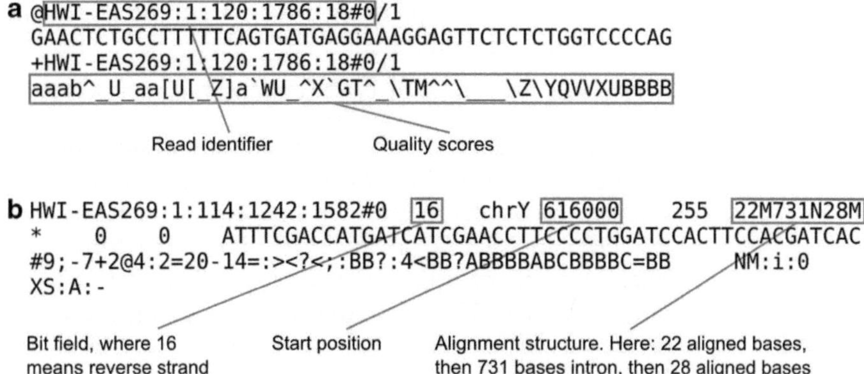

Fig. 1. Commonly used file formats for sequence reads and aligned reads. (**a**) A read in FASTQ file format (Solexa1.3+ flavor). (**b**) An aligned read in SAM file format. Selected entries have been annotated; see refs. 18 and 40 for details.

core facilities and service providers often comes in the FASTQ format (Fig. 1a). A FASTQ file contains the sequence name, nucleotides, and associated quality scores. The FASTQ format comes in a few flavors, differing in the encoding of the quality scores. Files in the SRA use the convention from Sanger sequencing, with Phred quality scores (14), whereas different versions of Illumina's software produce their own versions of the format (15). Some alignment programs can handle the conversion from these to the Sanger FASTQ format internally, otherwise tools within, e.g., Galaxy, Biopython, or Bioperl (16–18) can be used.

For Applied Biosystem's SOLiD machines, data is often provided in two separate files: one CSFASTA file and one QUAL file. The QUAL file contains the quality scores per base. The CSFASTA format differs from FASTA files in that sequences are in color space, an encoding where each digit (0–3) represents two adjacent bases in a degenerate way. To understand color space encoding, look at the following sequence:

T021331103300230232200230103322211233

The T is the last base of the adapter. The first 0 could be AA, CC, GG, or TT; thus the base after this T must be a T. 2 corresponds to GA, AG, TC, or CT, so the next base is a C. Together with the other two mappings (one corresponds to CA/AC/GT/TG and three to TA/AT/CG/GC), the sequence becomes:

TTCATACAATAAAGCCTAGAAAGCCAATAGACAGCG

However, if a sequencing error turned the tenth color into a 1, the sequence would be:

TTCATACAATGGGATTCGAGGGATTGGCGAGTGATA

This sequence would have too many mismatches to map to the genome. Instead of conversion, SOLiD reads are mapped in color space, so that they can be mapped despite sequencing errors.

Data in SRA is downloaded in FASTQ format, including SOLiD data, for which the sequence line of FASTQ file is in color space (see Note 1). However, SRA recommends uploads in sequence read format (SRF) for Illumina and SOLiD data (19). There are conversion tools to SRF format for both SOLiD (solid2srf) (20) and Illumina (illumina2srf) that comes with the Staden IO (21) and sequenceread (22) packages.

2.2. Aligning Reads Toward Genome and Transcriptome

Sequence alignment is the first step in the analysis of a new RNA-Seq data. Although one could directly map reads to databases of expressed transcripts, the most common approach has been to first map reads to the genome and then compare the alignments with known transcript annotations. A multitude of alignment tools for short read data exist and we refer readers to a recent review for a more thorough discussion (23). Data from Illumina's machine has few substitution errors per read and virtually no insertion or deletion (indel) errors (24). Thus, it can be mapped efficiently by, for example, Bowtie (25) and its junction-mapping extension by Tophat (26) that can handle up to three mismatches per sequence and no indels. Aligning SOLiD reads is however more computationally expensive and requires alignment software that works in color space. SOLiD data has more substitution errors per read in color space, so the mapping benefits from software that allow relaxed mismatch criteria, such as PerM (27). In addition to command line programs, these software and others are available through the Web-service Galaxy (28, 29). Other software use variations of the Needleman–Wunsch algorithm, e.g., Novoalign (30), BFAST (31), and Mosaik (32), allowing them to handle indels. This makes alignments more tolerant to DNA polymorphisms, as well as the indel errors that are common in Helicos' technology (33), at the cost of processor time. Aligning reads containing adapter sequence requires additional processing (see Note 2) and we have seen libraries where as many as 40% of the reads contained adapter sequence.

2.2.1. Aligning Reads to Exon–Exon Junctions

Some reads will overlap an exon–exon junction, that is, the position where an intron has been excised. These "junction reads" will not map directly to the genome. For de novo discovery of junctions, reads can be divided into multiple parts, which are aligned separately. This approach can only map a fraction of junction reads however. Another approach is to generate a sequence database that junction reads can map to. It can be applied by hand to any short read alignment program, by creating a library where each new "chromosome" corresponds to an exon–exon junction. After aligning reads to the genome and junction library, you will need to convert the junction-mapping reads to SAM or BED format for downstream analyses. If the read length is L and at least M nucleotides are required to map to each side of the junction (anchor length), then you extract $L - M$ bp for each exon end and the total sequence of the exon–exon junction becomes $2L - 2M$ bp. It is advisable to use at least four base pairs on each exon (2, 3) and it should be more than the number of mismatches/indels allowed. Both these approaches are used by Tophat (34), for the latter it can either be fed intron coordinates or try to find them itself from the positions of read clusters (putative exons).

2.2.2. De Novo Splice Junction Discovery

De novo junction discovery reduces accuracy compared to using a library of known exon–exon junctions, and longer anchor lengths are required to keep sequencing errors from causing false positives. We feel that using Tophat, provided with a library of known junctions, gives a fair trade-off between sensitivity and ability to find junctions outside current gene annotation for Illumina reads. For this, first specify a set of known junctions in "Tophat format." Each line of this file contains zero-based genomic coordinates of the upstream exon end and downstream exon start, for example, the last intron of the ACTB gene on hg19 assembly should be provided as:

```
chr7    5567521    5567634    -
```

One way to generate these is with the UCSC genome browser's table browser. Here, choose output in BED format (35), use, e.g. the knownGene table, click submit, and then specify that regions should be introns. You will have to subtract 1 from each start position in the file the browser produces, since Tophat requires a slightly different format than the one produced by table browser. In addition to the junctions you have specified, Tophat will by default try to find novel junctions. You also need to build a genome index with Bowtie's bowtie-build, or download one from its homepage (36). If your genome index files are called hg19index.ebwt.1, etc., you run Tophat from the command line with:

```
tophat -o outputdir -j junctionlibrary.bed hg19index reads.fastq
```

The resulting alignment will be found in outputdir/accepted_hits.sam (or .bam in more recent versions). Multimapping reads are listed multiple times and the NH flag in the SAM files can be used to identify uniquely mapping reads.

2.2.3. An Alternative Strategy for Read Mapping

For other alignment tools than Tophat, a library of sequences corresponding to splice junctions can be supplied. This strategy is useful where a higher tolerance for mismatches can be an advantage such as for SOLiD data. We provide such junction files at our Web site (37) for mouse and human, together with a python program to work with them. Assuming your reads are human and you want a minimum anchor length of 8, do the following to align with PerM:

1. Install PerM (38) and python 2.5/2.6/2.7 (39).
2. Download hg19junctions_100.fa and junctions2sam.py from our site (37).
3. Prepare a plain text file called, e.g., hg19files.txt listing the FASTA files, one per line:

   hg19_junctions_100.fa

   hg19/chr1.fa

   hg19/chr2.fa

   etc.

   where hg19 is the folder for the genome. Do not include chromosome files with a "hap" suffix unless you plan to handle multimapping reads, as these files do not represent unique loci. The same can be true for unplaced contigs (files with "chrUn" prefix or "random" suffix).
4. Assuming reads and quality scores are in reads_F3.csfasta and reads_F3_QV.qual, run PerM from the command line with:

   ```
   PerM hg19files.txt reads_F3.csfasta -v 5 -o alignment.sam
   ```

   Here, -v 5 means up to five mismatches.
5. The resulting alignment file cannot be used directly as the junctions reads will not have chromosome and position in the correct fields. Rather they will have names of junctions in the chromosome field. This will be true for all alignment tools without built-in junction support (i.e. all but Tophat). To use our conversion tool to correct these fields, run:

   ```
   python junctions2sam.py alignment.sam output.sam 100 --minanchor 8
   ```

   The –minanchor 8 option removes junction reads that do not map with at least eight bases to each exon. Without the option, the junction library would have been needed to be trimmed. The "100" refers to the anchor length in hg19_junctions_100.fa.

2.2.4. A Standard File Format to Store Sequence Alignment Data

The SAM file format produced in these examples can be specified as the output format for most alignment tools, instead of their native output formats. The SAM format allows storing different types of alignments such as junction reads (Fig. 1b) and paired-end reads. It has a binary version format called BAM, where files are smaller. SAM and BAM files can be interconverted using samtools (40). During the conversion, the BAM file can be sorted and indexed for some downstream tools, such as the visualization tool Integrative Genomics Viewer (IGV) (described below).

Conversion of a SAM file to BAM file followed by BAM file sorting and indexing can be done as follows by using human genome assembly hg19 as the reference genome:

1. Download chromosome sequence for hg19 (hg19.fa) from UCSC Genome Browser (41).
2. Run following commands on the command line:

```
samtools faidx hg19.fa                    # indexing of genome fasta file (first time only)
samtools view -bt hg19.fa.fai example.sam -o example_tmp # sam to bam conversion
samtools sort example_tmp example                        # bam sorting
samtools index example.bam                               # bam indexing
```

2.3. Visualization of RNA-Seq Data

Visualization of RNA-Seq data provides rapid assessment of data such as the signal-to-noise level by sequence coverage of exons in relation to introns and intergenic regions. It also shows possible limitations with current gene annotations, since clumps of sequences often map outside annotated 3′UTR regions (10, 11, 42). Although Web-based visualization is possible in the UCSC browser (under Genomes->add custom tracks) or Ensembl, this suffers from long uploading times for big data sets. It is still the easiest alternative for users who have relatively small data sets.

Desktop tools for visualization of RNA-Seq data are faster and more interactive (e.g., IGV (43)). IGV is convenient to use because it can read many file formats, including SAM, BAM, and BED, and supports visualization of additional types of data, such as microarray data and evolutionary conservation. All tools can export vector-based formats (e.g., EPS or PDF) that are suitable for creating illustrations for publication; example output is shown in Fig. 2.

2.4. Transcript Quantification

After mapping reads to a reference genome, one can proceed to estimate the gene expression levels. Due to the initial RNA fragmentation step, longer transcripts will contribute to more

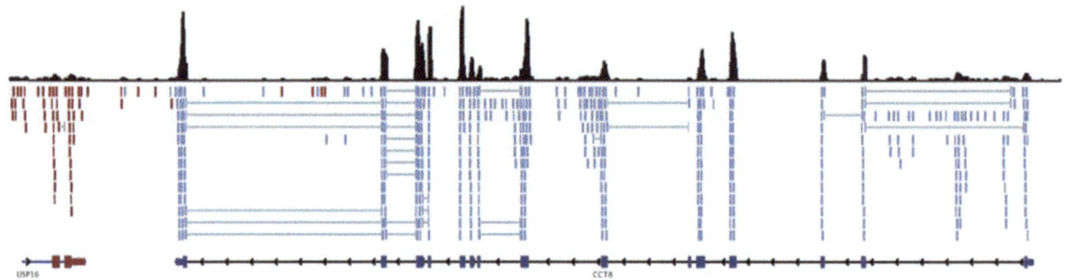

Fig. 2. Visualization of RNA-Seq data. Visualization of strand-specific RNA-Seq data in IGV. Reads mapping to the forward strand reads are colored *red*, and reads mapping to the reverse strand are colored *blue*.

fragments and are more likely to be sequenced. Therefore, the read counts should be normalized by transcript length in addition to the sequence depth when quantifying transcripts. A widely used expression level metric that normalizes for both these effects is reads per kilobase and million mappable reads (RPKM) (9). To estimate the expression level of a gene, the RPKM is calculated as:

$$\mathrm{RPKM} = R \times \frac{10^3}{L} \times \frac{10^6}{N}, \quad (1)$$

where R is the number of reads mapping to the gene annotation, L is the length of the gene structure in nucleotides, and N is the total number of sequence reads mapped to the genome. Although the calculation is common and trivial, there are certain issues that need careful consideration. Since the expression estimate for each gene is normalized by its annotated length and it is known that mRNA isoforms differ between cell types and tissues, the correct length to use is often not known. Furthermore, the lengths of 3′UTRs can differ by as much as a few kilobases between different kinds of cells (44, 45) and we recently found that it is more accurate to exclude 3′UTRs from gene models when calculating RPKM expression levels (46). Another issue arises in the normalization by sequence depth (N), since the types of RNAs present in the sequence data will differ depending upon RNA-Seq protocol used. It is inadvisable to use the total number of mapped reads when, e.g., comparing polyA$^+$ enriched data to data generated by ribosomal RNA reduction techniques since the latter data will contain many non-polyadenylated RNAs so that the total fraction of mRNA reads are lower and expression levels would be underestimated. An approach we have tried is to normalize by the number of reads mapping to exons of protein-encoding transcripts, this appears to help. A third issue is the estimation of transcript isoform expressions where multiple isoforms overlap. Although multiple tools exist (11, 46, 47), it is unclear how well they perform.

Finally, reads that map to multiple genomic locations present a problem, and tools differ in how they deal with these. If multi-mapping reads are discarded, then gene annotation lengths

(*L* in equation 1) become the number of uniquely mappable positions. This approach is efficient and accurate for most of the transcriptome, although a drawback is that recently duplicated paralogue genes will have few uniquely mappable positions and could therefore escape quantification. Another option is to first map uniquely mapping reads and then randomly assigning the multimapping reads to genomic locations based on the density of surrounding uniquely mapping reads (9). Here there is instead a risk that such paralogues are falsely detected as expressed, since paralogues not distinguishable with uniquely mapping reads will get roughly equal number of reads and similar expression levels. The latter approach can also lead to false-positive calls of differential expression, as small biases found in the unique positions could be reinforced through the proportional sampling of a much larger amount of multimapping reads.

2.4.1. Transcript Quantification Using rpkmforgenes

We have developed a script for RPKM estimation that is flexible to most of the issues discussed above, e.g., it can be run with only parts of gene annotations, calculate the uniquely mappable positions, and handle multiple inputs and normalization procedures (46) (rpkmforgenes (37)). To use it to quantify gene expression levels from the SAM files generated by Tophat, do the following:

1. Download a gene annotation file such as refGene.txt from ref. 41.
2. Install python 2.5, 2.6, or 2.7 (39) and numpy (48).
3. To use information about which human genome coordinates are mappable, download bigWigSummary (49, 50) and wgEncodeCrgMapabilityAlign50mer.bw.gz (51) (assuming your reads are ~50 bp – other files exist for other lengths) to the same folder as rpkmforgenes.py. If this information cannot be used, skip the -u option in the next step.
4. From the command line, run:

```
python rpkmforgenes.py -i accepted_hits.sam -a refGene.txt -o expressionlevels.txt -u wgEncodeCrgMapabilityAlign50mer.bw.gz -readcount
```

The -readcount option adds the number of reads, which is useful for calling differential expression.

The resulting gene expression values rarely have over twofold errors at a sequence depth of a few million reads, and at 20 million reads, half the values are within 5% accuracy (Fig. 3a). It is primarily lowly expressed genes that have uncertain expression values and for genes expressed above ten RPKM, the vast majority are accurately quantified with only five million mappable reads (Fig. 3b).

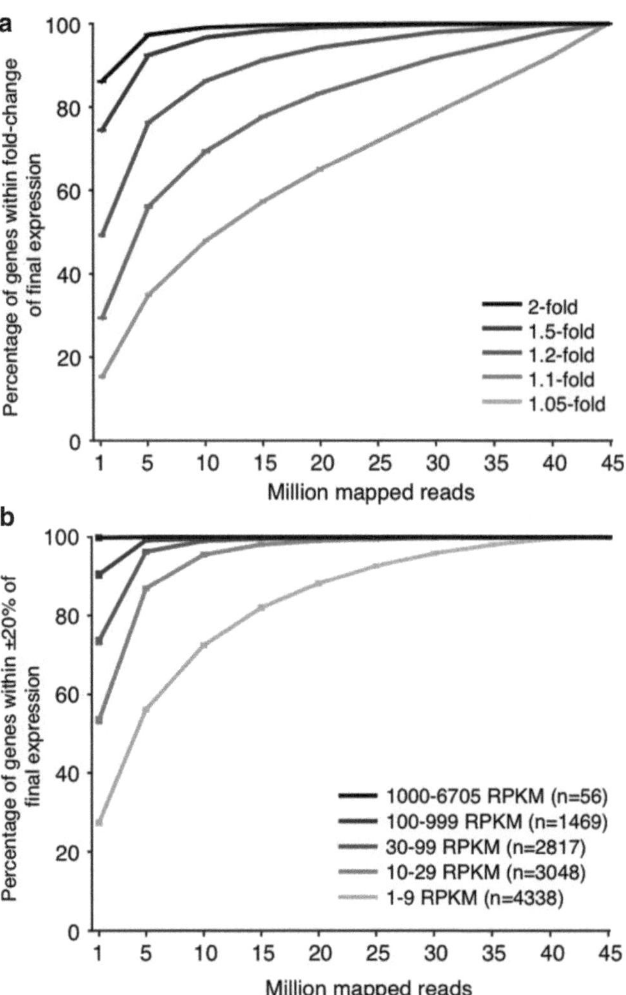

Fig. 3. Robustness of expression levels depending on sequencing depth. The robustness of expression levels was investigated by calculating expressions from randomly drawn subsets of reads and comparing with the final value using all 45 million reads (as a proxy for the real expression value). (**a**) The fraction of genes that are within specified fold-change interval from the final expression level at different sequence depths, for all genes expressed over one RPKM. (**b**) The fraction of genes at different sequence depths that are within ±20% of the final expression value that was estimated using all 45 million mappable reads. Genes have been grouped according to final RPKM expression level. The different sequence depths were obtained by selecting random subsets of mapped reads and the results are presented as mean and 95% confidence intervals.

2.5. Differential Expression

Most gene expression experiments include a comparison between different conditions (e.g., normal versus disease cells) to find differentially expressed genes. As with microarrays, we face a similar problem in that we measure the expression of thousands of genes and we only have a low number of biological replicates (often ≤3). In RNA-Seq experiments, there is little use of technical replicates, since background is lower and the variance better modeled (52). As in all experimental systems, however, the biological variation necessitates biological replicates to determine whether the observed differences are consistently found and to estimate the variance in the expression of genes (see Note 3). Improvements in the identification of differentially expressed genes have been made in both microarray and RNA-Seq analyses through a better understanding of the variance. Learning from the improvements in microarray data analyses, reviewed in ref. 53, it is clear that borrowing the variance from other genes help to better

Gene	Sample 1	Sample 2	Sample 3	Sample 4
Gene 1	7	10	7	9
Gene 2	6	8	2	8
Gene 3	4	4	1	4
Gene 4	5	8	3	4
Gene 5	1	2	1	7

Fig. 4. Read format for DESeq in R/Bioconductor. The tab-delimited file format should contain a header row with "Gene" followed by sample names. Each gene is represented as a gene name or identifier followed by the number of reads observed in each sample.

estimate the variation in read counts for a gene and condition. This overcomes a common problem with an underestimation of variance when based on a low number of observations. Recent tools such as edgeR (54) or DESeq (55) consider negative binomial distribution for read counts per region, overcoming an initial over-dispersion problem experienced when using only a Poisson model to fit the variance, e.g., in ref. 52. As in microarray analyses, many tests are being applied in parallel and one needs to correct for this multiple hypothesis testing. Benjamini–Hochberg correction is often performed to filter for a set of differentially expressed genes that have a certain false discovery rate (FDR).

Here we show how one proceeds to the estimation of differentially expressed genes using DESeq in the R/Bioconductor environment (see also Note 3):

1. Prepare a tab-delimited table of read counts (not RPKMs) to load to DESeq, with the following layout (Fig. 4).
2. Inside an R terminal, run the following commands:

```
source("http://bioconductor.org/biocLite.R") # Install DESeq (only done first time)

biocLite("DESeq")                    # Install DESeq (only done first time)

library("DESeq")                     # Load DESeq module

countsTable <- read.delim("countdata.txt", header=TRUE, stringsAsFactors=TRUE

) # Load data

rownames( countsTable ) <- countsTable$Gene     # Save gene names

countsTable <- countsTable[ , -1 ]   # Erase gene names from data table

conds = c("T","T","N","N")           # Give the condition for each input count file

cds <- newCountDataSet( countsTable, conds )    # Create a data frame object

cds <- estimateSizeFactors( cds )          # Estimate size factors for normalization

cds <- estimateVarianceFunctions( cds )    # Calculate variance estimates for

variance shrinkage

res <- nbinomTest( cds, "T", "N")          # Perform differential expression test

write.table(res,"result.txt", sep = "\t")  # Write resulting rows to a text file
```

2.6. Background Estimation and RNA-Seq Sensitivity

RNA-seq sensitivity depends on sequencing depth, however, only a few million reads are needed for detecting expressed transcripts with a sensitivity below a transcript/cell (46) (and see Note 2). Unexpressed regions will contain an even distribution of reads (42), so a single read that maps to a transcript is not enough to call detection. One way to estimate background is the following:

1. Find the distribution of transcript lengths in your annotation of choice.
2. Spread regions with these lengths across the genome.
3. Remove regions that overlap evidence of transcription (such as ESTs – coordinates for these can be found, e.g., at the download section of the UCSC genome browser).
4. Calculate RPKM values (e.g., Equation 1) for these regions, giving you a background distribution.

The simplest solution after this is to set the 95th percentile of the background distribution as your threshold of detection. A mathematically more complicated solution is to compare with observed gene expression values to derive the point where FDR balances false negative rate (46).

Sometimes you can find RNA-seq *too* sensitive, picking out transcripts from the background that are so rare that they must come from small subpopulations or contaminating cell types. As RPKM roughly equals transcripts/cell for hepatocyte-sized cells (9), a threshold on the order of one RPKM is reasonable. Less guesswork is required if a spike-in was added to the RNA sample, as RPKM values can then be converted to transcripts/cell. For example, say that 100 pg of a spike-in RNA which is 1 kb long is added to ten million cells, and you calculate an expression value of 30 RPKM for it. Assuming a molecular weight of 5×10^{-22} g/nucleotide, 30 RPKM corresponds to:

$$\frac{100 \times 10^{-12}(\text{g})}{5 \times 10^{-22}(\text{g/nt}) \times 1 \times 10^{3}(\text{nt/transcript}) \times 10 \times 10^{6}(\text{cells})} = 20(\text{transcripts/cell}). \tag{2}$$

With several spike-in RNAs, a line may be fitted by linear regression. If the numbers of transcripts per cell are A_1, A_2, ... A_n and the expression values are B_1, B_2, ... B_n, then the slope of such a line, which is the number of transcripts per cell and RPKM, will be:

$$\frac{\sum_i A_i B_i}{\sum_i B_i^2}. \tag{3}$$

2.7. De Novo Transcript Identification

Another application of RNA-Seq is high-resolution de novo reconstruction of transcripts. Two recent tools, Scripture (10) and Cufflinks (11), have been developed for transcript identification in organisms with a reference genome. They both require sequence reads mapped to the genome together with splice junctions as input (in SAM or BAM format) for the prediction of transcripts. Shallow sequencing will however lead to very fragmented transcripts for many genes, due to low coverage of exon–exon boundaries and junctions. Paired-end sequence reads are particularly useful for transcript identification, since the pairing enables many exons to be joined without direct exon–exon junction evidence (10). An alternative approach would be to first assemble RNA-Seq reads and then map the assembled contigs to a reference genome. This latter approach performs worse on lowly expressed genes that do not have sufficient coverage to be assembled.

This tutorial focuses on Scripture that predicts transcripts in two steps. First, the genome is segmented based on coverage into expressed islands. Then exon–exon junctions (and paired-end reads) join the expressed islands into multiexon transcripts. The analysis is done per chromosome and require in addition to the input SAM/BAM file, a chromosome file in FASTA, and a tab-delimited file with chromosome lengths. The BAM file needs to be sorted and indexed (see Subheading 2).

For each chromosome run the following command (here shown for chr19).

```
java --jar scripture.jar --alignment testdata.bam --out testdatascripture --

sizeFile CHRSIZE_FILE --chr chr19 --chrSequence chr19.fa -pairedEnd

testdata.paired.bam
```

where CHRSIZE_FILE is a tab-delimited file with lines containing each chromosome and its number of bases, and BAM index files must be present at the same folder as BAM files. For complete documentation, please see Scripture Web page (56). The resulting transcript predictions are in BED format and can be compared with existing annotations (e.g., RefSeq, Ensembl or UCSC known-genes) as well as those identified in recent RNA-Seq studies (10, 11, 42) to connect the discovered regions with known transcripts and tell apart the ones resembling novel transcription units.

3. Notes

1. The color space FASTQ format, which is sometimes called CSFASTQ, can differ depending on source. In files downloaded from SRA, the format has the same sequence line as in CSFASTA

format – a base letter followed by color space digits – and a quality score line the same length as the sequence, where the base has been given the quality score 0. However, some alignment tools use different formats: BFAST (31) requires the quality score for the base to be omitted and MAQ (59) requires both this quality score and the base in the sequence line to be omitted. Both tools provide commands to create such files from CSFASTA and QUAL files.

2. Sometimes sequence reads extend into adapter sequence, this can happen for example with Illumina's current strand-specific protocol as it leads to short insert sizes. These reads will not map to the genome unless the adapter sequence is removed. Many packages include code for adapter trimming that converts a FASTQ file with raw reads to a FASTQ file with reads of different lengths. Although many alignment programs (e.g., Bowtie) can handle mixed lengths, it gets harder to map splice junctions. Tophat cannot handle reads of different lengths, and one cannot simply present a precompiled junction library to a mapper such as Bowtie, since one cannot ensure a uniform anchor length in the junctions for reads of different lengths. Instead we favor a simple procedure where all reads are trimmed at fixed position (say, 30 nucleotides from the 3′ end) and then mapped with Tophat. This procedure is repeated using a few different cutting positions and each set is independently mapped. Finally, a downstream script compares alignments from the separate mappings and picks the longest possible alignment per read.

3. Often the experimental design is a trade-off between sequencing depth, the number of experimental conditions, and biological replicates. As in all biological experiments, the only way to tackle biological variation is to collect biological replicates. In RNA-Seq experiments, one has the ability to reduce the sequencing depth on each individual sample using sample barcoding and then have the ability to both determine the reproducibility in each replicate as well as to combine all biological replicates for a more sensitive comparison across conditions.

4. R is an open-source statistical package (57). Bioconductor (58) provides tools for the analysis of high-throughput data using the R language. Upgrade to a new version of R if DESeq has problem installing. DESeq can also give error if supplied with too few genes.

5. The sequence depth used will affect the downstream analysis options. A deep sequencing, e.g., a recent 160 million reads per condition (10), enables the complete reconstruction of the majority of all expressed protein-coding and noncoding transcripts and enables a sensitive analysis for alternative splicing and mRNA isoform expressions. Many studies have used depths around 20–40 million read sequences that is well suited

for quantification of alternative splicing and isoforms but will not have the coverage needed for complete reconstruction of sample transcripts. Using less depths in the range of 1–10 million reads is still very accurate for the quantifications of genes or transcripts but will not have the power to evaluate as many alternatively spliced events. Improvements in high-throughput sequencing (e.g., HiSeq) and efficient sample bar-coding now enable 96 samples to be run in a cost-effective manner with a depth of approximately 10 M reads per sample.

References

1. Wang Z, Gerstein M, Snyder M (2009) RNA-Seq: a revolutionary tool for transcriptomics. Nat Rev Genet 10:57–63
2. Wang ET, Sandberg R, Luo S et al (2008) Alternative isoform regulation in human tissue transcriptomes. Nature 456:470–476
3. Pan Q, Shai O, Lee L et al (2008) Deep surveying of alternative splicing complexity in the human transcriptome by high-throughput sequencing. Nat Genet 40:1413–1415
4. Yoder-Himes DR, Chain PSG, Zhu Y et al (2009) Mapping the Burkholderia cenocepacia niche response via high-throughput sequencing. Proc Natl Acad Sci USA 106:3976–3981
5. Armour CD, Castle JC, Chen R et al (2009) Digital transcriptome profiling using selective hexamer priming for cDNA synthesis. Nat Methods 6:647–649
6. Core LJ, Waterfall JJ and Lis JT (2008) Nascent RNA sequencing reveals widespread pausing and divergent initiation at human promoters. Science 322:1845–1848
7. Ingolia NT, Ghaemmaghami S, Newman JRS et al (2009) Genome-wide analysis in vivo of translation with nucleotide resolution using ribosome profiling. Science 324:218–223
8. Metzker ML (2010) Sequencing technologies – the next generation. Nat Rev Genet 11:31–46
9. Mortazavi A, Williams BA, McCue K et al (2008) Mapping and quantifying mammalian transcriptomes by RNA-Seq. Nat Methods 5:621–628
10. Guttman M, Garber M, Levin JZ et al (2010) Ab initio reconstruction of cell type-specific transcriptomes in mouse reveals the conserved multi-exonic structure of lincRNAs. Nat Biotechnol 28:503–510
11. Trapnell C, Williams BA, Pertea G et al (2010) Transcript assembly and quantification by RNA-Seq reveals unannotated transcripts and isoform switching during cell differentiation. Nat Biotechnol 28:511–515
12. Sequence Read Archive. http://www.ncbi.nlm.nih.gov/sra.
13. Gene Expression Omnibus. http://www.ncbi.nlm.nih.gov/geo.
14. Ewing B, Hillier L, Wendl MC et al (1998) Base-calling of automated sequencer traces using phred I accuracy assessment. Genome Res 8:175–185
15. Cock PJA, Fields CJ, Goto N et al (2010) The Sanger FASTQ file format for sequences with quality scores, and the Solexa/Illumina FASTQ variants. Nucleic Acids Res 38:1767–1771
16. Giardine B, Riemer C, Hardison RC et al (2005) Galaxy: a platform for interactive large-scale genome analysis. Genome Res 15:1451–1455
17. Stajich JE, Block D, Boulez K et al (2002) The Bioperl toolkit: Perl modules for the life sciences. Genome Res 12:1611–1618
18. Cock PJA, Antao T, Chang JT et al (2009) Biopython: freely available Python tools for computational molecular biology and bioinformatics. Bioinformatics 25:1422–1423
19. NCBI (2010) Sequence Read Archive Submission Guidelines. http://www.ncbi.nlm.nih.gov/Traces/sra/static/SRA_Submission_Guidelines.pdf. Accessed 2 Nov 2010
20. SOLiD Sequence Read Format package. http://solidsoftwaretools.com/gf/project/srf/
21. Staden IO module. http://staden.sourceforge.net/
22. Sequenceread package http://sourceforge.net/projects/sequenceread/
23. Pepke S, Wold B, Mortazavi A (2009) Computation for ChIP-seq and RNA-seq studies. Nat Methods 6:S22-S32
24. Dohm JC, Lottaz C, Borodina T et al (2008) Substantial biases in ultra-short read data sets from high-throughput DNA sequencing. Nucleic Acids Res 36:e105
25. Langmead B, Trapnell C, Pop M et al (2009) Ultrafast and memory-efficient alignment of

short DNA sequences to the human genome. Genome Biol 10:R25
26. Trapnell C, Pachter L and Salzberg SL (2009) TopHat: discovering splice junctions with RNA-Seq. Bioinformatics 25:1105–1111
27. Chen Y, Souaiaia T and Chen T (2009) PerM: efficient mapping of short sequencing reads with periodic full sensitive spaced seeds. Bioinformatics 25:2514–2521
28. Galaxy. http://g2.bx.psu.edu
29. Galaxy Experimental Features. http://test.g2.bx.psu.edu
30. Novoalign. http://www.novocraft.com
31. Homer N, Merriman B, Nelson SF (2009) BFAST: an alignment tool for large scale genome resequencing. PLoS ONE 4:e7767
32. Mosaik. http://bioinformatics.bc.edu/marthlab/Mosaik
33. Ozsolak F, Platt AR, Jones DR et al (2009) Direct RNA sequencing. Nature 461:814–818
34. Tophat. http://tophat.cbcb.umd.edu/index.html
35. UCSC Genome Browser FAQ File Formats. http://genome.ucsc.edu/FAQ/FAQformathtml#format1
36. Bowtie. http://bowtie-bio.sourceforge.net
37. RNA-Seq files at sandberg lab homepage. http://sandberg.cmb.ki.se/rnaseq/
38. PerM. http://code.google.com/p/perm/
39. Python. http://www.python.org
40. Li H, Handsaker B, Wysoker A et al (2009) The Sequence Alignment/Map format and SAMtools. Bioinformatics 25:2078–2079
41. UCSC Genome Browser Downloads. http://hgdownload.cse.ucsc.edu/downloads.html
42. van Bakel H, Nislow C, Blencowe BJ et al (2010) Most "dark matter" transcripts are associated with known genes. PLoS Biol 8: e1000371
43. Integrative Genome Browser. http://www.broadinstitute.org/igv
44. Sandberg R, Neilson JR, Sarma A et al (2008) Proliferating cells express mRNAs with shortened 3′ untranslated regions and fewer microRNA target sites. Science 320:1643–7
45. Neilson JR and Sandberg R (2010) Heterogeneity in mammalian RNA 3′ end formation. Exp Cell Res 316:1357–1364
46. Ramsköld D, Wang ET, Burge CB et al (2009) An abundance of ubiquitously expressed genes revealed by tissue transcriptome sequence data. PLoS Comput Biol 5:e1000598
47. Montgomery SB, Sammeth M, Gutierrez-Arcelus M et al (2010) Transcriptome genetics using second generation sequencing in a Caucasian population. Nature 464:773–777
48. NumPy. http://numpy.scipy.org
49. Kent WJ, Zweig AS, Barber G et al (2010) BigWig and BigBed: enabling browsing of large distributed datasets. Bioinformatics 26:2204–2207
50. UCSC stand-alone bioinformatic programs. http://hgdownload.cse.ucsc.edu/admin/exe/linux.x86_64/
51. UCSC Mappability Data. http://hgdownload.cse.ucsc.edu/goldenPath/hg19/encodeDCC/wgEncodeMapability/
52. Marioni JC, Mason CE, Mane SM et al (2008) RNA-seq: an assessment of technical reproducibility and comparison with gene expression arrays. Genome Res 18:1509–1517
53. Allison DB, Cui X, Page GP et al (2006) Microarray data analysis: from disarray to consolidation and consensus. Nat Rev Genet 7:55–65
54. Robinson MD, McCarthy DJ and Smyth GK (2010) edgeR: a Bioconductor package for differential expression analysis of digital gene expression data. Bioinformatics 26:139–140
55. Anders S, Huber W (2010) Differential expression analysis for sequence count data. Genome Biol 11:R106
56. Scripture. http://www.broadinstitute.org/software/scripture
57. R, http://www.r-project.org/
58. Bioconductor, http://www.bioconductor.org/
59. Li H, Ruan J, Durbin R (2008) Mapping short DNA sequencing reads and calling variants using mapping quality scores. Genome Res 18:1851–1858

Chapter 18

Analyzing ChIP-seq Data: Preprocessing, Normalization, Differential Identification, and Binding Pattern Characterization

Cenny Taslim, Kun Huang, Tim Huang, and Shili Lin

Abstract

Chromatin immunoprecipitation followed by sequencing (ChIP-seq) is a high-throughput antibody-based method to study genome-wide protein–DNA binding interactions. ChIP-seq technology allows scientist to obtain more accurate data providing genome-wide coverage with less starting material and in shorter time compared to older ChIP-chip experiments. Herein we describe a step-by-step guideline in analyzing ChIP-seq data including data preprocessing, nonlinear normalization to enable comparison between different samples and experiments, statistical-based method to identify differential binding sites using mixture modeling and local false discovery rates (fdrs), and binding pattern characterization. In addition, we provide a sample analysis of ChIP-seq data using the steps provided in the guideline.

Key words: ChIP-seq, Finite mixture model, Model-based classification, Nonlinear normalization, Differential analysis

1. Introduction

How proteins interact with DNA, the genomic locations where they bind to DNA, and their influence on the genes regulation have remained the topic of interests in the scientific community. By studying protein–DNA interactions, scientists are hopeful that they will be able to understand the mechanism of how certain genes can be activated while the others are repressed or remain inactive. The consequence of activation/repression/inactive will in turn affect the production of specific proteins. Since proteins play important roles for various cell functions, understanding protein–DNA relations is essential in helping scientists elucidate complex biological systems and discover treatment for many diseases.

Junbai Wang et al. (eds.), *Next Generation Microarray Bioinformatics: Methods and Protocols*, Methods in Molecular Biology, vol. 802, DOI 10.1007/978-1-61779-400-1_18, © Springer Science+Business Media, LLC 2012

There are several methods commonly used for analyzing specific protein–DNA interactions. One of the newer methods is ChIP-seq, an antibody-based chromatin immunoprecipitation followed by massively parallel DNA sequencing technology (also known as next-generation sequencing technology or NGS). ChIP-seq is quickly replacing ChIP-chip as the preferred approach for generating high-throughput accurate global binding map for any protein of interest. Both ChIP-seq and ChIP-chip goes through the same ChIP steps where cells are treated with formaldehyde to cross-link the protein–DNA complexes. The DNA is then sheared by a process called sonication into short sequences about ~500–1000 base-pair (bp). Next, an antibody is added to pull down regions that interact with the specific protein that one wants to study. This step filters out DNA fragments that are not bound to the protein of interest. The next step is where it differs between ChIP-chip and ChIP-seq experiments. In ChIP-chip, the fragments are PCR-amplified to obtain adequate amount of DNA and applied to a microarray (chip) spotted with sequence probes that cover the genomic regions of interest. Fragments that find their complementary sequence probes on the array will be hybridized. Thus, in ChIP-chip experiment, one needs to predetermine their regions of interest and "place" them onto the array. On the other hand, in ChIP-seq experiment, the entire DNA fragments are processed and their sequences are read. These sequences are then mapped to a reference genome to determine their location. Figure 1 shows a simplified workflow of ChIP-seq and ChIP-chip and the different final steps. Both ChIP-seq and ChIP-chip experiments require image analysis steps either to determine their probe binding intensities (DNA fragment abundance) or to read out their sequences (base calling). Some of the advantages of using ChIP-seq versus ChIP-chip include: higher quality data with lower background noise which is partly due to the need of cross hybridization for ChIP-chip, higher specificity (ChIP-chip array is restricted to a fixed number of probes), and lower cost (ChIP-seq experiments require less starting material to cover the same genomic region). Interested readers can find more information regarding ChIP-seq in refs. 1 and 2.

In a single run, ChIP-seq experiment can produce tens of millions of short DNA fragments that range in size between 500 and 1,000 bp long. Each fragment is then sequenced by reading a short sequence on each end (usually 35 bp or longer, newer illumina genome analyzer can sequence up to 100–150+ bp) leading to millions of short reads (referred to as tags). Sequencing can be done as single-end or paired-end reads. In single-end reads, each strand is read from one end only (the direction depends on whether it is a reverse or forward strand) while in paired-end each strand is read from both ends in opposite directions. Because of the way the sequences are read, some literatures either extend the reads

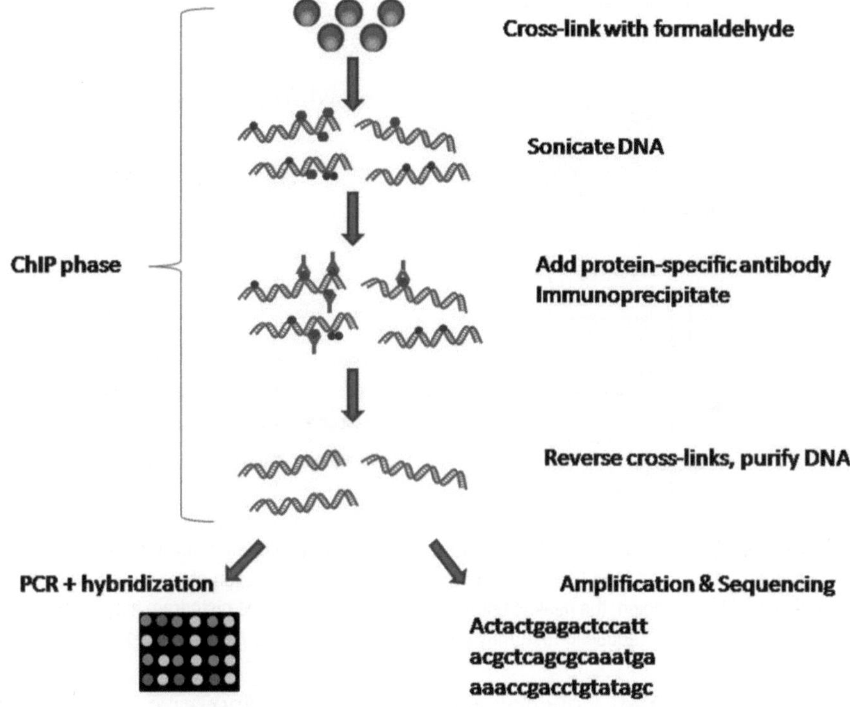

Fig. 1. Schematic of the ChIP-seq and ChIP-chip workflow. First the cells are treated with formaldehyde to cross-link the protein of interest to the DNA it binds to in vivo. Then the DNA is sheared by sonication and the protein–DNA complex is selected using antibody and by immunoprecipitation. Reverse cross-links is done to remove the protein and DNA is purified. For ChIP-chip, the fragments continue on to be cross hybridized. In ChIP-seq, they go through the sequencing process.

or shift the reads to cover the actual binding sites (see Note 1). In the sample analysis provided in this chapter, since the RNA polymerase II (Pol II) tends to bind throughout the promoter and along the body of the activated genes, it is unnecessary to shift or extend the fragments to cover the actual binding sites. Once all the tags are sequenced, they are aligned back to a reference genome to determine their genomic location. To prevent bias in the repeated genomic regions, usually only tags that are mapped to unique locations are retained. Preprocessing of ChIP-seq usually includes dividing the entire genome into *w*-bp regions and counting the number of short sequence tags that intersect with the binned region. The peaks of the binned regions signify the putative protein binding sites (where the protein of interest binds to the DNA). Figure 2 shows an example visualization of binned Pol II ChIP-seq data in MCF7, a breast cancer cell line.

Even though ChIP-seq data has been shown to have less error compared to ChIP-chip, they are still prone to biases due to variable quality of antibodies, nonspecific protein binding, material differences, and errors associated with procedures such as DNA library preparation, tags amplification, base calling, image processing,

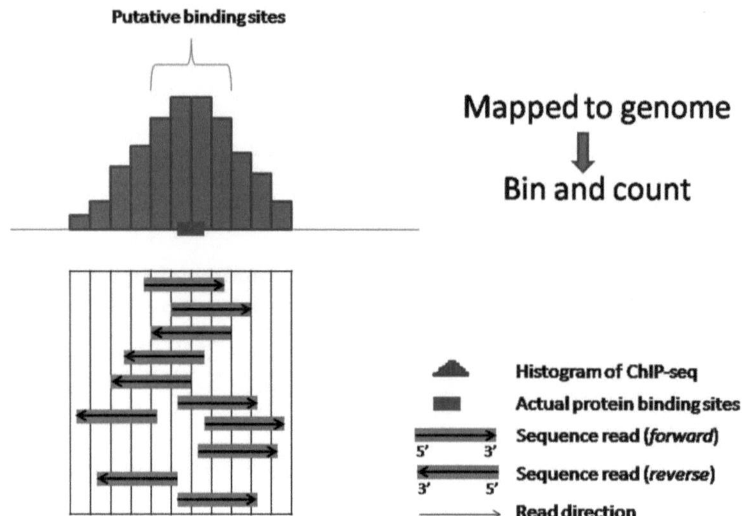

Fig. 2. An example visualization of the binned data with respect to the actual Pol II binding sites from ChIP-seq data. The single-end sequences are read from 5′ end or 3′ end depending on the direction of the strand. Note that since Pol II tends to bind throughout larger region, the peak is unimodal. For other protein, the histogram may be bimodal and hence some shifting or extension of the sequence read may be needed to identify the actual binding sites.

and sequence alignment. Thus, innovative computational and statistical approaches are still required to separate biological signal from noise. One of the challenges is data normalization which is critical when comparing results across multiple samples. Normalization is certainly needed to adjust for any systematic bias that is not associated with any biological conditions. Under ideal, error-free environment where every signal is instigated by its underlying biological systems, even a difference of one tag in a certain region can be attributed to a change in the conditions of the samples. However, various source of variability that is out of the experimenter's control can lead to differences that are not associated with any biological signal. Hence, normalization is critical to eliminate such biases and enable fair comparison among different experiments.

Our goal is to provide a general guideline to analyze ChIP-seq data including preprocessing, nonlinear data normalization, model-based differential analysis, and cluster analysis to characterize binding patterns. Figure 3 shows the flow chart of the analysis methods.

2. Methods

Given a library of short sequence reads from ChIP-seq experiment, the following steps are performed to analyze the data. We illustrate the process using the data generated from the Illumina Genome Analyzer platform, it nevertheless is applicable to data generated

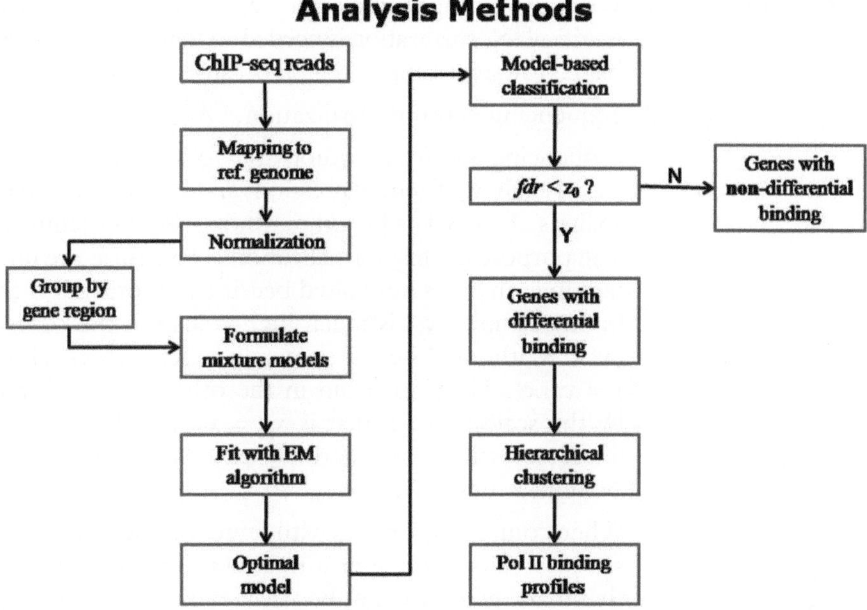

Fig. 3. Flow chart of the ChIP-seq analysis. The main steps of the methods to analyze ChIP-seq data including preprocessing are summarized in this figure.

from other sequencing platforms such as the Life Technology SOLiD sequencer.

2.1. Data Preprocessing

1. Determining genomic location of tags:
   (a) ELAND module within the Illumina Genome Analyzer Pipeline Software (Illumina, Inc., San Diego, CA) is used to align these tags back to a reference genome, allowing for a few mismatches.
   (b) After mapping, each tag will have its residing chromosome, starting and ending location. Depending on the software used, there may be a quality score associated with each base calling.
2. Filtering and quality control:
   (a) Filter out tags that are mapped to multiple locations.
   (b) Tags with low-quality score is filtered out internally in the Illumina pipeline.
   (c) Additional filtration maybe done as well. See Note 2.
3. Dividing genome into bins:
   (a) To reduce data complexity, the genome is divided into nonoverlapping w-bp regions (commonly called bins). The number of tags that overlap with each bin is then counted. We define x_{ij} as the sum of counts of tags that intersect with bin i in sample j.
   (b) Alternatively, one can use overlapping windows; see Note 3.

2.2. Normalization

1. When comparing multiple samples/experiments, normalization is critical. Normalization is needed so that the enrichment is not biased toward a sample/region because of systematic errors.
2. Sequencing depth normalization.

   Sequencing depth is a method used for normalization in SAGE (serial analysis of gene expression) and has been adapted for the analysis of NGS data by some authors; *See*, for example, ref. 10. The purpose of this normalization is to ensure the number of tags in each bin is not biased because the total number of tags in one sample (x_1) is much higher than in the other sample (x_2). Without lose of generality, let $x_1 > x_2$ and define $s = x_1/x_2$. Then, each bin in the other sample is multiplied by the scale factor s, that is $x'_{i2} = s \times x_{i2}$. This is a (scaling) linear normalization, where x_{i2} is the tag count in bin i.
3. Nonlinear normalization.

   When comparing samples with stages of disease progression or samples before and after a treatment in which it is expected that many genes will not be affected, nonlinear normalization may be used. The nonlinear normalization is done in two stages. In the first stage, the data is normalized with respect to the mean. In the second stage, the data is normalized with respect to the variance.

   (a) Mean-normalization:

   $$\hat{y}_i = \text{loess}\left((x_{i2} - x_{i1}) \sim \left(\frac{x_{i2} + x_{i1}}{2}\right)\right),$$

   $$D_{i.\text{mean}}(x_{i2} - x_{i1})\hat{y}_i, \quad (1)$$

   where $\hat{y}_i$ is the fitted value from regressing the difference on the mean counts using loess (locally weighted regression) proposed by Cleveland (3), and x_{i2} and x_{i1} are tag counts (may be after sequencing depth normalization) in bin i for control and treatment libraries, respectively. In this analysis, we assume no replicates are available. See Note 4 if replicates are available. This normalization step will find nonlinear systematic error and adjust them so the mean difference of unaffected genes becomes zero. $D_{i.\text{mean}}$ is the mean-normalized difference between reference and treatment libraries in bin i.

   (b) We choose to use the binding quantity for each sample directly (i.e., difference counts) rather than transforming it and using log-ratios for several reasons. First, it enables us to distinguish sites which have the same log-ratios but with vastly different magnitude. Furthermore, in ChIP-seq experiment, zeros indicate our protein of interest does not bind to the specific region. If we take log-ratios, these zero counts will be filtered out. In addition to those reasons, using difference counts will also help minimize problem with unbounded variance when fitting a mixture model; see Note 5.

(c) Wean-variance normalization:

$$\hat{z}_i = \text{loess}\left(|D_{i.\text{mean}}| \sim \left(\frac{x_{i2} + x_{i1}}{2}\right)\right),$$

$$D_{i.\text{var}} = \frac{D_{i.\text{mean}}}{\hat{z}_i}, \tag{2}$$

where $\hat{z}_i$ is the fitted value from regressing the absolute of mean-normalized difference on the mean counts. This step will find nonlinear and nonconstant variability in each region and adjust them so the spread is more constant throughout the genome. $D_{i.\text{var}}$ is the mean and variance normalized difference counts in bin i.

(d) For more detailed information including the motivation on nonlinear normalization for ChIP-seq analysis, the reader may refer to ref. 4.

4. Grouping tags into meaningful regions.

   (a) To study how the changes in the binding sites affect specific region of interest, we can sum the tags into grouped regions as follows:

$$R_g = \sum_{i \in I_g} D_{i.\text{var}}, \tag{3}$$

   where $D_{i.\text{var}}$ is the normalized difference in bin i as defined above. I_g is the index set specifying the bins belonging to group g. Thus, R_g is the sum of normalized tag-counts difference in region g for a total of G groups.

5. Although we did not scale our data based on the length of the groups, it may be a good idea to do further scaling normalization. See Note 6.

2.3. Differential Analysis: Modeling

1. With the normalized difference of grouped region (R_g) as input, we are now ready to perform statistical analysis. To determine whether there is a significant change in the tag counts of region g, we fit a mixture of exponential-normal component on R_g and apply a model-based classification. Assume that the data come from three groups, i.e., positive differential (genes that show increased bindings after treatment), negative differential (genes that have lower counts after treatment), and non-differential (those that do not change).

2. These three groups are assumed to follow certain distributions:

   (a) Positive differential: an exponential distribution.

(b) Negative differential: the mirror image of exponential.

(c) Nondifferential: a combination of one or more normal distribution.

(d) See Note 7 for special cases.

3. The choice of these distributions is based on observation that the characteristics of these distribution match well with the biological data (5).

4. The modeling are done by fitting a mixture of exponential (a special case of gamma) and normal components. This model is called GNG (Gamma-Normal^k-Gamma) which is described in ref. 5 and used in the analysis of ChIP-seq (4). The superscript k indicates the number of normal component in the mixture which will be estimated. Model fitted by GNG is as follows:

$$f(R_g;\psi)=\sum_{k=1}^{K}(\gamma_k\phi\{R_g,\mu_k,\sigma_k^2\})+\pi_1E_1(-R_g\times I\{R_g<-\xi_1\},\beta_1)$$
$$+\pi_2E_2(R_g\times I\{R_g>\xi_2\},\beta_2), \quad (4)$$

where ψ is a vector of unknown parameters of the mixture distribution. The first component $\sum_{k=1}^{K}(\gamma_k\varphi\{R_g,\mu_k,\sigma_k^2\})$ is a mixture of k normal component, where $\varphi\{.\}$ denotes the normal density function with mean μ_k and variance σ_k^2. Parameters γ_k indicate the proportion of each of the k normal components.

5. E_2 and E_1 each refers to an exponential component with π_2 and π_1 denoting their proportions and beta parameters β_2 and β_1, respectively. $I\{.\}$ is an indicator function that equals to 1 when the condition is satisfied and 0 otherwise; $\xi_2,\xi_1>0$ are the location parameters that are assumed to be known. In practice, we can set $\xi_1=|\max(R_g<0)|$ and $\xi_2=|\min(R_g>0)|$.

6. EM algorithm is used to find the optimal parameters by calculating the conditional expectation and then maximizing the likelihood function. See Note 5.

7. Akaike information criteria (AIC) (6), a commonly used method for model selection, is used to select k, the order of the mixture component that best represents the data.

2.4. Differential Analysis: Model-Based Classification

1. The best model selected by EM algorithm provides a model-based classification approach. Using this model, we can classify regions as differential and nondifferential binding sites.

2. Local false discovery rate (fdr) proposed by Efron (7) will be used to classify each binding sites based on the GNG model.

$$\mathrm{fdr}(R_g) = \frac{f(R_g; \psi_0)}{f(R_g; \psi_0) + f(R_g; \psi_1)}, \tag{5}$$

where $f(R_g; \psi_0)$ is the function of the k normal components and $f(R_g; \psi_1)$ is the function of the exponential components.

3. Ultimately, one can adjust the number of significantly different sites by setting the fdr value that they are comfortable with.

2.5. Binding Pattern Characterization

1. To further investigate the importance of protein binding profiles, one can perform clustering on the genes binding patterns which show significant changes.
2. Genes' lengths are standardized to enable genome-wide profiling.
3. The binding profiles for each gene are interpolated with optimum interpolator designed using direct form II transposed filter (8). As a result of this interpolation, all genes have the same length artificially.
4. Hierarchical clustering is then performed to group genes based on their binding profiles.

2.6. A Sample Analysis

In this section, we show a sample ChIP-seq analysis applying the above methodologies. Details on where to download the sample data and the software are provided in Subheading 2.7. The protein that we are interested in is RNA polymerase II (Pol II) and we are comparing MCF7, a normal breast cancer cell line before and after 17 β-estradiol (E2) treatments. We define MCF7 as the control sample and MCF7 + E2 as the treatment sample. The first part of the analysis is to discover genes that are associated with significant Pol II binding changes after E2 treatment. Because it is expected that the E2 treatment on cancer cell does not affect a large proportion of human genome, the above nonlinear normalization can be applied. See Note 8. Finally, significant genes are clustered to characterize their binding profiles.

2.6.1. Data Preprocessing

1. Sequence reads are generated by Illumina Genome Analyzer II. Reads are mapped to reference genome using ELAND provided by Illumina, allowing for up to two mismatches per tag.
2. Only reads that map to one unique location are used in the analysis. The total number of uniquely mapped reads (also known as sequence depth) for MCF7 sample is 6,439,640 and for MCF7 + E2 is 6,784,134. Table 1 shows details of the mapping result.
3. Nonoverlapping bins of size 1 kbp are used to divide the genome. 1 kbp is chosen to balance between data dimension and resolution. Thus, we set window size, $w = 1{,}000$ (bp).

Table 1
Reads of Pol II ChIP-seq data

Samples	Number of reads	Unique map	Multiple location	No match
MCF7	8,192,512	6,439,640 (79%)	1,092,519 (13%)	660,353 (8%)
MCF7 + E2	8,899,564	6,784,134 (76%)	1,233,574 (14%)	881,415 (10%)

The number of reads gives the raw counts from Solexa Genome Analyzer. Those under unique map are the reads that are used in our analysis. Those that are not uniquely map are either mapped to multiple loci or there is no match in the genome even allowing for two bases mismatches

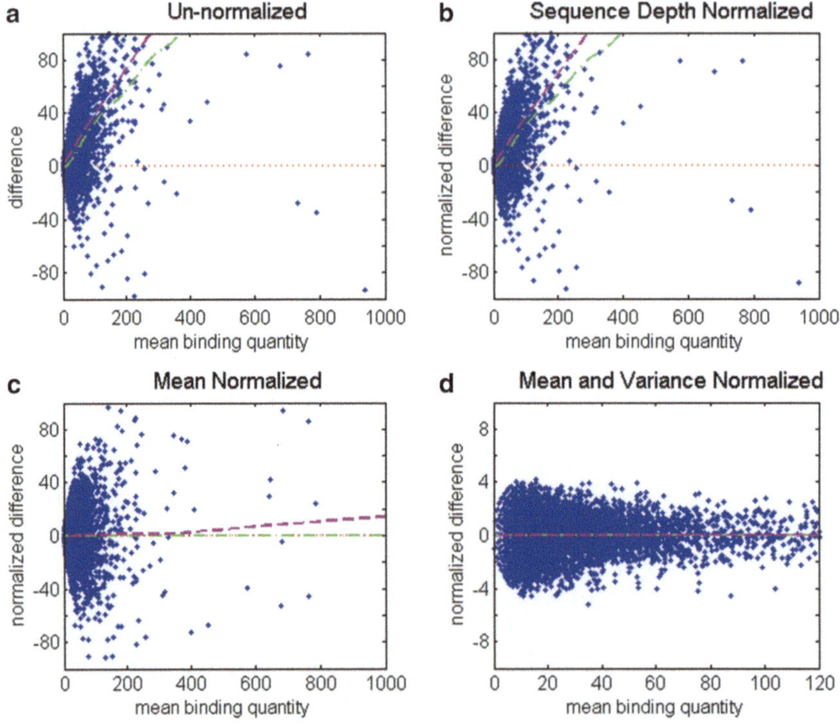

Fig. 4. Normalization process. The effects of the different normalization for chromosome 1 in MCF7 sample are shown. (**a**) The unnormalized data shows biases toward positive difference and nonconstant variance. (**b**) Data normalized using sequencing depth. (**c**) Data after normalization with respect to mean. (**d**) Data after two-stage normalization (with respect to mean and variance). *Dot-dashed* (*green*) *line* is the average of the difference counts estimated using loess regression. *Dashed* (*magenta*) *line* is the average absolute variance estimated using loess. *Dot* (*red*) *line* indicates the zero difference.

2.6.2. Sequencing Depth and Nonlinear Normalization Detailed in Subheading 2 Is Applied

1. We define MCF7 sample as the reference ($j = 1$) and MCF7 + E2 data as the treatment ($j = 2$). Figure 4a (raw data) shows that a large proportion of regions in treatment sample have Pol II binding that are higher than the control sample (indicated by the green dot-dashed line, estimated mean difference $D_{i.\text{mean}}$ in (1), which is always above zero). Sequencing depth normalization is commonly used for normalizing ChIP-seq

data. This normalization method scales the data to make the total sequence reads the same for both samples. As shown in Fig. 4b, since the total number of reads in control versus treatment sample is about the same, normalization based on sequencing depth has little effect. Figure 4b depicts the data after applying sequence depth (linear) normalization which still show biases toward positive difference and unequal variance, hence it is not sufficient as a normalization method. Figure 4c, d show the effect of the nonlinear normalization. In our application, we use a span of 60% and 0.1 to calculate loess estimate of the mean and variance, respectively. Since E2 treatment should only affect a small proportion of binding sites, i.e., most regions should have zero difference, normalization with respect to mean is applied to correct for this bias. Figure 4c shows the data after the mean adjustment. In addition, since the spread of the region increases with the mean as shown in Fig. 4c (indicated by the magenta dashed line, $D_{i.\mathrm{var}}$ in (2)), we apply normalization with respect to variance. Figure 4d shows that the data after mean and variance normalization is spread more evenly around zero (difference) which indicate the systematic error caused by unequal variance and bias toward positive difference has been corrected.

2. Grouping. In our application, we are interested in the Pol II binding quantities changes in the gene regions. Thus, after normalization, we summed tags count differences that fall into gene region based on RefSeq database (9). Hence, in Equation 3 above, I_g is the index of bins that overlap with gene region g and R_g is the sum of normalized tag-counts difference in gene region g for all 18,364 genes. The number of genes is small enough for a whole genome analysis.

2.6.3. Differential Analysis: Modeling

1. We fit GNG on the normalized difference R_g for all $g = 1,\ldots,$ G genes (genome-wide). In Fig. 5, the fit of the best model superimposed on the histogram is plotted in panel a, which shows the model fits the data quite well. The individual component of the best GNG model with two normal components is shown in Fig. 5b. The QQ plot of the normalized data versus the GNG mixture in Fig. 5c, where most of the points scatter tightly around the straight line, further substantiates that the model provides a good fit for the data. The EM algorithm was re-initialized with 1,125 random starting points to prevent it from getting stuck in the local optimum. The EM algorithm is set to stop when the maximum iteration exceeding 2,000 or when the improvement on the likelihood functions is less than 10^{-16}.

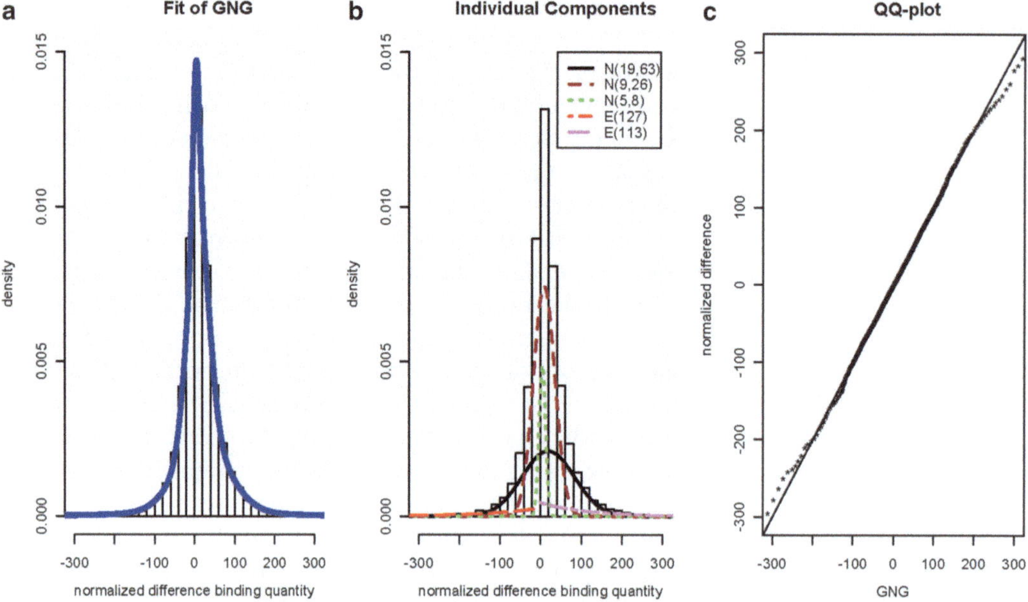

Fig. 5. The goodness of fit of the optimal GNG mixture to ChIP-seq data. (**a**) The fit of the best model imposed on the histogram of the normalized data (**b**) Plot of the individual components of the best GNG model. The best mixture has three normal components with parameters: $(\mu_1 = 5, \sigma_1 = 8)$, $(\mu_2 = 9, \sigma_2 = 26)$, and $(\mu_3 = 19, \sigma_3 = 63)$ represented by *dot* (*green*), *brown* (*dashed*), and *solid* (*black*) *lines*, respectively. The parameters for each of the exponential components are $\beta_1 = 127$ and $\beta_1 = 113$ represented by (*dot-dashed*) *red* and (*long-dash*) *magenta lines*, respectively. (**c**) QQ plot of the data versus the GNG model. All together these plots show that the optimal GNG model estimated by EM algorithm provide a good fit to the data.

2.6.4. Differential Analysis: Classification

1. Genes which have local fdr less than 0.1 are called to be significant. Using this setting, we find 448 genes to be associated with differential Pol II binding quantities in MCF7 versus MCF7 + E2 where around 60% of them are associated with increased bindings.
2. This finding is consistent with previous breast cancer study where the treatment of E2 appears to make more genes to be upregulated. Furthermore, we find PGR and GREB1 to be associated with significant increase of Pol II bindings (after E2 treatment) which are also found to be ER target genes that are upregulated in refs. 10 and 11.
3. A functional analysis on the genes associated with increased Pol II bindings is done using Ingenuity Pathway Analysis (17) (see Note 9) and shown in Fig. 6. The top network functions associated with these genes are cancer, cellular growth and proliferation, and hematological disease. Our finding thus suggests a regulation of nervous system development, cellular growth and proliferation, and cellular development in E2-induced breast cancer cells.

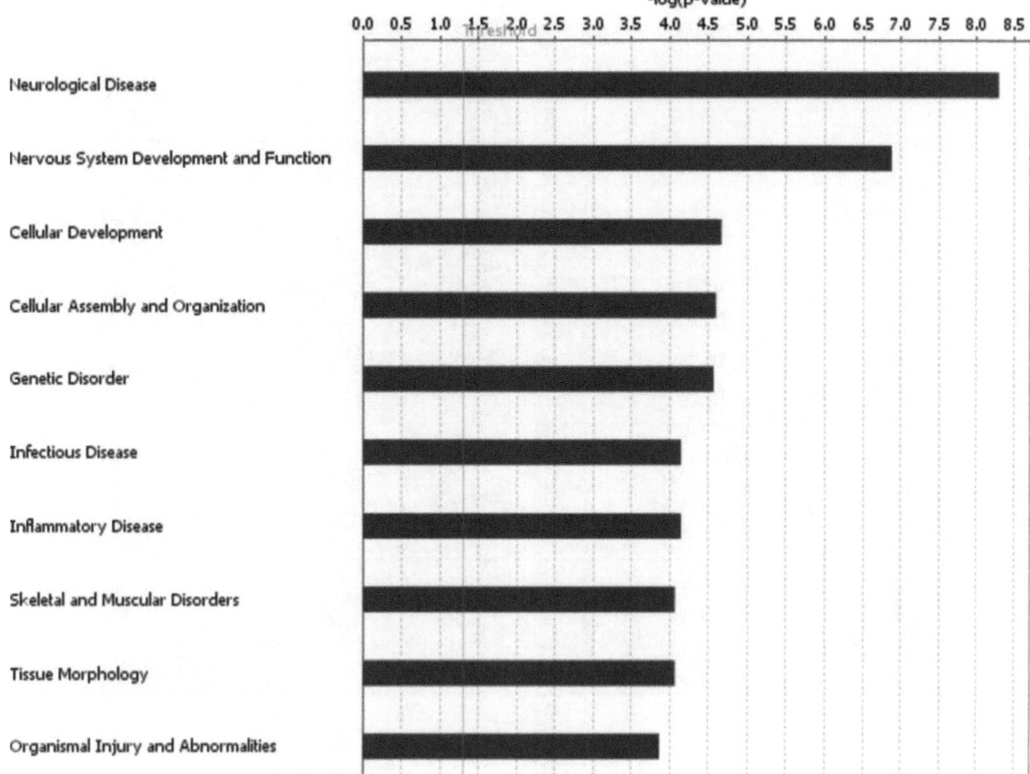

Fig. 6. The top ten functional groups identified by IPA. Analysis is done on the 264 genes which are found to show significant increase of Pol II binding in E2-induced MCF7. The *bar* indicates the minus log10 of the *p*-values calculated using Fisher's exact test. The *threshold line* indicates $p = 0.05$.

2.6.5. In Order to Characterize Pol II Binding Profiles of the Significant Genes Found in Previous Step, We Perform Hierarchical Clustering on These Regions

1. First, we filter out all the tags associated with introns retaining only those falling into exons regions. We did this filtration because the protein we are studying mainly acts on the exons regions.
2. Pearson correlation is used as the similarity distance in the hierarchical clustering procedure.
3. Binding profiles for each of the genes is interpolated to artificially make all genes length to be the same.
4. We find distinct clusters of genes with high Pol II binding sites at 5′ end (yellow, cluster 1) and genes with high Pol II binding quantity at 3′ end (blue, cluster 2), see Fig. 7.
5. Interestingly, there are more genes associated with high Pol II binding sites at 5′-end in MCF7 after E2 treatment.
6. This seems to indicate that different biological conditions (specifically treatment of E2) not only lead to changes in the Pol II binding quantity but it can also induce modification in the Pol II dynamics and patterns.

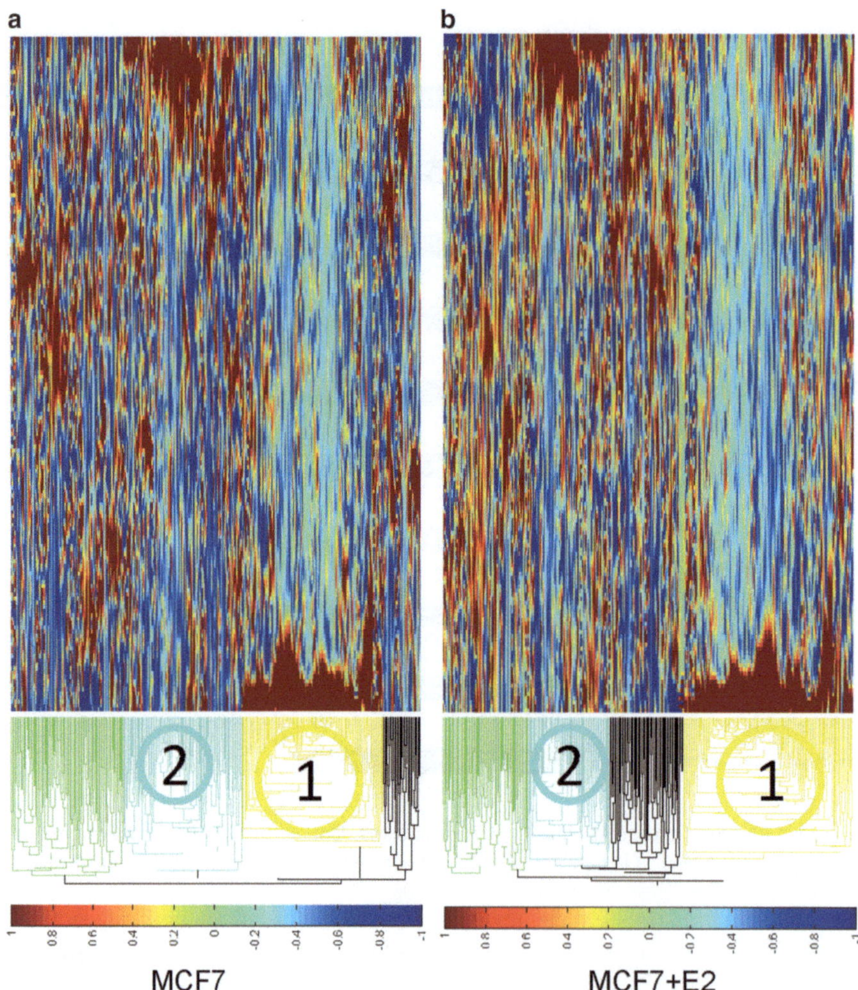

Fig. 7. Clustering of Pol II binding profiles in genes with significant changes in MCF7 after being treated with E2. Each column represent the Pol II binding profiles in each gene. Cluster 1 shows genes that are associated with high Pol II binding at the 5′ end and cluster 2 shows genes that are associated with high Pol II binding quantity at the 3′ end. (**a**) Binding profiles in MCF7; (**b**) binding profiles in MCF7 after E2 treatment. This indicate that E2 stimulation on MCF7 cell line not only change the Pol II binding quantity but it also modify its binding dynamics.

2.7. Software

The model fitting (GNG) is implemented as an R-package and is publicly available (21). The data used in the sample analysis is also downloadable from the same Web site.

3. Notes

1. Because the sequencing process cannot read the sequence of the entire tag length, some literature extends the sequenced tags to *x*-bp length and others shift each tag *d*-bp along the direction it was read in an attempt to cover the actual protein binding sites. For example, Rozowsky et al. (12) extend each

mapped tag in the 3′ direction to the average length of DNA fragments (~200 bp) and Kharchenko (13) shift the tags relative to each other. In our sample analysis, since Pol II tends to bind throughout the promoter and the body regions of a regulated gene, it is unnecessary to do shifting or extension. Readers should consider doing extension or shifting for any other protein binding analysis.

2. By combining number of mismatches with QC values of each base, one may be able to filter out low-quality/high mismatch reads from the analysis. On the other hand, one can also include more sequence reads with reasonable number of mismatches that are associated with high-quality score.
3. Instead of a fixed bin, some literature, for example, Jothi et al. (14) use a sliding window of size w where each consecutive window overlapped by $w/2$.
4. The methodology outlined here focus on analyzing ChIP-seq data without any replicates. When replicates are available, the same methodology can be applied by treating each replicates as individual independent samples or by taking the average of the replicates.
5. By allowing more than one normal components and not restricting them to have constant variances, the EM algorithm can have spurious solution where the variance becomes closer to zero and the model achieve artificially higher likelihood. We advise readers to use difference counts which would have a larger range than log-ratios in the modeling to minimize this problem. Re-initializing EM with multiple starting points will also help minimize this problem and prevent it from being trapped in a local optimum. For more information regarding the unboundedness problem of the likelihood function, see ref. 15.
6. A scaling normalization method known as RPKM (reads per kilobase per million mapped), proposed in ref. 16, is commonly used for ChIP-seq because of its simplicity. The main goal of this normalization is to scale all counts based on the length of the region and the total number of sequence reads. Although we did not apply this in our sample analysis, it may be a good idea to further scale our normalized data to minimize bias due to genes length and sequence depth. In this case, we can apply it on our normalized data as follows

$$y_g = \frac{R_g}{L_g \times \mathrm{SD}} \times 10^3 \times 10^6, \quad g = 1, \ldots, G;$$

where R_g is the number of loess-normalized tags in region g of a set of G regions, L_g is the gene length (in bp) of region g, and SD is the loess-normalized sequence depth (the total number of tags after loess normalization).

7. In the special situation where a normal component have either a large variance (say > 2IQR) or a large mean (say > 1.5 IQR), then such normal components should also be classified as differential components.
8. The nonlinear normalization described above is applicable when comparing samples in which the majority of genes do not show significant changes in treatment versus control samples. This assumption is satisfied for application in which the difference between the samples (i.e., effects of a drug treatment) is not expected to influence a large proportion of binding sites.
9. IPA is proprietary. There are free programs that provide similar information such as KEGG (18), GO (19), WebGestalt (20).

Acknowledgments

This work was partially supported by the National Science Foundation grant DMS-1042946, the NCI ICBP grant U54CA113001, the PhRMA Foundation Research Starter Grant in Informatics and the Ohio State University Comprehensive Cancer Center.

References

1. Johnson DS, Mortazavi A, Myers R et al (2007) Genome-Wide Mapping of in Vivo Protein-DNA Interactions. Science 316: 1441–1442
2. Liu E, Pott S, Huss M (2010) Q&A: ChIP-seq technologies and the study of gene regulation. BMC Biology 8: 56
3. Cleveland WS (1988) Locally-Weighted Regression: An Approach to Regression Analysis by Local Fitting. J. Am. Stat. Assoc. 85: 596–610
4. Taslim C, Wu J, Yan P et al (2009) Comparative study on ChIP-seq data: normalization and binding pattern characterization. Bioinformatics 25: 2334–2340
5. Khalili A, Huang T, Lin S (2009) A robust unified approach to analyzing methylation and gene expression data. Computational Statistics and Data Analysis 53: 1701–1710
6. Akaike H (1973) Information Theory and an Extension of the Maximum Likelihood Principle. In *International Symposium on Information Theory, 2nd, Tsahkadsor, Armenian SSR*: 267–281.
7. Efron B (2004) Large-Scale Simultaneous Hypothesis Testing: The Choice of a Null Hypothesis. Journal of the American Statistical Association 99: 96–104
8. Oetken G, Parks T, Schussler H (1975) New results in the design of digital interpolators. IEEE Transactions on Acoustics, Speech and Signal Processing [see also IEEE Transactions on Signal Processing] 23: 301–309
9. Pruitt KD, Tatusova T, Maglott DR (2007) NCBI reference sequences (RefSeq): a curated non-redundant sequence database of genomes, transcripts and proteins, Nucleic Acids Research 35: D61–65
10. Lin CY, Strom A, Vega V et al (2004) Discovery of estrogen receptor alpha target genes and response elements in breast tumor cells. Genome Biology 5, R66
11. Feng W, Liu Y, Wu J et al (2008) A Poisson mixture model to identify changes in RNA polymerase II binding quantity using high-throughput sequencing technology. BMC Genomics 9: S23

12. Rozowsky J, Euskirchen G, Auerbach RK et al (2009) PeakSeq enables systematic scoring of ChIP-seq experiments relative to controls. Nat Biotech 27: 66–75
13. Kharchenko PV, Tolstorukov MY, Park PJ (2008) Design and analysis of ChIP-seq experiments for DNA-binding proteins. Nature biotechnology 26: 1351–1359
14. Jothi R, Cuddapah S, Barski A et al (2008) Genome-wide identification of in vivo protein-DNA binding sites from ChIP-Seq data. Nucl. Acids Res. 36: 5221–5231
15. McLachlan G, Peel D (2000) Finite Mixture Models. Wiley-Interscience, New York
16. Mortazavi A, Williams BA, McCue K et al (2008) Mapping and quantifying mammalian transcriptomes by RNA-Seq. Nat Meth 5:621–628
17. The networks and functional analyses were generated through the use of Ingenuity Pathways Analysis (Ingenuity® Systems), see http://www.ingenuity.com
18. KEGG pathway analysis, see http://www.genome.jp/kegg/
19. Gene Ontology website, see http://www.geneontology.org/
20. WEB-based GEne SeT AnaLysis Toolkit, see http://bioinfo.vanderbilt.edu/webgestalt/
21. Software and datasets used can be downloaded, see http://www.stat.osu.edu/~statgen/SOFTWARE/GNG/

Chapter 19

Identifying Differential Histone Modification Sites from ChIP-seq Data

Han Xu and Wing-Kin Sung

Abstract

Epigenetic modifications are critical to gene regulations and genome functions. Among different epigenetic modifications, it is of great interest to study the differential histone modification sites (DHMSs), which contribute to the epigenetic dynamics and the gene regulations among various cell-types or environmental responses. ChIP-seq is a robust and comprehensive approach to capture the histone modifications at the whole genome scale. By comparing two histone modification ChIP-seq libraries, the DHMSs are potentially identifiable. With this aim, we proposed an approach called ChIPDiff for the genome-wide comparison of histone modification sites identified by ChIP-seq (Xu, Wei, Lin et al., Bioinformatics 24:2344–2349, 2008). The approach employs a hidden Markov model (HMM) to infer the states of histone modification changes at each genomic location. We evaluated the performance of ChIPDiff by comparing the H3K27me3 modification sites between mouse embryonic stem cell (ESC) and neural progenitor cell (NPC). We demonstrated that the H3K27me3 DHMSs identified by our approach are of high sensitivity, specificity, and technical reproducibility. ChIPDiff was further applied to uncover the differential H3K4me3 and H3K36me3 sites between different cell states. The result showed significant correlation between the histone modification states and the gene expression levels.

Key words: ChIP-seq, Epigenetic modification, Differential histone modification site, ChIPDiff, Hidden Markov model

1. Introduction

Eukaryotic DNA is packaged into a chromatin structure consisting of repeating nucleosomes by wrapping DNA around histones. The histones are subject to a large number of posttranslational modifications such as methylation, acetylation, phosphorylation, and ubiquitination. The histone modifications are implicated in influencing gene expression and genome function. Considerable evidence suggests several histone methylation types play crucial roles in biological processes (1). A well-known example is the repression

Junbai Wang et al. (eds.), *Next Generation Microarray Bioinformatics: Methods and Protocols*, Methods in Molecular Biology, vol. 802, DOI 10.1007/978-1-61779-400-1_19,

of development regulators by trimethylation of histone H3 lysine 27 (H3K27me3 or K27) in mammalian embryonic stem cell (ESC) to maintain stemness and cell puripotency (2, 3). Some epigenetic stem cell signature of K27 is also found to be cancer-specific (4). Moreover, the tri- and dimethylation of H3 lysine 9 are implicated in silencing the tumor suppressor genes in cancer cells (5). In the light of this, the specific genomic locations with differential intensity of histone modifications, which are called differential histone modification sites (DHMSs) in this chapter, are of great interest in the comparative study among various cell-types, stages, or environmental response.

The histone modification signals can be captured by chromatin immunoprecipitation (ChIP), in which an antibody is used to enrich DNA fragments from modification sites. Several ChIP-based techniques, including ChIP-chip, ChIP-PET, and ChIP-SAGE, have been developed in the past decade for the study of histone modification or transcription factor binding in large genomic regions (6–8). With the recent advances of ultra-high throughput sequencing technologies such as Illumina/Solexa GA sequencing and ABI SOLiD sequencing, ChIP-seq is becoming one of the main approaches since it has high coverage, high resolution, and low cost, as demonstrated in several published works (9–11). The basic idea of ChIP-seq is to read the sequence of one end of a ChIP-enriched DNA fragment, followed by mapping the short read called *tag* to the genome assembly in order to find the genomic location of the fragment. Millions of tags sequenced from a ChIP library are mapped and form a genome-wide profile. Regions with enriched number of ChIP fragments are potential histone modification sites or transcription factor binding sites.

Inspired by the success of ChIP-seq in identifying histone modification sites in a single library, we asked if the DHMSs could be identified by computationally comparing two ChIP-seq libraries generated from different cell-types or experimental conditions. Mikkelsen et al. (12) mapped the H3K4me3 (K4) and K27 sites in mouse ESC, neural progenitor cell (NPC), and embryonic fibroblast (MEF) and compared the occurrence of modification sites in promoter regions across three cell-types. A limitation of their study is that the modification sites are compared qualitatively but not quantitatively. An example demonstrating this limitation is the regulation of Klf4 by K4, which is known to be positively correlated to gene expression. The Klf4 promoter was flagged as "with K4" in both ESC and NPC by qualitative analysis hence it could not explain the upregulation of Klf4 in ESC. On the other hand, quantitative comparison indicated the intensity of K4 in Klf4 promoter is more than fivefold higher in ESC than in NPC (Fig. 1), consistent with the observation of expression change.

Triggered by the idea from microarray analysis (14), a simple solution to the problem of quantitative comparison is to partition

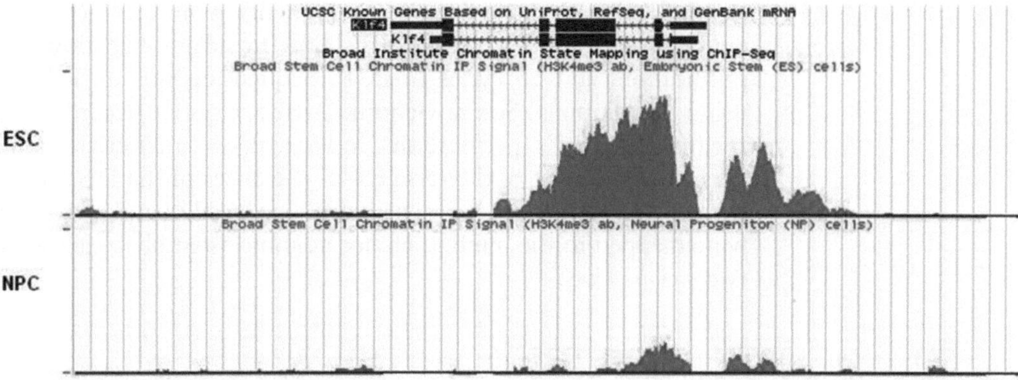

Fig. 1. Quantitative comparison of H3K4me3 intensity at Klf4 promoter between ESC and NPC. The intensity shown in the figure was normalized against the sequencing depth of ChIP-seq libraries. Image generated using UCSC Genome Browser (13).

the genome into bins and to compute the fold-change of the number of ChIP fragments in each bin. However, fold-change approach is sensitive to the technical variation caused by random sampling of ChIP fragments. In this chapter, we propose an approach called ChIPDiff to improve the fold-change approach by taking into account the correlation between consecutive bins (15, 16). We modeled the correlation in a hidden Markov model (HMM) (17), in which the transmission probabilities were automatically trained in an unsupervised manner, followed by the inference of the states of histone modification changes using the trained HMM parameters. We evaluated the performance of ChIP-Diff using the H3K27me3 libraries prepared in ESC and NPC (12). We demonstrated that our method outperforms the previous qualitative analysis, as well as the fold-change approach, in sensitivity, specificity, and reproducibility. We further applied ChIPDiff to H3K4me3 (K4) and H3K36me3 (K36) for the discovery of DHMSs on these two types of histone modifications and studied their potential biological roles in stem cell differentiation. Several interesting biological discoveries were achieved in the study.

2. Materials

In our study, we employed the histone modification ChIP-seq libraries in mouse ESCs and NPCs, which were published by Mikkelsen et al. (12, 18). The ESC libraries were prepared on murine V6.5 ES cells (129SvJae3C57BL/6; male), and the NPCs were cultured as described by Conti et al. (17) and Bernstein et al. (3).

In the ChIP experiment, three different antibodies were used to enrich the ChIP-DNA, corresponding to H3K4me3, H3K36me3, and H3K27me3, respectively (19). Sequencing libraries were generated from 1 to 10 ng of ChIP-DNA by adaptor ligation, gel purification, and 18 cycles of PCR. Sequencing was carried out using the Illumina/Solexa Genome Analyzer system according to the manufacturer's specifications. In average, ~10 million successful tags, which consist of the terminal 27–36 bases of the DNA fragments, were sequenced for each library. The first 27 bases in the tags were mapped to the mm8 reference genome assembly by allowing two mismatches.

3. Methods

3.1. Quantitative Comparison of Modification Intensity by Fold-Change

Tags in the raw data generated from a ChIP-seq experiment were mapped onto the genome to obtain their positions and orientations. Due to the PCR process in ChIP-seq experiments, multiple tags may be derived from a single ChIP fragment. To remove the redundancy, tags mapped to the same position with the same orientation were treated as a single copy (see Note 1). In ChIP-seq protocol, a tag is retrieved by sequencing one end of the ChIP fragment, of which the median length is around 200 bp (9, 20). To approximate the center of the corresponding ChIP fragment, we shifted the tag position by 100 bp toward its orientation. The whole genome was partitioned into 1 kbp bins and the number of centers of ChIP fragments was counted in each bin (see Note 2). After the above preprocessing procedure, a profile of ChIP fragment counts was generated. Given two ChIP-seq libraries L_1 and L_2, and considering a genome with m bins, the profiles of L_1 and L_2 are represented as $X_1 = \{x_{1,1}, x_{1,2}, \ldots, x_{1,m}\}$ and $X_2 = \{x_{2,1}, x_{2,2}, \ldots, x_{2,m}\}$, respectively, where $x_{i,j}$ is the fragment count at the jth bin in L_i.

Histone modifications exhibit a variety of kinetics and stoichiometries (21). For a ChIP-seq experiment, we define the modification intensity at the ith bin in library L_j to be the probability of an arbitrary ChIP fragment captured from the ith bin in the ChIP process, denoted $p_{j,i}$. We define a DHMS as a bin in which the ratio of intensities between L_1 and L_2 is larger than τ(L_1-enriched DHMS) or smaller than $1/\tau$ (L_2-enriched DHMS), where τ is a predetermined threshold, and $\tau \geqslant 1.0$. A simple solution for identifying DHMSs is to estimate the fold-change of expected intensity (preferably in term of log-ratio) from the ChIP fragment counts, as follows:

$$lr_i = \log\left[\frac{\alpha + x_{1,i}}{\alpha + x_{2,i}} \times \frac{m\alpha + n_2}{m\alpha + n_1}\right], \tag{1}$$

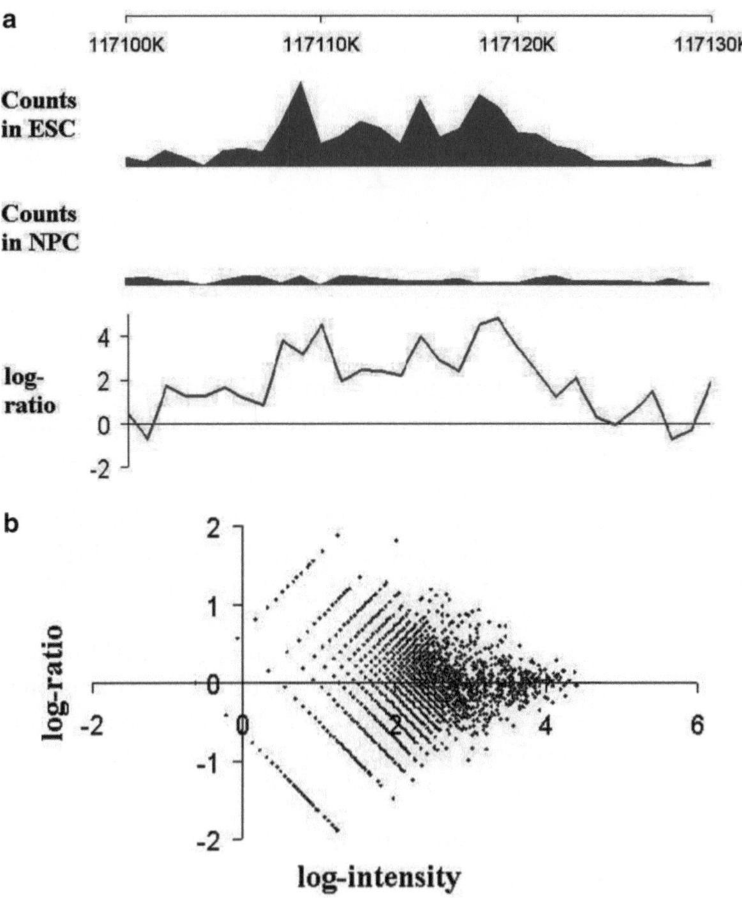

Fig. 2. Comparison of ChIP-seq libraries based on fold-change. (**a**) An example of the log-ratio estimation of H3K27me3 intensity between mouse ESC and NPC. Bin size set to be 1 k; displayed genomic region range from chr14:117,100,000 to 117,130,000; data retrieved from Mikkelson et al.'s dataset (12); (**b**) an RI-plot for chromosome 19 in K27 data.

where α was a small constant introduced as a pseudocount to avoid zero denominator in the ratio, and $n_1$and n_2 are the sequencing depths of L_1 and L_2, respectively. By such, the log-ratio of intensity was normalized against the sequencing depths (see Note 3).

An example of the log-ratio estimation is shown in Fig. 2a. A drawback of the fold-change approach is that it is prone to the technical variation caused by random sampling. Figure 2b shows an RI-plot (14) to depict the variation of the log-ratio dependent on the intensity. When the intensity is relatively small, the variation of log-ratio becomes too high, which may result in considerable false positives.

3.2. An HMM-Based Approach to Identifying DHMSs

Histone modifications usually occur in continuous regions that span a few hundreds or even thousands of nucleosides. Hence, one may expect strong correlation between consecutive bins in the measurements of intensity changes. This argument is supported

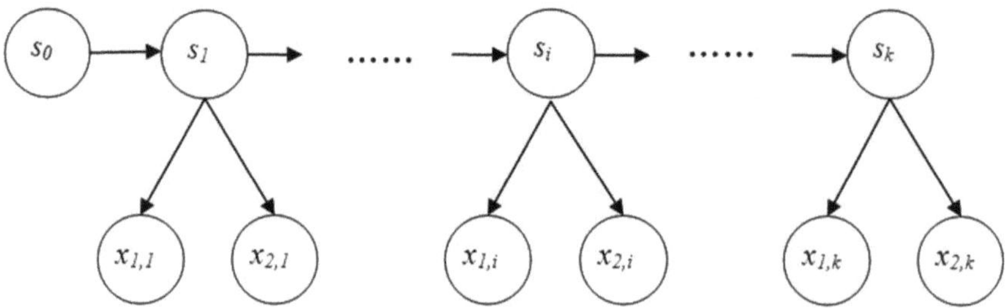

Fig. 3. The graphic representation of the HMM used in ChIPDiff.

by our observations from ChIP-seq profile. As an example, the log-ratio profile in Fig. 1a has an autocorrelation of 0.84. In ChIP-chip data analysis, Li et al. have designed an HMM to model the correlation of signals between consecutive probes and successfully applied it for the identification of p53 binding sites (22), suggesting the potential ability of HMM for identifying DHMSs in our study. Here, we propose a HMM-based approach called ChIPDiff to solve the problem.

The graphic representation of the HMM used in ChIPDiff is shown in Fig. 3. We denote s_i be the state of histone modification change at the ith bin ($i = 1, 2, \ldots, k$). Based on the definition of DHMS in Subheading 3.1, the state s_i takes one of the following three values:

- α_0: Nondifferential site, if $1/\tau \leqslant p_{1,i}/p_{2,i} \leqslant \tau$.
- α_1: L_1-enriched DHMS, if $p_{1,i}/p_{2,i} > \tau$.
- α_2: L_2-enriched DHMS, if $p_{1,i}/p_{2,i} < 1/\tau$.

In ChIPDiff, the HMM was trained by Baum–Welch algorithm (23), which takes expectation maximization (EM) steps to iteratively estimate the parameters of the HMM from hidden states in an unsupervised manner. Forward–backward algorithm was then employed to estimate the probability distributions of the states in each bin. Bins with posterior probability larger than a confidence threshold ρ ($0 < \rho < 1$) for $s_i = \alpha_1$ or $s_i = \alpha_2$ were identified as DHMSs. Consecutive DHMSs with no gap between them were merged into DHMS regions.

3.3. Evaluation on H3K27me3 Data

H3K27me3 was selected for the evaluation since its DHMSs in highly conserved noncoding elements (HCNEs) have been implicated in literature (3). Moreover, K27 preferentially marks gene regions and functions as a repressor, which facilitated our indirect validation using expression data. We compared the K27 ESC library and NPC library with ChIPDiff, in which the fold-change threshold τ was set to be 3.0 and the confidence threshold ρ was set to be 0.95. The HMM was trained with 10,000 randomly selected histone modification regions. 26,230 bins were identified to be DHMSs, corresponding to 4,722 continuous regions. Among them, 3,833 (81.2%) regions are ESC enriched and 889 (18.8%) are NPC enriched, implying a global trend of K27 depletion upon cell differentiation.

We first assessed the capability of ChIPDiff in identifying the biologically significant DHMSs, i.e., sensitivity. Bernstein et al. have shown that K27 is enriched in HCNEs in ESC, repressing a number of development regulators to maintain the stemness of the cell (3). These histone marks are depleted in diverse differentiated cells. From HCNEs, we selected 223 genes of which the expressions were studied by Mikkelson et al. (12). Since K27 functions as a gene repressor, we reasoned that some of those HCNE genes marked by K27 in ESC will be upregulated in NPC, and DHMSs should be identified at these genes. As expected, a subset containing 30 genes were determined to be upregulated with the criterion of more than fourfold. Among them, 24 (80%) are marked by DHMSs identified by ChIPDiff in promoter region ±1 kb from transcription start site (TSS). By contrast, only 37 (19.2%) out of the 193 genes that are not upregulated in NPC are marked by DHMSs.

To test the specificity of ChIPDiff result, we need to estimate the fraction of falsely identified DHMS regions that are not cell-specific. For this purpose, we partitioned each library into two technical replicates: $L_{\mathrm{ESC,rep1}}$ and $L_{\mathrm{ESC,rep2}}$ for ESC, $L_{\mathrm{NPC,rep1}}$ and $L_{\mathrm{NPC,rep2}}$ for NPC. The replicates consist of tags retrieved from different lanes in ChIP-seq experiments, with similar sequencing depth. Two new libraries were generated by merging the tags in $L_{\mathrm{ESC,rep1}}$ and $L_{\mathrm{NPC,rep1}}$, $L_{\mathrm{ESC,rep2}}$ and $L_{\mathrm{NPC,rep2}}$, respectively. Since the replicates are of similar sequencing depth, the difference between these two libraries should not be cell-specific and only reflect the technical variations in the experiments. Comparing these non-cell-specific controls, nine differential regions were identified by ChIPDiff. Hence, we approximated a false-positive rate of 0.19% (9/4,722) for the DHMS regions identified in cell-specific comparison.

We also tested the reproducibility by conducting two independent passes of cell-specific comparison: $L_{\mathrm{ESC,rep1}}$ vs. $L_{\mathrm{NPC,rep1}}$, and $L_{\mathrm{ESC,rep2}}$ vs. $L_{\mathrm{NPC,rep2}}$. To measure the reproducibility, we defined a score as the ratio of the number of DHMSs identified in both passes to the average number of DHMSs in individual pass. From the test, we obtained a reproducibility score of 57.4% for ChIPDiff. Note that the reproducibility is conditional on the sequencing depth of the replicates, which ranges from three to four million tags in our assessment.

To compare the performance among different methods, we repeated the sensitivity, specificity, and reproducibility tests for fold-change and qualitative method. In qualitative method, K27

Table 1
Comparison of the performance of ChIPDiff, fold-change approach, and qualitative method

	ChIPDiff	Fold-change	Qualitative method
Number of DHMS regions in cell-specific comparison	4,722	4,958	4,790
FPR estimated from non-cell-specific control (%)	0.19	10.8	52.3
Detection rate on HCNE DHMSs (%)	80.0	63.3	73.3
Reproducibility score (%)	57.4	23.4	43.8

The results are based on H3K27me3 data. FPR refers to false-positive rate

modification sites were identified for ESC and NPC individually using the binning approach proposed by Mikkelson et al. (12), and bins marked as K27 site in only one cell-type were identified to be DHMSs. Consecutive DHMSs were merged into DHMS regions as well. For a fair comparison, the thresholds were adjusted to allow similar number of DHMS regions to be identified for all three methods (the numbers are not identical because the thresholds take discrete values). The evaluation results are summarized in Table 1. ChIPDiff outperformed the other two methods in all three aspects. Fold-change approach and qualitative method had much higher false-positive rates, indicating these methods are sensitive to technical variation and experimental bias (see Note 4).

3.4. Application to H3K4me3 and H3K36me3 Data

We extended our study to trimethylations on K4 and K36. Both histone modification types positively correlate with gene expression but in different manner. Guenther et al. revealed K4 marks the active promoters where the transcription of the genes is initiated, while K36 occupies the gene region as a hallmark of elongation (24). Our previous study also showed that K4, together with K27, establishes distinct genomic domains of active and inactive chromatin structures in human ESC (25). Thus, it attracted our interest to study the DHMSs of these histone modifications between ESC and NPC. Moreover, K4 sites usually appear in punctated pattern sharply around TSSs in ChIP-seq profile, while K36 sites appear in a much broader pattern, providing a comprehensive test-bed for evaluating the adaptability of our approach to diverse histone modification types.

Table 2
A summary of DHMSs identified from H3K4me3 and H3K36me3 libraries

	H3K4me3		H3K36me3	
	ESC enriched	NPC enriched	ESC enriched	NPC enriched
Number of DHMS bins	32,384	3,742	15,111	16,719
Number of DHMS regions	12,976	1,768	1,158	1,228
Number of RefSeq genes marked	3,877	211	747	417

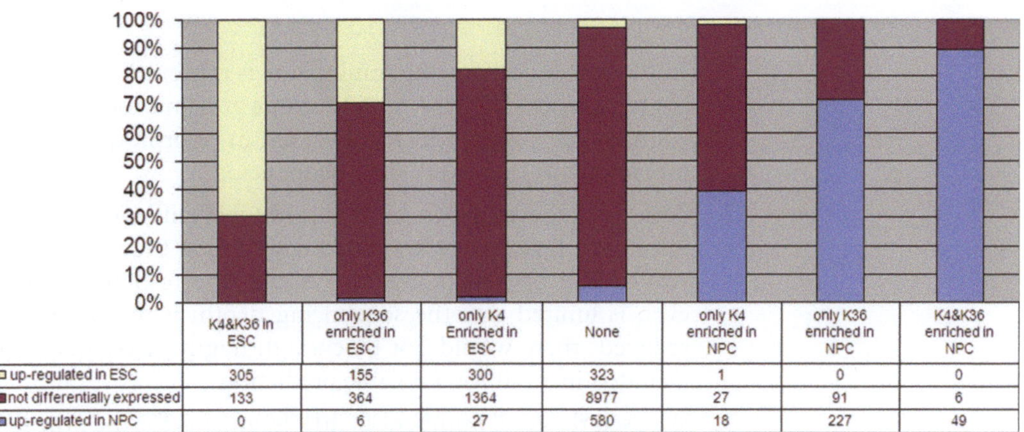

	K4&K36 in ESC	only K36 enriched in ESC	only K4 Enriched in ESC	None	only K4 enriched in NPC	only K36 enriched in NPC	K4&K36 enriched in NPC
up-regulated in ESC	305	155	300	323	1	0	0
not differentially expressed	133	364	1364	8977	27	91	6
up-regulated in NPC	0	6	27	580	18	227	49

Fig. 4. Combinatorial effect of H3K4me3 and H3K36me3 on gene expression between ESC and NPC. Up- and down-regulations were determined by the criterion of fourfold change in microarray expression data.

We processed the libraries with the same ChIPDiff configurations as mentioned in Subheading 3.3. The results are summarized in Table 2. Consecutive bins identified as DHMSs were merged into regions. Strikingly, the number of ESC-enriched K4 DHMSs is much larger than NPC-enriched ones. Considering such imbalance was also observed for K27, we hypothesized that it may be associated with the bivalent chromatin structure marked by K4 and K27 (14). In further analysis, we found 1,961 (51.2%), out of 3,833 ESC-enriched K27 DHMS regions overlap with ESC-enriched K4 DHMSs. In contrast, K36 and K27 seemed to be mutually exclusive: only 8 (0.21%) of these 3,833 regions overlap with ESC-enriched K36 DHMSs.

To study the correlation between DHMSs and gene expression, we annotated the RefSeq genes with DHMS regions and expression data published by Mikkelson et al. RefSeq genes were retrieved from UCSC database (26). To remove the redundancy, the longest ORF was selected for gene annotation if multiple transcripts are mapped to the same gene, which resulted in 18,795 unique genes in total. As shown in Fig. 4, K4 and K36 co-regulate the gene expression with strong significance. This observation is consistent with the conclusion by Guenther et al. (26). Among 1,085 genes upregulated in ESC, 791 (72.9%) are associated with ESC-enriched K4 or K36 DHMSs, suggesting the gene expression is potentially predictable from DHMSs. Notably, two key transcription factors in ESC, Nanog and Oct4, are marked by DHMSs of both K4 and K36, implying the critical roles played by these histone modification marks in ESC by interfering the transcription regulatory network.

4. Notes

1. In the preprocessing step, multiple tags retrieved from different fragments and mapped to the same genomic location were counted only once, which may result in error in quantitative measurement upon very deep sequencing.
2. The bin size was set to be 1 kbp in ChIPDiff, of which the resolution is relatively low when considering the nucleosome size of 200 bp (including the linker). The resolution, however, is limited due the sequencing depth; if the bin size is reduced, there would not be enough fragment counts to be included in a bin for a reliable prediction.
3. We used the total number of ChIP fragments for the normalization against sequencing depth. This normalization procedure is subject to the noise level of ChIP experiment. As an alternative, qPCR measurements (27) on a few "control" sites may provide a better way for normalization. In addition, recently we developed an approach called CCAT (Control based ChIP-seq Analysis Tool) to estimate the signal-to-noise ratio of a ChIP-seq library based on an input control library (28).
4. The specificity, sensitivity, and repeatability of our approach were evaluated based on technical replicates or a limited list of DHMSs inferred from biological knowledge and gene expression. There might be an argument on whether these data provide a "golden" standard for the evaluation. In fact, such a "golden" standard is very difficult to define for most biological data.

References

1. Martin C, Zhang Y (2005) The diverse functions of histone lysine methylation. Nature Rev Mol Cell Biol 6:838–849
2. Boyer LA, Plath K, Zeitlinger J et al (2006) Polycomb complexes repress developmental regulators in murine embryonic stem cells. Nature 441:349–353
3. Bernstein BE, Mikkelsen TS, Xie X et al (2006) A bivalent chromatin structure marks key developmental genes in embryonic stem cells. Cell 125:315–326
4. Widschwendter M, Fiegl H, Egle D et al (2007) Epigenetic stem cell signature in cancer. Nature Genet 39:157–158
5. McGarvey KM, Fahrner JA, Greene E et al (2006) Silenced tumor suppressor genes reactivated by DNA demthylation do not return to a fully euchromatic chromatin state. Cancer Res 66:3541–3549
6. Impey S, McCorkle SR, Cha-Molstad H et al (2004) Defining the CREB regulon: a genome-wide analysis of transcription factor regulatory regions. Cell 119:1041–1054
7. Wei CL, Wu Q, Vega VB et al (2006) A global mapping of p53 transcription factor binding sites in the human genome. Cell 124:207–219
8. Kim TH, Ren B (2006) Genome-wide analysis of protein-DNA interactions. Annu Rev Genom Hum Genet 7:81–102
9. Barski A, Cuddapah S, Cui K et al (2007) High-resolution profiling of histone methylations in the human genome. Cell 129:823–837
10. Johnson DS, Mortazavi A, Myers RM et al (2007) Genome-wide mapping of in vivo protein-DNA interactions. Science 316:1497–1502
11. Mardis ER (2007) ChIP-seq: welcome to the new frontier. Nature Methods 4:613–614
12. Mikkelsen TS, Ku M, Jaffe DB et al (2007) Genome-wide maps of chromatin state in pluripotent and lineage-committed cells. Nature 448:553–560
13. http://genome.ucsc.edu/
14. Quackenbush J. (2002) Microarray data normalization and transformation. Nature Genet 32:496–501
15. Xu H, Wei CL, Lin F et al. (2008) An HMM approach to genome-wide identification of differential histone modification sites from ChIP-seq data. Bioinformatics 24:2344–2349
16. http://cmb.gis.a-star.edu.sg/ChIPSeq/paperChIPDiff.htm
17. Conti L, Pollard SM, Gorba T et al (2005) Niche-independent symmetrical self-renewal of a mammalian tissue stem cell. PLoS Biol 3:e283
18. http://www.broad.mit.edu/seq_platform/chip
19. Bernstein BE, Kamal M, Lindblad-Toh K et al (2005) Genomic maps and comparative analysis of histone modifications in human and mouse. Cell 120:169–181
20. Robertson G, Hirst M, Bainbridge M et al (2007) Genome-wide profiles of STAT1 DNA association using chromatin immunoprecipitation and massively parallel sequencing. Nature Methods 4:651–657
21. Gan Q, Yoshida T, McDonald OG et al (2007) Concise review: epigenetic mechanism contribute to pluripotency and cell lineage determination of embryonic stem cells. Stem Cell 25:2–9
22. Li W, Meyer CA, Liu XS (2005) A hidden Markov model for analyzing ChIP-chip experiments on genome tiling arrays and its application to p53 binding sequences. Bioinformatics (ISMB2005) 21 Suppl 1:i274-i282
23. Welch LR (2003) Hidden Markov Models and the Baum-Welch Algorithm. IEEE Information Theory Society Newsletter 53:1–1
24. Guenther MG, Levine SS, Boyer LA et al (2007) A chromatin landmark and transcription initiation at most promoters in human cells. Cell 130:77–88
25. Zhao XD, Han X, Chew JL et al (2007) Whole-genome mapping of histone H3 Lys4 and 27 trimethylations reveals distinct genomic compartments in human embryonic stem cells. Cell Stem Cell 1:286–298
26. Pruitt KD, Tatusova T, Maglott DR (2005) NCBI Reference Sequence (RefSeq): a curated non-redundant sequence database of genomes, transcripts and proteins. Nucleic Acids Res 33:D501–504
27. Ding C, Cantor CR (2004) Quantitative analysis of nucleic acids – the last few years of progress. J of Biochem and Mol Bio 37:1–10
28. Xu H, Handoko L, Wei X et al (2010) A Signal-noise Model for Significance Analysis of ChIP-seq with Negative Control. Bioinformatics 26:1199–1204

Chapter 20

ChIP-Seq Data Analysis: Identification of Protein–DNA Binding Sites with SISSRs Peak-Finder

Leelavati Narlikar and Raja Jothi

Abstract

Protein–DNA interactions play key roles in determining gene-expression programs during cellular development and differentiation. Chromatin immunoprecipitation (ChIP) is the most widely used assay for probing such interactions. With recent advances in sequencing technology, ChIP-Seq, an approach that combines ChIP and next-generation parallel sequencing is fast becoming the method of choice for mapping protein–DNA interactions on a genome-wide scale. Here, we briefly review the ChIP-Seq approach for mapping protein–DNA interactions and describe the use of the SISSRs peak-finder, a software tool for precise identification of protein–DNA binding sites from sequencing data generated using ChIP-Seq.

Key words: ChIP-Seq, SISSRs, Protein–DNA interaction, Binding sites, Transcription factor, Next-generation sequencing, Genomics

1. Introduction

DNA-binding proteins are essential for the proper functioning of several cellular processes such as transcriptional regulation, which is primarily mediated by interactions between proteins called transcription factors and specific regions on the DNA. These interactions play key roles in determining gene-expression programs during development, differentiation, proliferation, and lineage-specification (1–5). Besides regulating transcription, DNA-binding proteins are essential for DNA replication (6), DNA repair (7), and chromosomal stability (8). Identification of regions targeted by such proteins is therefore crucial for a better understanding of these cellular processes.

Originally developed to investigate protein–DNA binding at a Drosophila locus (9), chromatin immunoprecipitation (ChIP) has become the most widely used assay for determining DNA regions

Junbai Wang et al. (eds.), *Next Generation Microarray Bioinformatics: Methods and Protocols*, Methods in Molecular Biology, vol. 802, DOI 10.1007/978-1-61779-400-1_20,

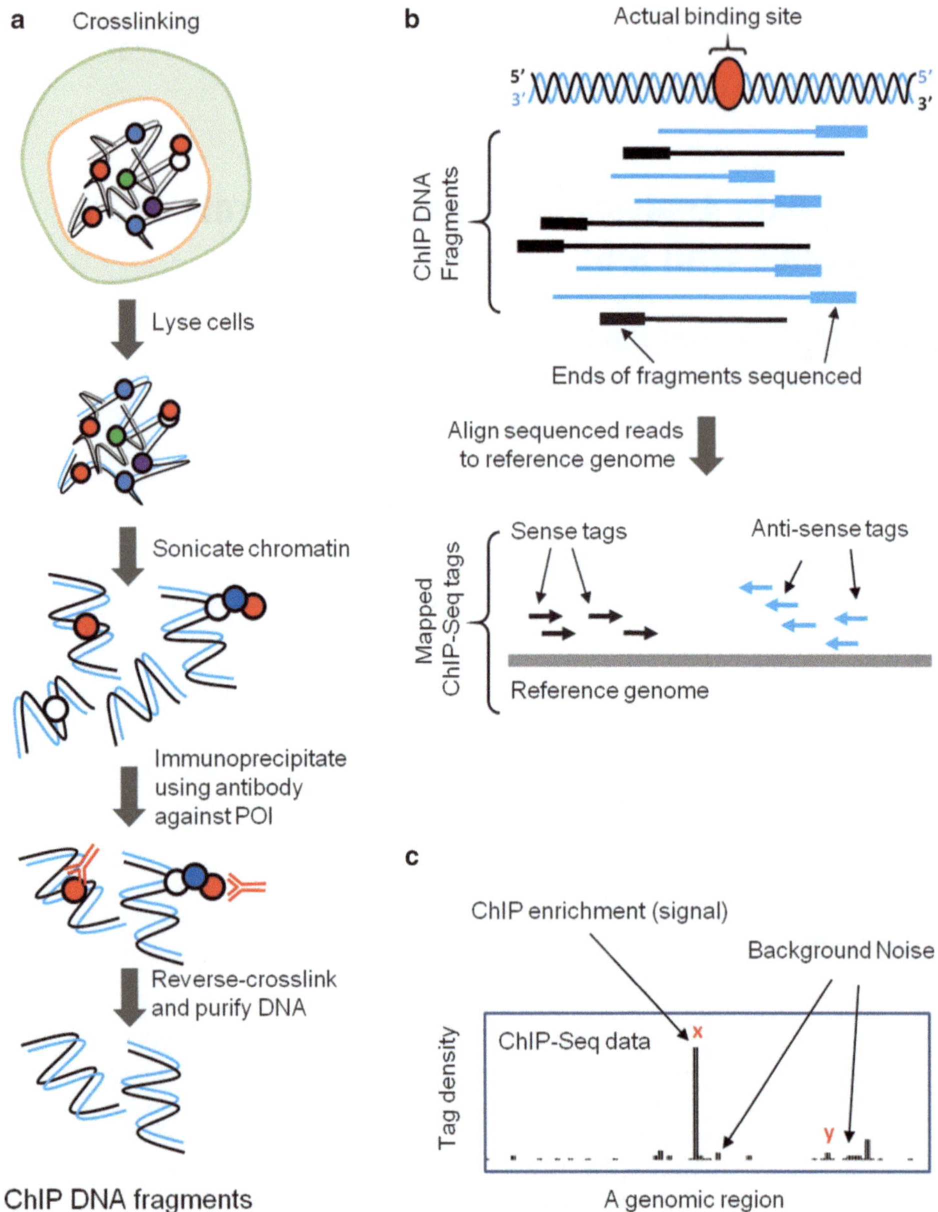

Fig. 1. ChIP-Seq experiment and data. (**a**) Steps involved in chromatin immunoprecipitation (ChIP). Proteins are represented as *circles*. The antibody used in the immunoprecipitation step is represented as a Y-shaped structure. (**b**) Ends of DNA fragments obtained from ChIP are sequenced and aligned back to the reference genome (*arrows* represent the sequenced portion of the ChIP DNA fragment). (**c**) Tags mapped to a genomic region are visualized as a histogram of tag density. Regions with signal and noise are marked with *x* and *y*, respectively.

bound by the protein of interest (POI) in vivo. In this assay, protein–DNA and protein–protein interactions are first crosslinked by treating living cells with formaldehyde (Fig. 1a). This crosslinking step can be omitted in case of proteins such as histones that stably bind DNA. Next, the crosslinked cells are lysed and then sonicated – a process in which ultrasonic waves are used to shear the chromatin into short fragments of desired length (~0.2–0.5 kb). The sheared chromatin is then immunoprecipitated with a specific antibody against the POI. The antibody may not necessarily target only direct POI–DNA complexes but also those complexes where the POI is indirectly bound to the DNA via its interaction with another protein or protein complex (Fig. 1a). The immunoprecipitated protein–DNA crosslinks are reversed, and the DNA is purified for downstream assays designed to characterize the sequences bound by the POI.

Traditionally, PCR or quantitative/real-time PCR (qPCR) with primers designed to probe regions of interest are used to detect and quantify ChIP-derived DNA in relation to a control input DNA, which is obtained the same way as the ChIP DNA but without the immunoprecipitation step. Although ChIP-qPCR still remains the gold-standard assay for quantifying specific protein–DNA interactions, the necessity to design primers for every region to be probed makes it ill-suited for profiling protein–DNA interactions on a large scale. ChIP-chip (10), an approach that combines ChIP with DNA microarrays, was the most widely used technique for mapping protein–DNA interactions on a global scale until recently (11, 12). Advances in sequencing technology have enabled millions of short DNA fragments to be sequenced within a day or two in a cost-effective manner. These sequences can then be aligned back to the reference genome to determine the source of origin. This is exploited in ChIP-Seq (13–17), where ChIP is combined with next-generation massively parallel sequencing technology to identify DNA regions bound by the POI. Its superior coverage and resolution have resulted in ChIP-Seq replacing ChIP-chip as the method of choice. Readers are referred to ref. 18, 19 for a detailed review on ChIP-Seq.

In ChIP-Seq, ChIP-derived DNA fragments are directly sequenced on a next-generation sequencing platform. Although the length of ChIP DNA fragments can range anywhere between a few hundred and a few thousand nucleotides, sequencing just ~25–75 nucleotides from the ends of the DNA fragments is sufficient to align/map the fragments back to unique locations in the reference genome (Fig. 1b). Bowtie (20), MAQ (21), and ELAND from Illumina are popular tools for aligning short sequence reads back to the reference genome. During the alignment process, reads that map to multiple locations in the reference genome are discarded and only those reads that map to unique genomic locations are retained. Such reads are commonly referred to as tags. Henceforth, "reads" and "tags" are used interchangeably.

The first step in interpreting a ChIP-Seq dataset involves identifying regions bound by (or associated with) the POI using the mapped tags. Hereafter, we will refer to these regions as binding sites/regions. Regions with higher tag densities compared to the background "noise" are typically good binding site candidates (site *x* compared to site *y* in Fig. 1c). In theory, only the regions bound by the POI are expected to have tags associated with them since these would be the regions immunoprecipitated and sequenced (Fig. 1a, b). In practice, however, sequencing errors can cause some of the incorrectly sequenced reads to get mapped to regions that were not immunoprecipitated, resulting in background noise tags at these regions (Fig. 1c; see Note 1). Noise in the data could also be due to biological reasons, primarily stemming from antibodies that are not specific to the POI. For instance, nonspecific antibodies targeting additional proteins can result in ChIP-derived DNA fragments that bind one of these proteins and not the POI. Since this type of noise is difficult to detect postsequencing, pre-ChIP experiments are typically performed to confirm antibody specificity.

Issues outlined above highlight the need for a systematic approach for the precise identification of binding sites from ChIP-Seq data. Such an approach must not only identify regions bound by the POI but also filter out false-positive regions by evaluating the test dataset (obtained from ChIP DNA) against a control dataset obtained from input DNA or IgG ChIP (see Note 2). In this chapter, we describe a widely used method called SISSRs (22), a peak-finder that leverages the direction of ChIP-Seq tags (mapped to sense/antisense strands) to identify binding sites at a high resolution, typically within few tens of base pairs. We provide a detailed description of the SISSRs software application tool and instructions for using it effectively to identify protein–DNA binding sites from data generated using ChIP-Seq.

2. Methods

2.1. SISSRs Algorithm

SISSRs, short for *S*ite *I*dentification from *S*hort *S*equence *R*ead*s*, is a peak-finder algorithm that uses the direction and density of mapped ChIP-Seq tags along with the average length (F) of sequenced DNA fragments to identify protein–DNA binding sites (see Note 3; Fig. 2a). If the user does not know the average fragment length of the ChIP DNA, SISSRs can estimate F from the tags within the dataset (see ref. 22 for details). SISSRs begins by scanning regions mapped with sequence tags in the test data using a sliding window of size w nucleotides with consecutive windows overlapping by $w/2$. For a region i spanned by the sliding window, a measure called "net-tag count" (c_i) is computed by subtracting the number of tags mapped to the antisense strand of i (antisense tags) from the number of tags mapped to the sense strand of i (sense tags). As the window slides along, whenever the

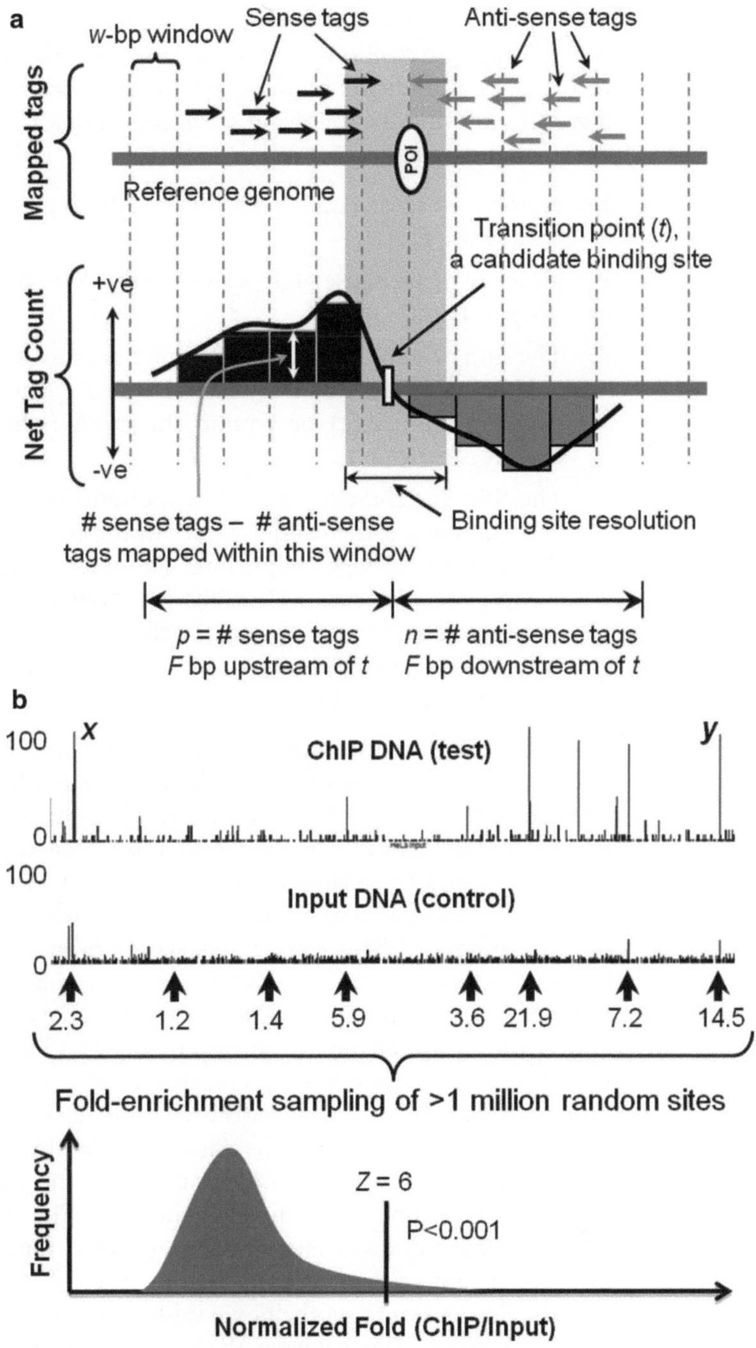

Fig. 2. SISSRs algorithm. (**a**) Typical distribution of tags mapped to sense and antisense strands of a region ChIP-seq using an antibody against the protein of interest (POI), and a schematic showing candidate binding site identification using the direction and density of tags mapped to sense and antisense strands. (**b**) Illustration of how candidate binding sites identified from a test dataset are evaluated against the control dataset to determine the true binding sites. Distribution of fold-enrichment, defined as ratio of the number of tags within a $2F$ bp long region in the test dataset to that within the same region in the control dataset, computed for over one million random sites is used to determine the empirical p-values for candidate binding sites. Only those candidate sites with fold-enrichment value greater than or equal to the smallest fold threshold Z (with p-value not greater than the user-set threshold) are reported as true binding sites. For $Z = 6$, candidate site y with 14.5-fold enrichment will be reported as a true binding site, whereas site x with a similar ChIP signal but with a smaller fold enrichment over the control (2.3-fold) will not be reported as a true site.

net-tag count transitions from a positive to a negative value, the corresponding transition point marked by genomic coordinate t is recorded as a candidate binding site. Only those candidate binding sites satisfying the following set of conditions are retained and designated as true binding sites.

1. Number of sense tags (p) within the F bp region upstream of t is at least E.
2. Number of antisense tags (n) within the F bp downstream of t is at least E.
3. The sum of p and n is at least R, which is estimated based on a user-defined false discovery rate (FDR) D (when no control dataset is available) or e-value threshold (when a control dataset is provided).
4. The fold-enrichment, defined as the ratio of the number of tags supporting the candidate site in the test data ($p + n$) to the number of tags supporting the exact same site in the control data, is at least Z, which is determined based on an empirical distribution of fold-enrichment values of at least a million randomly selected sites and a chosen p-value threshold (Fig. 2b; see Note 4).

Condition 4 applies only when a control dataset is available to evaluate the enrichment of tags supporting the binding site in the test versus the control. When no control dataset is available, the background tag distribution is modeled using a Poisson distribution.

E is set to 2 by default and can be changed by the user. The value of R is estimated as follows. The FDR is defined as the ratio of the number of $2F$-bp long regions with V or more tags that the background model indicates should occur by chance (e^V) to the number observed in the real data. If no control dataset is available, R is equal to the smallest V corresponding to FDR $< D$, otherwise R is equal to the largest V such that $e^V < e$. The expected number of tags (λ) within a window of length $2F$ bp is given by $2F$ times the number of tags in the dataset divided by the mappable genome length M (which is roughly 0.8 times the actual genome length for the human and mouse genomes). The probability of observing a binding site supported by at least R tags by chance is given by a sum of Poisson probabilities as $1 - \sum_{n=0}^{R-1} (e^{-\lambda}\lambda^n)/n!$ SISSRs allows users to set their own values for all of the parameters discussed above. This provides the users the leverage to control sensitivity, specificity, resolution, and noise subtraction.

Identified binding sites are reported by their chromosomal coordinates (e.g., chr1:123450–123490). The resolution of each reported binding site is essentially the distance between the sense tag immediately upstream of the identified site and the antisense tag immediately downstream of this site (Fig. 2a; see Note 5). For additional details on the SISSRs algorithm, the reader may refer to ref. 22.

2.2. Identification of Protein–DNA Binding Sites Using SISSRs

This section gives detailed instructions for installing and using SISSRs on a ChIP-Seq dataset.

2.2.1. Getting and Installing SISSRs

A perl implementation of the SISSRs peak-finding algorithm is freely available at refs. 23, 24. Users with Linux operating system (or most UNIX systems, including Mac OS X) typically have an installation of perl. Users with other operating systems can download the latest version of perl for free using ref. 25. After downloading the SISSRs zipped archive, users should save the extracted sissrs.pl executable either onto their working directory (to run it from the working directory) or to a directory containing executables (to enable execution of sissrs.pl from anywhere within the home directory).

2.2.2. Preparing the Input Data Files

SISSRs takes as input data file(s) containing genomic coordinates of the mapped reads or tags in BED file format (26). In BED file format, each line contains six tab-separated terms as follows:

```
chr1    1234501    1234536    U0    1    +
chr2    9876540    9876585    U0    1    -
chr7    4567825    4567860    U0    1    -
chr3    1472585    1472620    U0    1    +
...
...
```

The first term denotes the chromosome, and the second and third terms denote the chromosomal start and end coordinates of the mapped read, respectively. The sixth term denotes the DNA strand onto which the read was mapped (+ and – for sense and antisense strand, respectively). The fourth and the fifth terms are not used by SISSRs.

2.2.3. Running SISSRs

Typing the name of the executable (sissrs.pl or ./sissrs.pl or perl sissrsl.pl) on the command line displays the help menu. A simple execution of the SISSRs application on a ChIP-Seq dataset (without a control dataset) requires three parameters outlined below with optional parameters discussed next.

-i The name of the input file containing the mapped tags in BED file format.

-o The name of the file onto which the output from SISSRs will be stored.

-s Size or length of the reference genome (number of bases/ nucleotides) onto which the sequenced reads were mapped. For example, 3080436051 for the human genome (hg18 assembly). If analyzing data for a specific chromosome (or a set of chromosomes), then this would be the length of that chromosome (or sum of the lengths of those chromosomes).

If a control dataset is available, option -b, described below, should be used (see Note 2). Various other options available on SISSRs application are listed below. Some of these parameters are preset to default values, which the users can reset to their desired values. Users are recommended to set the -a option, which controls false positives due to amplification or sequencing biases.

-a Setting this option allows only one read per genomic coordinate to be retained even if multiple reads align to the same coordinate, thus effectively minimizing the effects of sequencing and/or PCR amplification bias. During PCR amplification, certain DNA fragments may be amplified into several orders of magnitude in a biased fashion, which after sequencing and mapping will show up as regions enriched with inordinate number of tags. To avoid calling these pseudo-enriched regions as binding sites, we strongly recommend using this option when running SISSRs.

-F Average length of the DNA fragments from ChIP. Typically, DNA fragments of certain length are size-selected for sequencing. Set *F* to this length (integer), if it is known. The individual performing the ChIP experiment and size-selection usually has a good estimate of the average length of sequenced DNA fragments. If this information is not available, this parameter can be left unset in which case SISSRs estimates this measure from the tags in the dataset (also check option -L below; see ref. 22 for details on length estimation).
Default: estimated from tags.

-D FDR if random background model based on Poisson probabilities needs to be used as control. This parameter is relevant only when a control data (e.g., input DNA or nonspecific IgG control) is not provided using the -b option.
Default: 0.001.

-b The name of the file containing the control data (e.g., input DNA or nonspecific IgG control; see Note 2). This file should be in the BED format. The tags in this file are used as a negative control. Subheading 2.2 contains a detailed description of how SISSRs uses the control data to minimize the number of false positives. Users may use -e and -p options (see below) to set the *e*-value and *p*-value thresholds to control sensitivity and specificity, respectively. If no control data is available, SISSRs uses a random background model based on Poisson probabilities (in which case, use option -D to set the FDR).

-e *e*-Value threshold. It is the expected number of enriched regions (based on Poisson probabilities) in a similar-sized dataset. The value entered for this parameter is used to estimate the minimum number of reads (R) necessary to identify candidate binding sites. This option controls sensitivity (the -p option explained below controls specificity), and is ignored if -b option is not used (no control data).
Default: 10.

-p *p*-Value threshold. For a given F value (average DNA fragment length), the fold/ChIP enrichment for a candidate binding site is the ratio of the number of tags supporting the site, which is $p + n$ (Fig. 2a), to the number of tags supporting the same site in the control dataset. This fold enrichment is normalized with respect to the number of tags in both the test and the control datasets. To assess the statistical significance of the observed fold enrichment (the probability that the observed fold enrichment is by chance), an empirical distribution of fold enrichments from at least one million random sites, spanning the set of all chromosomes in the test dataset, is used to estimate the *p*-value for each candidate binding site. Only those sites with *p*-values not over the *p*-value threshold are reported as true binding sites. This option controls specificity (the -e option explained above controls sensitivity), and is ignored if -b option is not used (no control data).
Default: 0.001.

-m Fraction of genome (0.0–1.0) mappable by reads. Typically, not all sequenced reads map to unique genomic locations. Portions of the genome containing repetitive elements, which account for roughly 20% of the genome, are not mappable. The value entered for this parameter is used to estimate Poisson probabilities.
Default: 0.8.

-w Size of the scanning window (must be an even number >1), which is one of the parameters that attempts to control for noise in the data. The scanning window slides so that there is a 50% overlap between two consecutive window positions. As a result, the resolution of the identified binding sites (t in Fig. 2a) is $w/2$. For example, for $w = 20$, each binding site in the output file (with default -c option) will have a starting and ending coordinate with 1 and 0 in the Units position, respectively (e.g., 1234561–1234620). A larger window size reduces the influence of nonspecific reads and thus false positives at the cost of resolution. A smaller window size provides for increased resolution but may increase the number of false

positives if the data is noisy (contains a high number of nonspecific reads). In other words, smaller window size makes for higher sensitivity possibly at the cost of lower specificity, and larger window size makes for higher specificity possibly at the cost of lower sensitivity. The amount of background noise in the data is an important factor one needs to consider before setting a value for -w.
Default: 20.

-E Threshold for the number of tags mapped within F bp upstream or downstream of the center of the inferred binding site (t in Fig. 2a). This is one of the parameters that controls for specificity to a small degree. The higher the E, the more specific (and slightly less sensitive) SISSRs will be, and vice versa.
Default: 2 (assuming that the data file contains ~5–10 million reads; the user may consider increasing this value if the total number of reads is much larger).

-L Upper-bound on the DNA fragment length. It is the approximate length/size of the longest DNA fragment that was sequenced. This value is one of the critical parameters used during the estimation of average DNA fragment length. The individual who performed the ChIP and size-selection of the DNA fragments before sequencing should have a good estimate on of the upper-bound for the DNA fragment length.
Default: 500 (assuming that DNA fragments of length <500 bp were size-selected).

-q The name of the file containing genomic regions in simple three-column tab-separated format (chr start-coordinate end-coordinate). Reads falling within these regions will not be considered for the analysis.

-t If this option is set, each binding site is reported as a single genomic coordinate representing the center of the inferred binding site (t in Fig. 2a). If this option is not selected, SISSRs uses the -c option (see below).

-r If this option is set, SISSRs, instead of reporting each binding site as a single genomic coordinate (representing the center t of the inferred binding site; e.g., chr1 12345), each binding site is reported as an X-bp binding region, where X represents the resolution of the identified site (Fig. 2a). X varies for each binding site depending upon the availability of tags supporting the site. If this option is not selected, SISSRs uses the -c option as default (see below).

-c This option is same as the -r option, except that it reports binding sites that are clustered within F-bp of each other as a single binding region by merging those sites. As a result, the number of binding sites reported using this option could be typically fewer than that reported using the -r option. For each binding region reported in the output file, the entry in the "NumTags" column indicates the number of tags supporting the strongest binding site in the reported binding region. The -c option is the recommended option especially if *w* is set to smaller values (ten or less).

Default: This is the default option, which SISSRs is used to report binding sites.

-u If this option is set, SISSRs also reports binding sites supported only by reads mapped to either sense or antisense strand. This option will recover binding sites whose sense or antisense reads were not mapped for some reason, e.g., the actual binding site lies right next to a repetitive region in which case reads aligning to the repetitive side were not mapped because they also align to other region(s) in the genome (see ref. 22 for details).

-x If this option is set, the summary and the progress report are not displayed on the terminal during the execution of the application.

2.3. Examples

Example 1: A simple example with no control dataset:

./sissrs.pl -i ctcf.bed -s 3080436051 -o ctcf.sissrs

SISSRs identify binding sites based on the reads in the test data file ctcf.bed. Since no control data file was provided (−b option), the default background model based on Poisson probabilities and the default FDR (0.001) will be used to determine statistically significant number of tags (R in Fig. 2) necessary to identify binding sites. SISSRs automatically use the default values for other parameters.

Example 2: Using the -a option, which considers only one read per genomic position:

./sissrs.pl -i ctcf.bed -s 3080436051 -o ctcf.sissrs -a

This is same as Example 1, except that only one read per genomic position is kept even if multiple reads get mapped to the same gnomic position.

Example 3: Using a control dataset:

./sissrs.pl -i ctcf.bed -s 3080436051 -o ctcf.sissrs -b control.bed -a

This is same as Example 2, except that a background control file is used as negative control (replacing the default random model based on Poisson probabilities). Default values are used for other parameters including the -e and -p parameters, which assume the default values 10 and 0.001, respectively.

Example 4: Ignoring reads that fall within certain genomic regions:

./sissrs.pl -i ctcf.bed -s 3080436051 -o ctcf.sissrs -b control.bed -a -q repeatsFile.txt

This is same as Example 3, except that the input reads that fall within the genome regions listed in the repeatsFile.txt will be ignored during the analysis. Effectively, this may reduce the number of binding sites reported compared to that reported in the case of Example 3.

Example 5: General run with no control data (relevant options listed using separate square brackets [ ]):
./sissrs.pl -i ctcf.bed -s 3080436051 -o ctcf.sissrs [−a] [−F 200] [−D 0.001] [−m 0.8] [−w 20] [−E 2] [−L 500] [−q repeatsFile.txt] [−t]/[−r]/[−c] [−u] [−x]

Example 6: General run with a control dataset (relevant options listed using separate square brackets [ ]):
./sissrs.pl -i ctcf.bed -s 3080436051 -o ctcf.sissrs [−a] [−F 200] [−b bg.bed] [−e 10] [−p 0.001] [−m 0.8] [−w 20] [−E 2] [−L 500] [−q repeatsFile.txt] [−t]/[−r]/[−c] [−u] [−x]

2.4. SISSRs Output, Interpretation, and Downstream Analyses

The results from a SISSRs run are stored under the file name that was provided by the user with the -o parameter. This output file contains the summary of the test and control datasets, the list of command line and estimated parameters which SISSRs used to process the data, and the list of binding sites identified using the statistical thresholds chosen by the user. A typical SISSRs output is shown in Fig. 3. Each identified binding site is listed as a genomic region along with the number of tags supporting that site. If a background control data was used, fold enrichment over the control data along with a *p*-value accompanies each reported site.

The first term denotes the chromosome on which the binding site resides. The second and the third terms denote the chromosomal start and end coordinates of the binding site, respectively. The fourth term "NumTags" denotes the number of tags supporting the identified binding site, which is equal to $p + n$ in Fig. 2a. The fifth and the sixth terms "Fold" and "*p*-value," respectively, are reported only if a background control data was used. Fold denotes fold-enrichment, which is the ratio of NumTags to the number of tags supporting the exact same site in the background control data (see Note 6). While computing the fold enrichment, the number of tags supporting the binding site in the test and control data is normalized by the total number of tags in the test and control data. The *p*-value denotes the probability that one would expect to see this fold-enrichment between the test and the control data just by chance, which is computed based on the empirical distribution of fold-enrichment values for one million or more random sites (Fig. 2b). Only those binding sites with fold-enrichment *p*-value less than or equal to the *p*-value threshold (set by the user using the -p option) are reported in the results file.

Typical downstream analyses of SISSRs-reported binding sites include de novo motif analysis to identify the consensus sequence within the identified binding sites/regions. De novo motif analysis is an unbiased search for a consensus sequence motif present within the identified binding sites (Fig. 4; see Note 7). Software tools such as PRIORITY (27), MEME (28), and GADEM (29)

```
==========================================================================
SISSRs: A tool to identify binding sites from ChIP-Seq data
==========================================================================
SISSRs version 1.4 (Release date: Mon, 24 November 2008)

...
...

==========================================================================
COMMAND LINE SUMMARY & ESTIMATED PARAMETERS
==========================================================================
Data file (i)                            : hg18-test.bed
Number of tags in the data file          : 17977403
Number of tags selected for analysis     : 17381505 (96.69%)
Tags mapped to sense strand              : 8690727 (50.00%)
Tags mapped to anti-sense strand         : 8690778 (50.00%)
Background model (Negative Control)      : hh18-control-input.bed
Number of tags in the control file       : 18655942
Genome length (s)                        : 3080436051
Fraction of genome mappable by reads (m): 0.80
Effective Genome length  (s*m)           : 2464348840
E-value (e)                              : 10
P-value (p)                              : 0.001
Scanning window size (w)                 : 20
Average DNA fragment length (f)          : 202
Minimum number of 'directional' tags
  required on each side of the inferred
  binding site (E)                       : 2
Keep one tag per genomic position (a)    : YES
Also reports binding sites supported
  only by reads mapped to either sense
  or anti-sense strand (u)               : NO
==========================================================================



==========================================================================
BINDING SITES
==========================================================================
Tags necessary to identify binding sites: 15 (Fold >= 6.00)
(at least E=2 tags on each side)

Chr     cStart     cEnd       NumTags  Fold    p-value
---     ------     ----       -------  ----    -------
chr1    6397351    6397391    91       19.53   7.0e-06
chr1    9164791    9165031    109      29.25   2.0e-06
chr1    11890651   11890951   98       17.53   1.0e-05
chr1    11892391   11892491   40       10.73   5.0e-05
...
```

Fig. 3. A typical SISSRs output file.

can be used to identify the consensus sequence, if any, present within identified sites (see Note 8). If the DNA binding preference for the POI is known, then the identified consensus sequence is expected to match the known binding sequence. Otherwise, the user needs to investigate at least two possible scenarios with regard

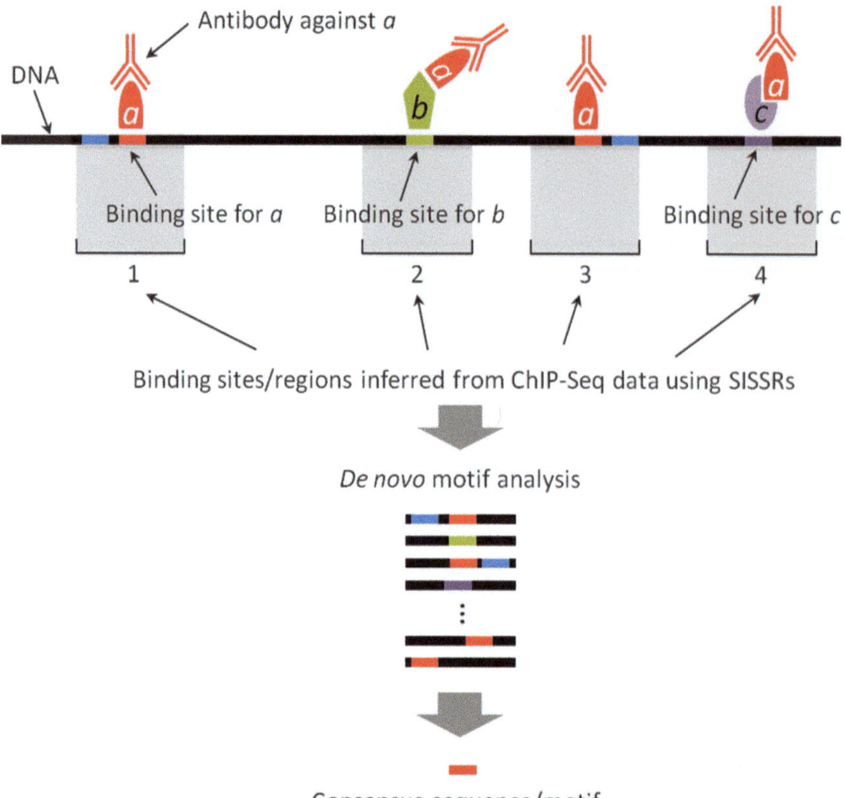

Fig. 4. De novo motif analysis for discovering consensus sequence motif within the identified binding sites.

to the novel consensus sequence: (a) the consensus sequence could characterize an undiscovered novel binding preference of the POI or (b) the POI binds DNA indirectly via another protein, in which case the identified consensus sequence would correspond to the binding preference of that protein.

Other analyses include determining the genomic distribution of identified binding sites in relation to genomic landmarks, and defining a list of genes targeted by the protein being profiled. For a given reference genome and a set of gene annotations, custom software can be written to determine the fraction of identified binding sites that fall within intronic/exonic regions, promoter regions (defined as a few kilo-bases upstream and/or downstream of transcription start sites of known genes), and other genomic landmarks of interest. Given that a binding site may or may not be functional, defining target genes based on the set of identified binding sites alone is not straightforward. But, in practice, genes that contain one of more identified binding sites within a few kilo-bases upstream or downstream of their transcription start sites are defined as targets of protein being profiled.

2.5. SISSRs Running Time

SISSRs running time primarily depends on whether or not a background control data is being used. When no background control data is used, the running time is typically few minutes. Most of this time is spent reading the data files. In general, it takes ~5 min for SISSRs to analyze a test dataset containing approximately ten million reads with default settings and no background control data. If a background control data is used, then SISSRs could take anywhere between ~10 and 30 min for a *p*-value threshold of 0.001, with the additional time spent sampling one million random sites to determine the empirical *p*-value distribution. Setting the *p*-value to smaller values will further increase the running time. Thus, it is recommended that the *p*-value is not set to extremely small values if running time is of primary concern (see Note 9).

3. Notes

1. A high noise-to-signal ratio raises a red flag on the sequencing quality, and it is a good practice to avoid datasets where signal and noise cannot be easily distinguished.
2. Many nucleosome-free (open chromatin) regions in the genome can bind proteins in a nonspecific manner and certain genomic regions are prone to biased amplification/sequencing. These biases in the test dataset can be neutralized to some extent by using a control dataset, which will help reduce the number of nonspecific binding sites inferred as true binding sites. Input DNA and IgG ChIP-derived DNA are the two commonly used controls. Input DNA is prepared the same way as the ChIP DNA without the immunoprecipitation step. IgG ChIP is performed with an antibody against IgG, which binds DNA in a nonspecific manner. If antibody specificity against the POI is not a concern, input DNA serves as a better control for amplification and sequencing bias compared to IgG ChIP DNA. Although not necessary, we strongly recommend using a control data when using SISSRs.
3. SISSRs was designed to identify protein–DNA interaction sites from ChIP-Seq datasets and is not suitable for analyzing histone modification data to identify regions enriched with a specific histone modification. ChIP-Seq data characterizing histone modifications in general have much broader footprints of signal of varying lengths (anywhere from few hundred to several thousand bases) compared to that for protein–DNA interaction sites, which is typically ~200 nucleotides (13). Distinguishing broader footprints of signal from the background noise requires accurate characterization

of boundaries demarcating signal and noise, a task that requires sequencing of the ChIP sample to near saturation. Since samples are rarely sequenced to near saturation, identification of regions with broad footprints of signal (e.g., histone modifications H3K4me1, H3K9me3, H3K27me3, and H3K26me3 (13)) is a relatively difficult task compared to protein–DNA binding sites. We do not recommend SISSRs for analyzing histone modification data in general, but it may be used to analyze histone modification data such as H3K4me3 or H3K9ac (that have ~200–500 bp footprints) with caution.

4. The statistics used to determine *Z* is highly dependent on how well saturated the control data is. If the control data does not contain sufficient reads (much less than what may be necessary), then using such a dataset as a control is as good as using no control. Thus, it is important to make sure that the control data contains sufficient number of reads. As a rule of thumb, for a genome of length *L* nucleotides and the average fragment length of *F* nucleotides, it is desirable that the control dataset contains at least about L/F tags to make reliable inferences.
5. The resolution of the reported binding site is dependent on the number of tags in the dataset. The larger the dataset (more tags), the higher the likelihood of identifying sites with better resolution. Typically, the average resolution of the reported sites is somewhere between 40 and 80 bp, but it could be as much as the length of the average ChIP fragment.
6. The value for ChIP fold-enrichment (when a control is used) or number of tags (when a control is not used) is a good indicator of protein–DNA binding affinity/stability (22). When comparing two or more binding sites, higher (lower) values for these measures can be interpreted as stronger (weaker) binding.
7. If one wishes to perform motif analysis on the DNA sequences corresponding to the reported binding sites, we recommend using the 200 nucleotide sequence centered on the reported binding site. Although the ~5–20 bp DNA sequence bound by a protein is highly likely to be present within the region reported as the binding site, it is quite possible that all or part of this binding sequence is just outside of the reported binding site. And, since the resolution of the reported sites are dependent on the tags that map near these sites, some of which could be noise, there is always a chance that a reported coordinate defining a binding site could be off by a few base

pairs. It is therefore good practice to consider using a 200 nucleotide sequence centered on the reported binding site.

8. Since ChIP using an antibody against POI captures genomic regions bound directly as well as indirectly by POI (Fig. 4), one cannot expect all of the reported binding sites for POI to contain the consensus binding sequence/motif. Thus, a lack of consensus sequence at a site cannot be interpreted as that site being a false-positive.
9. If running time is of concern, do not set the *p*-value ($-p$) to a number less than 0.0001 (0.001 is the default).

Acknowledgments

This work was supported by the Intramural Research Program of the National Institutes of Health, National Institute of Environmental Health Sciences (Project number ES102625–02 to R.J.).

References

1. Boyer LA, Lee TI, Cole MF et al (2005) Core transcriptional regulatory circuitry in human embryonic stem cells. Cell 122:947–956.
2. Chen X, Xu H, Yuan P et al (2008) Integration of external signaling pathways with the core transcriptional network in embryonic stem cells. Cell 133:1106–1117.
3. Ho L, Jothi R, Ronan JL et al (2009) An embryonic stem cell chromatin remodeling complex, esBAF, is an essential component of the core pluripotency transcriptional network. Proceedings of the National Academy of Sciences of the United States of America 106:5187–5191.
4. Molkentin JD (2000) The zinc finger-containing transcription factors GATA-4, -5, and −6. Ubiquitously expressed regulators of tissue-specific gene expression. J Biol Chem 275:38949–38952.
5. Hou C, Dale R, Dean A (2010) Cell type specificity of chromatin organization mediated by CTCF and cohesin. Proceedings of the National Academy of Sciences of the United States of America 107:3651–3656.
6. Rampakakis E, Gkogkas C, Di Paola D et al (2010) Replication initiation and DNA topology: The twisted life of the origin. J Cell Biochem 110:35–43.
7. Cohn MA, D'Andrea AD (2008) Chromatin recruitment of DNA repair proteins: lessons from the fanconi anemia and double-strand break repair pathways. Mol Cell 32:306–312.
8. Shivji MK, Venkitaraman AR (2004) DNA recombination, chromosomal stability and carcinogenesis: insights into the role of BRCA2. DNA Repair (Amst) 3:835–843.
9. Solomon MJ, Larsen PL, Varshavsky A (1988) Mapping protein-DNA interactions in vivo with formaldehyde: evidence that histone H4 is retained on a highly transcribed gene. Cell 53:937–947.
10. Ren B, Robert F, Wyrick JJ et al (2000) Genome-wide location and function of DNA binding proteins. Science 290:2306–2309.
11. Mardis ER (2007) ChIP-seq: welcome to the new frontier. Nat Methods 4:613–614.
12. Park PJ (2009) ChIP-seq: advantages and challenges of a maturing technology. Nat Rev Genet 10:669–680.
13. Barski A, Cuddapah S, Cui K et al (2007) High-resolution profiling of histone methylations in the human genome. Cell 129:823–837.
14. Johnson DS, Mortazavi A, Myers RM et al (2007) Genome-wide mapping of in vivo protein-DNA interactions. Science 316:1497–1502.
15. Robertson G, Hirst M, Bainbridge M et al (2007) Genome-wide profiles of STAT1 DNA association using chromatin

immunoprecipitation and massively parallel sequencing. Nat Methods 4:651–657.

16. Barski A, Jothi R, Cuddapah S et al (2009) Chromatin poises miRNA- and protein-coding genes for expression. Genome Research 19:1742–1751.
17. Cuddapah S, Jothi R, Schones DE et al (2009) Global analysis of the insulator binding protein CTCF in chromatin barrier regions reveals demarcation of active and repressive domains. Genome Research 19:24–32.
18. Barski A, Zhao K (2009) Genomic location analysis by ChIP-Seq. J Cell Biochem 107:11–18.
19. Cuddapah S, Barski A, Cui K et al (2009) Native chromatin preparation and Illumina/Solexa library construction. Cold Spring Harb Protoc 2009:pdb prot5237.
20. Langmead B, Trapnell C, Pop M et al (2009) Ultrafast and memory-efficient alignment of short DNA sequences to the human genome. Genome Biol 10:R25.
21. Li H, Ruan J, Durbin R (2008) Mapping short DNA sequencing reads and calling variants using mapping quality scores. Genome Research 18:1851–1858.
22. Jothi R, Cuddapah S, Barski A et al (2008) Genome-wide identification of in vivo protein-DNA binding sites from ChIP-Seq data. Nucleic Acids Research 36:5221–5231.
23. http://www.rajajothi.com.
24. http://dir.nhlbi.nih.gov/papers/lmi/epigenomes/sissrs/.
25. http://www.perl.org.
26. http://genome.ucsc.edu/FAQ/FAQformat#format1.
27. Narlikar L, Gordan R, Hartemink AJ (2007) A nucleosome-guided map of transcription factor binding sites in yeast. PLoS Comput Biol 3:e215.
28. Bailey TL, Elkan C (1994) Fitting a mixture model by expectation maximization to discover motifs in biopolymers. Proc Int Conf Intell Syst Mol Biol 2:28–36.
29. Li L (2009) GADEM: a genetic algorithm guided formation of spaced dyads coupled with an EM algorithm for motif discovery. J Comput Biol 16:317–329.

Chapter 21

Using ChIPMotifs for De Novo Motif Discovery of OCT4 and ZNF263 Based on ChIP-Based High-Throughput Experiments

Brian A. Kennedy, Xun Lan, Tim H.-M. Huang, Peggy J. Farnham, and Victor X. Jin

Abstract

DNA motifs are short sequences varying from 6 to 25 bp and can be highly variable and degenerated. One major approach for predicting transcription factor (TF) binding is using position weight matrix (PWM) to represent information content of regulatory sites; however, when used as the sole means of identifying binding sites suffers from the limited amount of training data available and a high rate of false-positive predictions. ChIPMotifs program is a de novo motif finding tool developed for ChIP-based high-throughput data, and W-ChIPMotifs is a Web application tool for ChIPMotifs. It composes various ab initio motif discovery tools such as MEME, MaMF, Weeder and optimizes the significance of the detected motifs by using bootstrap re-sampling error estimation and a Fisher test. Using these techniques, we determined a PWM for OCT4 which is similar to canonical OCT4 consensus sequence. In a separate study, we also use de novo motif discovery to suggest that ZNF263 binds to a 24-nt site that differs from the motif predicted by the zinc finger code in several positions.

Key words: Motif, ChIP, Position weight matrix, OCT4, ZNF263

1. Introduction

During the past decade, several computational approaches have been developed to study large and complex datasets generated from high-throughput technologies such as mRNA expression profiling (1, 2), ChIP-chip (3, 4), DamID (5), DNase-chip (6), and ChIP-PET (7). The computational algorithms behind these approaches include (1) statistically driven ab initio motif discovery methods such as hidden Markov models (8), Gibbs sampling (9), expectation-maximization (MEME (10)), exhaustive enumeration (Weeder (11)), and words enumeration with a positional weight matrix updating (12); and (2) prior-compiled position weight

Junbai Wang et al. (eds.), *Next Generation Microarray Bioinformatics: Methods and Protocols*,
Methods in Molecular Biology, vol. 802, DOI 10.1007/978-1-61779-400-1_21, © Springer Science+Business Media, LLC 2012

matrices (PWMs) library-based motifs detection methods such as MATCH (13) combined with the TRANSFAC database (14) and MSCAN (15) combined with the JASPAR database (16).

All of the above-motioned methods have been proven to be useful in detecting novel motifs and deciphering the logics of transcription regulatory networks; however, there are still several major challenges facing these de novo methods. First, TF binding sites are short and easily confused among the noise of larger sequences; second, variability in TF binding sites is not well understood; and third, many consensus binding sites are derived from a small set of in vitro experiments. Some of these challenges in identifying motifs can be minimized by using ChIP-chip data to derive a consensus binding site to which a factor is bound in vivo. Also, some of the issues concerned with background (control) sequences can be eliminated using a bootstrap re-sampling of the data.

The sequences identified from ChIP-based high-throughput techniques such as ChIP-chip (4, 17), ChIP-seq (18, 19), and ChIP-PET (7) are called "peaks," which are defined as significantly dense clusters in the sequence reads. Usually ranging from ~150 to ~1,500 bases, these peaks are currently considered to be highly reliable data sets for detecting the novel motif. Many computational tools including ours (20–24) have been recently developed to de novo find the motifs for the data generated from these techniques.

2. Methods

The flow chart in Fig. 1a demonstrated the general protocol used for de novo motif discovery. In which, sequences are ranked according to some metric external to this algorithm, and the top *k* sequences are selected for de novo motif detection. In the case of in vivo ChIP-based data, for which this protocol was originally developed, the criteria for selecting input sequences would be that the binding sites (sequences) were identified by a peak detection program and ranked based on a statistical measurement (a *p*-value or a false discovery rate). Binding sites (sequences) above an appropriate significance (such as $p < 0.05$) would be used as an input data in the following protocol.

2.1. General Protocol for De Novo Motif Discovery

1. Select the input data set of the top *k* sequences ordered by significance (see Note 1).
2. Process the input data set in Weeder (see Note 2).
3. Process the input data set in MEME.
4. Process the input data set in MaMF.

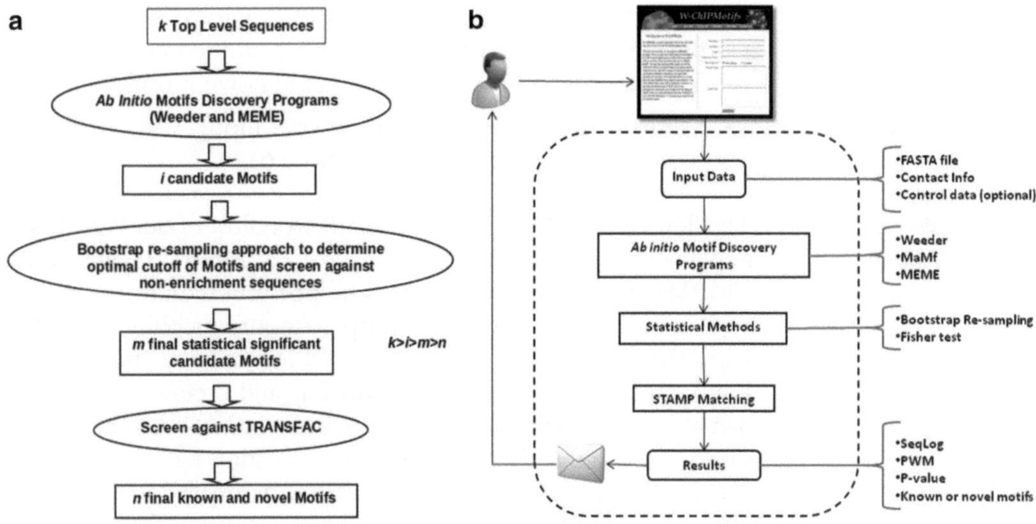

Fig. 1. Ab initio motif discovery workflows. (**a**) The general ChIPMotifs workflow from initial data selection to the final set of motif predictions. (**b**) The workflow of W-ChIPMotifs, in more detail, with the specific input file formats and types of data being processed at each stage. The overall workflow can be summarized as follows: (1) select input sequences; (2) run Weeder, MEME, and MaMF on the input; (3) use bootstrap re-sampling and the Fischer exact test to filter the output by quality; (4) use STAMP to predict a phylogenetic hierarchy of the results and identify matches to existing motifs.

5. The union of the output of these three programs is the set of candidate motifs, of size i.
6. Construct position weight matrices for each of the i candidate motifs.
7. Perform bootstrap re-sampling by randomizing each of the k sequences for 100 times, and generate a total of $100xk$ sequences. These sequences have same nucleotides' identities with original sequences but in different orders (see Note 3).
8. Scan these randomized sequences for each candidate motif (using the PWMs derived from step 6) starting at a minimal core score of 0.5 and a minimal PWM score of 0.5. This score is the sum of the weight for the nucleotide in the sequence being scored at a position i in the PWM, for each such i in the PWM, 1, ..., n where n is the length of the PWM (25).
9. Retrieve core scores and PWM scores at the Top X % percentile (one-tailed p-value is less than $X/100$).
10. Filter these i candidate motifs to those which meet any additional experimental constraints, if any (see Note 4).
11. Apply the Fisher test to measure the significance of the motifs using nonenrichment (or control) data (see Note 5).
12. Discard nonsignificant motifs, i.e., motifs with a significance of $p > 0.001$, to obtain a significant set of m putative motifs.

13. Feed this set of motifs and their PWMs to STAMP (26) for phylogenetic hierarchical clustering and comparison with TRANSFAC (14) and JASPAR (16) known motifs.
14. STAMP will output the final set of *n* motifs with significant similarity to known motifs (see Note 6).

2.2. Introduction to W-ChIPMotifs

The flow chart in Fig. 1b illustrates the workflow of our Web-based implementation of this algorithm, W-ChIPMotifs. Usage of W-ChIPMotifs web service is simple and does not require any knowledge of the underlying software (http://motif.bmi.ohio-state.edu/ChIPMotifs). There are three required inputs from the user: the DNA sequence data, contact information, and a transcription factor name. DNA sequences are required to be in the FASTA format. They can be uploaded either by selecting an existing file or by directly copying the data into the form. Results will be emailed to the address given in the contact information. The transcription factor name is used as a label in the results. Also, control data can be specified as an optional input, which is used to infer the statistical significance for detected motifs. In case of no control data input from users, we will use default control data sets where we randomly selected 5,000 promoter sequences per run from all human or mouse promoter sequences depending on the user selected species.

2.3. W-ChIPMotifs Workflow

1. Select the input data set of the top *k* sequences ordered by significance (see Note 1).
2. Provide these sequences in a FASTA format and contact information.
3. Optionally provide control data. If no control data is submitted, a default control data set is used composing 5,000 randomly selected promoter sequences from all promoter region sequences in the target species.
4. Process the input data set in Weeder (see Note 2).
5. Process the input data set in MEME.
6. Process the input data set in MaMF.
7. The union of the output of these two programs is the set of candidate motifs, of size *i*.
8. Construct position weight matrices for each of the *i* candidate motifs.
9. Perform bootstrap re-sampling by randomizing each of the user input's sequences for 100 times (see Note 3).
10. These randomized sequences are used for scanning the identified motifs (represented with PWMs, from step 8) at a minimal core score of 0.5 and a minimal PWM score of 0.5.

11. Retrieve core and PWM scores at the top 0.1, 0.5, and 1% percentiles.
12. Apply the Fisher test to measure the significance of each motif.
13. We also apply the Bonferroni correction by adjusting the p-value multiplying by the number of samples being input. If the adjusted p-value ended up greater than 1.0, it would be rounded down to 1.0 (see Note 5).
14. Discard nonsignificant motifs, i.e., motifs with a significance of $p > 0.001$, to obtain a significant set of m putative motifs.
15. Feed this set of motifs and their PWMs to STAMP for phylogenetic hierarchical clustering and comparison with TRANSFAC and JASPAR known motifs.
16. STAMP will output the final set of n motifs with significant similarity to known motifs.
17. The results from W-ChIPMotifs are composed of two files. The first file contains detected motifs with their SeqLOGOs, PWMs, core and PWM scores, p-values, and Bonferroni correction p-value at different percentile levels. The second file contains matched similar motifs from the STAMP tool. These files are in PDF format.

2.4. W-ChIPMotifs Implementation

W-ChIPMotifs is written in Perl, and uses a Web interface developed with PHP. Multiple scripts are used to produce output from the included motif discovery programs, parse this output, and apply statistical techniques. The sequence logos for the motifs are generated using the WEBLOGO tool. The open-source HTMLDOC program is used to convert these logos to PDF format (http://www.htmldoc.org/). A tree in Newick format is created with the DRAWTREE tool (see Note 7). The PHPGmailer package is used for sending results to the user from the W-ChIPMotifs email account.

2.5. Case Studies for De Novo Motif Discovery of OCT4 and ZNF263

We present two case studies in the application of these techniques (see the sample data at http://motif.bmi.ohio-state.edu/BookChIPMotifs). The study in OCT4 illustrates how in vivo ChIP sequence data can be used to computationally predict motifs ab initio. The ZNF263 research shows that computationally predicted motifs may differ from in vitro predicted motifs while still having high predictive capability, i.e., they can be used to identify sites on the genome which correlate with the genome wide in vivo experimental results.

2.6. In Vivo OCT4 Motif Discovery

Recent ChIP-chip studies have revealed that many in vivo binding sites have a weak match to the consensus sequence for the transcription factor being analyzed. Possible explanations for these observations include (a) the consensus site was derived from in vitro analyses and does not represent the preferred in vivo binding site and/or (b) the factor is recruited to a weak binding site via interaction with a protein that binds nearby. To investigate case (b), we performed the following analysis. Using OCT4 ChIP-chip data derived from genomic tiling arrays and the ChIPMotifs approach, we developed a refined OCT4 PWM. We then used the in vivo derived PWM and a ChIPModules approach to identify transcription factors co-localizing with OCT4 in a testicular germ cell tumor (Ntera2 cells). We found that the consensus binding site for SRY, a transcription factor critical for testis development, co-localizes with the OCT4 PWM. To further characterize the relationship between OCT4 and SRY binding sites, we used ChIP-chip analysis of human promoter microarrays, and found that 49% of the top ~1,000 OCT4 target promoters were also bound by SRY. This analysis represents the first identification of SRY target promoters. Our studies not only validate the ChIP-Motifs and ChIPModules combinatorial approach but also identify a possible new regulatory partner of OCT4.

2.7. Methods for OCT4 Data

1. Input a set of 154 in vivo OCT4 binding sequences into the Weeder and MEME programs (see Note 1).
2. Using these programs, we identified ten candidate motifs, each having a length of 8–12 bp.
3. We then constructed ten positional weight matrices for each candidate motif.
4. We randomized the sequences of each of the 154 OCT4 binding sequences 100 times to generate a set of 15,400 randomized sequences (see Note 8).
5. We then scanned these randomized sequences for each candidate motif (using the PWMs derived from Weeder and MEME) starting at a minimal core score of 0.5 and a minimal PWM score of 0.5.
6. We retrieved core scores and PWM scores at the Top 0.1% percentile (one-tailed p-value is less than 0.001).
7. We retrieved core scores and PWM scores at the Top 0.5% percentile (one-tailed p-value is less than 0.005, see Note 9).
8. We retrieved core scores and PWM scores at the Top 1% percentile (one-tailed p-value is less than 0.01).
9. Using these scores, we tested the 154 OCT4 binding regions (Dataset 1) and 499 regions that were not bound by OCT4 (defined as Dataset 2).
10. A Fisher test was applied and the p-value was used to define the significance measure for this data (see Note 5).

a

b

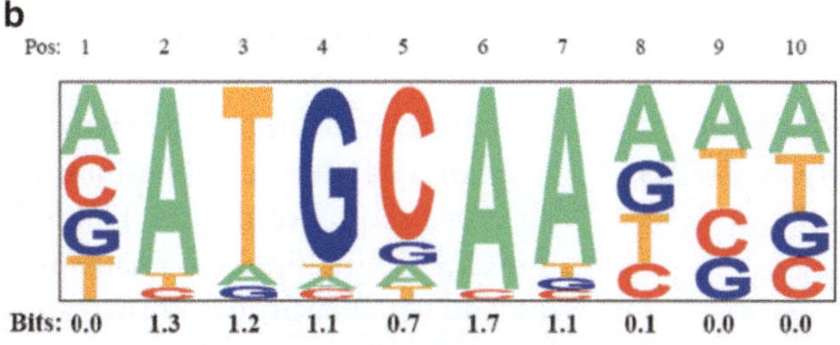

c	Nucleotide Frequency			
Position	A	C	G	T
1	331	247	217	207
2	**881**	**59**	**0**	**62**
3	**92**	**0**	**64**	**846**
4	**53**	**50**	**839**	**60**
5	**100**	**730**	**110**	**62**
6	**954**	**47**	**0**	**1**
7	833	44	57	68
8	363	173	237	229
9	301	231	198	272
10	321	201	209	271

Fig. 2. The OCT4 motif and position frequency matrix. The motif nATGCAAAnn, (**b**) which resembles the OCT4 consensus site, (**a**) of ATGCAAAT was identified. Importantly, our ChIPMotifs analysis provided not only a consensus site, but also a position frequency matrix, (**c**) (OCT4H_PWM) for in vivo OCT4 binding in the above table.

11. We filtered the set by keeping only those motifs that were found in the OCT4 binding sites, but not in the control Dataset 2, which were considered to be over-represented motifs.
12. These motifs have a confidence level at the Top 0.1% percentile and a Fisher test *p*-value less than 0.001. Thus, a *p*-value of 0.00026 for the OCT4H_PWM at the top 0.1% percentile with a core score of 0.88 and PWM score of 0.85 (see Note 10) is considered to be significant, nonsignificant motifs are discarded (Fig. 2).

2.8. The Results for OCT4 Data

The motif **NATGCAAA**NN which resembles the OCT4 consensus site of **ATGCAAA**T (Fig. 2a) was identified. We found that a 0.88 match to the core sequences (S_c) and a 0.85 match to the PWM (S_p) clearly distinguishes the OCT4 dataset from the control set (with a *p*-value of 0.00026) and demonstrates high specificity (eliminating 60% of the fragments in the negative control set) and high sensitivity (capturing ~70% of the binding sites). However, when using 0.88 (S_c) and 0.85 (S_p) criteria, 28.6% of the experimentally determined Oct4 binding regions still lack a match to the OCT4H_PWM.

2.9. In Vivo Motif Discovery for ZNF263

Recent in vitro studies (27) have shown that approximately half of a set of 104 mouse DNA-binding proteins recognized multiple different sequence motifs. Half of all human transcription factors use C2H2 zinc finger domains to specify site-specific DNA binding and yet very little is known about their role in gene regulation. Based on in vitro studies, a zinc finger code has been developed that predicts a binding motif for a particular zinc finger factor (ZNF). However, very few studies have performed genome-wide analyses of ZNF binding patterns, and thus, it is not clear if the binding code developed in vitro will be useful for identifying target genes of a particular ZNF. We performed genome-wide ChIP-seq for ZNF263, a C2H2 ZNF that contains nine finger domains, a KRAB repression domain, and a SCAN domain and identified more than 5,000 binding sites in K562 cells (28). Although ZNFs containing a KRAB domain are thought to function mainly as transcriptional repressors, many of the ZNF263 target genes are expressed at high levels. To address the biological role of ZNF263, we identified genes whose expression was altered by treatment of cells with ZNF263-specific small interfering RNAs. Our results suggest that ZNF263 can have both positive and negative effects on transcriptional regulation of its target genes.

2.10. Methods for ZNF263 Data

1. We identified a set of 1,473 binding sites in common in the two ChIP-seq experiments at the top 0.1% level to derive an in vivo binding motif for ZNF263 (see Note 1).
2. A set of ~24,000 human promoter sequences of 500 bp in length for each promoter from 1,000 bp upstream to the 5′ transcription start site were selected as a negative control data set.
3. Process the input data set in Weeder.
4. Process the input data set in MEME.
5. Process the input data set in MaMF.
6. The union of the output of these two programs is the set of candidate motifs.
7. Construct position weight matrices for each of the candidate motifs.
8. Perform bootstrap re-sampling by randomizing each of 1,473 sequences for 100 times.

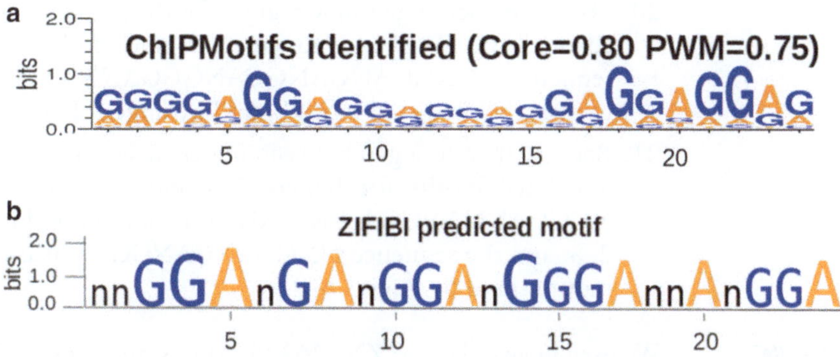

Fig. 3. Comparison of in vivo and in vitro predicted ZNF263 motifs. (**a**) A WebLogo representing the 24 nt experimentally in vivo derived ZNF263 binding site is shown. (**b**) A WebLogo representing the ZNF263 binding site in vitro predicted using the zinc finger code is shown. ZNFs bind in the C-terminal to N-terminal orientation; therefore, the first 12 nt in the motif are those predicted to be bound by fingers 9–6, and the second 12 nt in the motif are those predicted to be bound by fingers 5–2. For searching of the ZNF263 binding sites for the predicted motif, the sequence nnGGAnGAnGGAnGGGAn-nAnGGA was used as the motif bound by fingers 2–9; the sequence nGGGAnnAnGGA was used as the motif bound by fingers 2–5, and the sequence nnGGAnGAnGGA was used as the motif bound by fingers 6–9.

9. Scanned these randomized sequences for each candidate motif (using the PWMs derived from step 11) starting at a minimal core score of 0.5 and a minimal PWM score of 0.5.
10. Retrieve core scores and PWM scores at the Top 0.1% percentile (one-tailed p-value is less than 0.001).
11. Filter these candidate motifs to those which are over-represented in the input set compared to the negative control set.
12. Apply the Fisher test to measure the significance measure for the motifs (see Note 5).
13. Discard nonsignificant motifs, i.e., motifs with a significance of $p > 0.001$, to obtain a significant set of m putative motifs.
14. Feed this set of motifs and their PWMs to STAMP for phylogenetic hierarchical clustering and comparison with TRANSFAC and JASPAR known motifs.
15. STAMP will output the final set of n motifs with significant similarity to known motifs.
16. A de novo ZNF263 motif (Fig. 3a) is then determined.
17. For those ZNF263 binding sites without a good match to the first identified novel ZNF263 motif, ChIPMotifs were further run on these sites, and other known or novel motifs were then determined.
18. To obtain a motif predicted for ZNF263 by the zinc finger code, we used a prediction program ZIFIBI that predicts binding sites for zinc finger domains (see Note 11).
19. We merged the individual triplet predictions to obtain a predicted WebLogo for fingers 2–9 (Fig. 3b).

20. To search a set of genomic regions for the predicted motif, we adapted the WebLogo to create a nucleotide string; the sequence NNGGANGANGGANGGGANNANGGA was used as the predicted motif bound by fingers 2–9.
21. Because there is a gap between fingers 5 and 6, we also made individual motifs for fingers 2–5 and 6–9; the sequence NGGGANNANGGA was used as the motif bound by fingers 2–5, and the sequence NNGGANGANGGA was used as the motif bound by fingers 6–9.

2.11. The Results for ZNF263 Data

We used in vivo derived ZNF263 PWM to scan a set of 5,273 sites identified from the Top 0.5% level from two biological replicates in K562 cells (28). We found that 75% of the 5,273 sites contained a good match (Core/position weight matrix 0.80/0.75) to this motif. We next examined the distribution of this motif in the two largest categories of ZNF263 binding site locations, promoters, and introns. We found that 86% of the 5′ transcription start site category and 73% of the intragenic category contained this site. Therefore, it seems that ZNF263 is recruited to the intragenic sites using the same motif as used in the core promoter regions. Our results suggest that ZNF263 binds to a 24-nt site, Fig. 3a, that differs from the motif predicted by the zinc finger code in several positions. Interestingly, many of the ZNF263 binding sites are located within the transcribed region of the target gene.

3. Notes

1. It is important to use a large enough number of sequences to get statistically significant results from de novo motif discovery. Use at least ten different sequences; however, there are also technical concerns: MEME performs best with less than 2,000 input sequences.
2. The W-ChIPMotifs currently include three ab initio motif programs: MEME, MaMF, and Weeder. We will plan to add more programs in the next version of program.
3. In step 7 of Subheading 2.1, these randomized sequences no longer correspond to binding sites, but have the same nucleotide frequencies as the original binding sites and are therefore used as a negative control set for motif finding.
4. In step 10 of Subheading 2.1, for many experiments there will be no such additional constraints. See Subheading 2.7, step 11 for an example.

5. It is very important to use Bonferroni correction to adjust the *p*-value by multiplying by the number of samples being input in order to reduce inaccuracy from small sample sizes.
6. Common transcription factors with poorly specifies positional weight matrices may show up as matches from STAMP with poor but possibly acceptable *p*-values. Experience and background knowledge are important in interpreting these results.
7. "Newick format" is a common textual representation of a tree graph.
8. In step 4 of Subheading 2.7, these randomized sequences no longer correspond to binding sites, but have the same nucleotide frequencies as the original binding sites and are therefore used as a negative control set for motif finding.
9. In steps 6–8 of Subheading 2.7, allowing too many changes from the consensus motif results in the identification of OCT4 binding sites in the great majority of both datasets, whereas requiring a complete match to the consensus eliminates the majority of the true binding sites.
10. We compute any possible six consecutive nucleotides for the OCT4H_PWM and define the one with a maximum value as a core and the corresponding value as core score, while a sum of the OCT4H_PWM is considered as PWM score.
11. In step 18 of Subheading 2.10, this program predicted motifs for fingers 2–3–4, 3–4–5, 6–7–8, and 7–8–9.

References

1. Lockhart D, Dong H, Byrne MC et al (1996) Expression monitoring by hybridization to high-density oligonucleotide arrays. Nat Biotechnol 14:1675–1680
2. Schena M, Shalon D, Davis RW et al (1995) Quantitative monitoring of gene expression patterns with a complementary DNA microarray. Science 270:467–470
3. Iyer VR, Horak CE, Scafe CS et al (2001) Genomic binding sites of the yeast cell-cycle transcription factor SBF and MBF. Nature 409:533–538
4. Ren B, Robert F, Wyrick JJ et al (2000) Genome-wide location and function of DNA binding proteins. Science 290:2306–2309
5. Steensel B, Henikoff S (2000) Identification of in vivo DNA targets of chromatin proteins using tethered dam methyltransferase. Nat Biotechnol 18:424–428
6. Crawford GE, Davis S, Scacheri PC et al (2006) DNase-chip: a high-resolution method to identify DNase I hypersensitive sites using tiled microarrays. Nat Methods 3:503–509
7. Loh YH, Wu Q, Chew JL et al (2006) The Oct4 and Nanog transcription network regulates pluripotency in mouse embryonic stem cells. Nature Genet 38:431–440
8. Pedersen JT, Moult J (1996) Genetic algorithms for protein structure prediction. Curr Opin Struct Biol 6:227–231
9. Lawrence C, Altschul S, Boguski M et al (1993) Detecting subtle sequence signals: a Gibbs sampling strategy for multiple alignment. Science 262:208–214
10. Bailey TL, Elkan C (1995) The value of prior knowledge in discovering motifs with MEME. Proc Int Conf Intell Syst Mol Biol 3:21–29
11. Pavesi G, Mereghetti P, Mauri G et al (2004) Weeder Web: discovery of transcription factor binding sites in a set of sequences from co-regulated genes. Nucleic Acids Res 32: W199-203

12. Liu J, Stormo GD (2008) Context-dependent DNA recognition code for C2H2 zinc-finger transcription factors. Bioinformatics 24:1850–1857
13. Kel AE, Gossling E, Reuter I et al (2003) MATCH: A tool for searching transcription factor binding sites in DNA sequences. Nucleic Acids Res 31:3576–3579
14. Wingender E, Chen X, Hehl R et al (2000) TRANSFAC: an integrated system for gene expression regulation. Nucleic Acids Res 28:316–319
15. Alkema WB, Johansson O, Lagergren J et al (2004) MSCAN: identification of functional clusters of transcription factor binding sites. Nucleic Acids Res 32:W195-198
16. Sandelin A, Alkema W, Engstrom P et al (2004). JASPAR: an open-access database for eukaryotic transcription factor binding profiles. Nucleic Acids Res 32:D91-94
17. Weinmann AS, Yan PS, Oberley MJ et al (2002) Isolating human transcription factor targets by coupling chromatin immunoprecipitation and CpG island microarray analysis. Gene Dev 16:235–244
18. Barski A, Cuddapah S, Cui K et al (2007) High-resolution profiling of histone methylations in the human genome. Cell 129:823–837
19. Robertson G, Hirst M, Bainbridge M et al (2007) Genome-wide profiles of STAT1 DNA association using chromatin immunoprecipitation and massively parallel sequencing. Nat Methods 4:651–657
20. Ettwiller L, Paten B, Ramialison M et al (2007) Trawler: de novo regulatory motif discovery pipeline for chromatin immunoprecipitation. Nat Methods 4:563–565
21. Gordon DB, Nekludova L, McCallum et al (2005) TAMO: a flexible, object-oriented framework for analyzing transcriptional regulation using DNA-sequence motifs. Bioinformatics 21:3164–3165
22. Hong P, Liu XS, Zhou Q et al (2005) A boosting approach for motif modeling using ChIP-chip data. Bioinformatics 21:2636–2643
23. Jin VX, O'Geen H, Iyengar S et al (2007) Identification of an OCT4 and SRY regulatory module using integrated computational and experimental genomics approaches. Genome Res 17:807–817
24. Jin VX, Apostolos J, Nagisetty NS et al (2009) W-ChIPMotifs: a web application tool for de novo motif discovery from ChIP-based high-throughput data. Bioinformatics 25: 3191–3193
25. Jin VX, Leu YW, Liyanarachchi S et al (2004) Identifying estrogen receptor alpha target genes using integrated computational genomics and chromatin immunoprecipitation microarray. Nucleic Acids Res 32:6627–6635
26. Mahony S, Benos PV (2007) STAMP: a web tool for exploring DNA-binding motif similarities. Nucleic Acids Res 35:W253-258
27. Badis G, Berger MF, Philippakis AA et al (2009) Diversity and complexity in DNA recognition by transcription factors. Science 324:1720–1723
28. Frietze S, Lan X, Jin VX et al (2010) Genomic targets of the KRAB and SCAN domain-containing zinc finger protein 263 (ZNF263). J Biol Chem 285:1393–1403

Part V

Emerging Applications of Microarray and Next Generation Sequencing

Chapter 22

Hidden Markov Models for Controlling False Discovery Rate in Genome-Wide Association Analysis

Zhi Wei

Abstract

Genome-wide association studies (GWAS) have shown notable success in identifying susceptibility genetic variants of common and complex diseases. To date, the analytical methods of published GWAS have largely been limited to single single nucleotide polymorphism (SNP) or SNP–SNP pair analysis, coupled with multiplicity control using the Bonferroni procedure to control family wise error rate (FWER). However, since SNPs in typical GWAS are in linkage disequilibrium, simple Bonferonni correction is usually over conservative and therefore leads to a loss of efficiency. In addition, controlling FWER may be too stringent for GWAS where the number of SNPs to be tested is enormous. It is more desirable to control the false discovery rate (FDR). We introduce here a hidden Markov model (HMM)-based PLIS testing procedure for GWAS. It captures SNP dependency by an HMM, and based which, provides precise FDR control for identifying susceptibility loci.

Key words: Genome-wide association, SNP, Hidden Markov model, False discovery rate, EM algorithm, Multiple tests

1. Introduction

Genome-wide association studies (GWAS), interrogating the architecture of whole genomes by single nucleotide polymorphism (SNP), have shown notable success in identifying susceptibility genetic variants of common and complex diseases (1). Unlike traditional linkage and candidate gene association studies, GWAS have enabled human geneticists to examine a wide range of complex phenotypes, and have allowed the confirmation and replication of previously unsuspected susceptibility loci. GWAS typically test hundreds of thousands of markers simultaneously. To date, the analytical methods of published GWAS have largely been limited to single SNP or SNP–SNP pair analysis, coupled with multiplicity control using the Bonferroni procedure to control family wise error rate (FWER), the probability of having at least one false

Junbai Wang et al. (eds.), *Next Generation Microarray Bioinformatics: Methods and Protocols*,
Methods in Molecular Biology, vol. 802, DOI 10.1007/978-1-61779-400-1_22, © Springer Science+Business Media, LLC 2012

positive out of all loci claimed to be significant. However, since SNPs in typical GWAS are in linkage disequilibrium (LD), simple Bonferonni correction is usually over conservative and therefore leads to a loss of efficiency. Furthermore, the power of a FWER controlling procedure is greatly reduced as the number of tests increases. In GWAS, the number of SNPs is enormous and the number of susceptibility loci can be large for many complex traits and for common diseases, it is more desirable to control the false discovery rate (FDR) (2), the expected proportion of false positives among all loci claimed to be significant (3).

We have developed a hidden Markov model (HMM) to capture SNP local LD dependency, and based on which, proposed a FDR controlling procedure for identifying disease-associated SNPs (4). Under our model, the association inference at a particular SNP will theoretically combine information from all typed SNPs on the same chromosome, although the influence of these SNPs decreases with increasing distance from the locus of interest. The SNPs rankings based on our procedure are different from the rankings based on *p*-values of conventional single SNP association tests. We have shown that our HMM-based PLIS (pooled local index of significance) procedure has a significantly higher sensitivity of identifying susceptibility loci than conventional single SNP association tests (4). In addition, GWAS is often criticized for its poor reproducibility in that a large proportion of SNPs claimed to be significant in one GWAS are not significant in another GWAS for the same diseases. Compared to single SNP analysis, our procedure also yields better reproducibility of GWAS findings (4). We introduce here how to conduct genome-wide association analysis using our HMM-based PLIS testing procedure.

2. Materials

Case–control GWAS compare the DNA of two groups of participants: samples with the disease (cases) and *comparable* samples without (controls). Cases are readily obtained and can be efficiently genotyped and compared with control populations. The selection of controls should be careful because any systematic allele frequency differences between cases and controls can appear as disease association. Controls should be comparable with cases as much as possible, so that their DNA differences are not caused by the results of evolutionary or migratory history, gender differences, mating practices, or other independent processes, but are only coupled with differences in disease frequency (5). All DNA samples of the cohort (cases and controls) are genotyped for a large number of genome-wide SNPs using high-throughput SNP arrays, for example,

550,000 SNPs on the Illumina HumanHap550 array (Illumina, San Diego, CA, USA). A sample dataset and the program to implement the analysis introduced in this chapter can be downloaded from the author's Web site (6).

3. Methods

We use HMM to characterize the dependency among neighboring SNPs. In our HMM, each SNP has two hidden states: disease-associated or nondisease-associated, and the states of all SNPs along a chromosome are assumed to follow a Markov chain with a normal mixture model as the conditional density function for the observed genotypes.

Suppose there are n_1 cases and n_2 controls being genotyped over the m SNPs on a chromosome. We first conduct single SNP association tests for each SNP to assess the association between the allele frequencies and the disease status. We then transform the association significance p-values to z-values $Z(z_1, \ldots, z_m)$ for further analysis (as detailed in step 4 of Subheading 3.1). Let $\theta(\theta_1, \ldots, \theta_m)$ be the underlying states of the SNP sequence in the chromosome from the $5'$ end to the $3'$ end, where $\theta_i = 1$ indicates that SNP i is disease-associated and $\theta_i = 0$ is nondisease-associated. We assume that θ is distributed as a stationary Markov chain with transition probability $a_{ss'} = \Pr(\theta_i = s'|\theta_{i-1} = s)$ and the stationary distribution $\pi(1 - \pi_1, \pi_1)$, where π_1 represents the proportion of disease-associated SNPs. We model $f(z_i|\theta_i) \sim (1 - \theta_i)F_0 + \theta_i F_1$. We assume that for nondisease-associated SNPs, the z-value distribution is standard normal $F_0 = N(0, 1)$, and for disease-associated SNPs, the z-value distribution is a L-component normal mixture $F_1 = \sum_{l=1}^{L} w_l N(\mu_l, \sigma_l^2)$. The normal mixture model can approximate a large collection of distributions and has been widely used. When the number of components in the normal mixture L is known, the maximum likelihood estimate (MLE) of the HMM parameters can be obtained using the EM algorithm (7, 8). When L is unknown, we use the Bayesian information criterion (BIC) (9) to select an appropriate L.

After HMM model fitting using the EM algorithm, we can calculate for each SNP the local index significance (LIS) score, defined as $\text{LIS}_i = \text{Prob}(\theta_i = 0|z)$, the probability that a SNP is nondisease-associated given the observed data (z-values of all SNPs in the same chromosome). We will fit each chromosome by a separate and independent HMM and obtain the LIS statistics for all SNPs, which will be used by our PLIS procedure for selecting disease-associated SNPs with FDR control.

Table 1
Genotype counts and allele frequencies for a hypothetic SNP

Genotype	Count
AA	30
Aa	55
aa	15
Total	100
Allele	Frequency
A	0.575 (P_A)
a	0.425 (P_a)

The whole detailed procedure for genome-wide association analysis is outlined as follows.

3.1. Obtain SNP Association p-Values, Odds Ratios, and z-Values by Single SNP Analysis

1. We first perform a series of standard quality control procedures to eliminate problematic markers that are not good for association analysis. We remove any SNPs with minor allele frequency less than 1% or with genotype call rate smaller than 5%.
2. Hardy–Weinberg Disequilibrium (10, 11) may suggest genotyping errors or, in samples of affected individuals, an association between the marker and disease susceptibility. Therefore, we also exclude markers that fail the Hardy–Weinberg equilibrium (HWE) test in controls at a specified significance threshold 10^{-6}. The HWE test is performed using a simple χ^2 goodness-of-fit test (see Note 1), and for case–control samples in GWAS, this test will be based on *controls* only. Here is an example of how to do a HWE test. Suppose that a hypothetic SNP has the genotype counts and allele frequencies in the *control* samples as shown in Table 1. Under HWE, the expected genotype counts for AA, Aa, and aa are $(P_A^2, 2P_A P_a, P_a^2)\times$ Total count, respectively. We can calculate the χ^2value $= \sum_i (O_i - E_i)^2/E_i$ as shown in Table 2. Since we are testing HWE with two alleles, this test statistic has a "chi-square" distribution with 1 degree of freedom. It can be shown that under 1 degree of freedom chi-square distribution be $\Pr(\chi^2 \geq 23.928) \leq 10^{-6}$. Therefore, for any SNPs with $\chi^2 \geq 23.928$, we will exclude it for further analysis as it significantly deviates from HWE in controls. In the given example, the resultant χ^2 value $1.50 < 23.928$, implies no evidence for Hardy–Weinberg disequilibrium, so we will keep it.

Table 2
An example of χ^2 value calculation

Genotype	Observed	Expected	$\frac{(O-E)^2}{E}$
AA	30	33	0.27
Aa	55	49	0.73
aa	15	18	0.50
Total	100	100	1.50

Table 3
Allele counts in case and control for a hypothetic SNP

Allele	Control	Case	Total
A	115	80	195
a	85	40	125
Total	200	120	320

3. For the remaining SNPs that survive the above quality control, we calculate their disease association significant p-values using basic allelic test (χ^2 test with 1 degree of freedom, see Note 2) and odds ratio. Continuing with the previous hypothetic SNP example, suppose we have its *observed* allele counts in the case and control samples as shown in Table 3. As $P_A = 195/320 = 0.61$ and $P_a = 125/320 = 0.39$, if the two alleles A and a distribute the same in controls and cases, namely, they are not associated with sample status, then the expected counts for alleles A and a are $P_A \times 200 = 122$ and $P_a \times 200 = 78$, respectively, for controls; and $P_A \times 120 = 73.2$ and $P_a \times 120 = 46.8$, respectively, for cases. So we can calculate its χ^2 value as $(115-122)^2/122 + (85-78)^2/78 + (80-73.2)^2/73.2 + (40-4.8)^2/46.8 = 2.65$. By 1 degree freedom of χ^2 distribution, its association significance p-value will be $\Pr(\chi^2 \geq 2.65) = 0.104$. Its odds ratio (case–control) can be easily computed as $(80/40)/(115/85) = 1.48$.
4. Transform p-values to z-values using the following formula,

$$z = \begin{cases} \Phi^{-1}\left(1-\frac{p}{2}\right), \textit{oddsratio}>1, \\ \Phi^{-1}\left(\frac{p}{2}\right), \text{otherwise}, \end{cases}$$

where Φ indicates the standard normal cumulative distribution function. Continuing with the previous hypothetic SNP example with p-value 0.104 and odds 1.48, we have its z-value as $\Phi^{-1}(1-(0.104/2)) = 1.626$ (see Note 3).

3.2. HMM-Based PLIS Procedure for Identifying Disease-Associated Loci

Given the z-values from the previous single SNP analysis step, now we fit an HMM for each chromosome using an EM algorithm and apply the PLIS procedure for selecting disease-associated SNPs with FDR control. For each chromosome, arrange the z-values in the order of their corresponding SNPs' chromosome positions. Assume that there are L components in the normal mixture $F_1 = \sum_{l=1}^{L} w_l N(\mu_l, \sigma_l^2)$ for the disease-associated SNPs in each chromosome. The nominal FDR level we want to control is α. The HMM-based PLIS procedure is outlined as follows.

1. Initialize transition probabilities $a_{00} = 0.95$ and $a_{11} = 0.5$, and the stationary distribution $(1-\pi_1, \pi_1) = (1-10^{-5}, 10^{-5})$; each component $\sim N(\mu_i = 1.5 \times (i-1) - 1, 1)$, with weight $w_l = 1/L$, $i = 1, \ldots, L$ (see Note 4).
2. Iterate the E-step and the M-step until converged (see Note 5).
3. Calculate the BIC score for the converged model

$$\mathrm{BIC} = \log\left(\Pr\left(\widehat{\Psi_{\mathrm{L}}}|Z\right)\right) - \frac{\left|\widehat{\Psi_{\mathrm{L}}}\right|}{2}\log(m),$$

where $\Pr\left(\widehat{\Psi_{\mathrm{L}}}|Z\right)$ is the likelihood function, $\widehat{\Psi_{\mathrm{L}}}$ is the MLE of HMM parameters, and $\left|\widehat{\Psi_{\mathrm{L}}}\right|$ is the number of HMM parameters, and m is the number of SNPs in that fitted chromosome. We have $L \times 2$ parameters for the L normal components $N(\mu_l, \sigma_l^2)$, $(L-1)$ for their weights, 1 for the stationary distribution (π_1), and 2 for the transition probabilities $(a_{00}\mathrm{and}a_{11})$. So $\left|\widehat{\Psi_{\mathrm{L}}}\right| = L \times 2 + \mathrm{L} - 1 + 1 + 2 = 3L + 2$.
4. Repeat the above procedure for $L = 2, \ldots, 6$ (see Note 6).
5. Select L with the highest BIC score and the corresponding converged HMM model as the final model.
6. Calculate LIS statistics for each SNP based on the selected converged HMM model. The standard forward–backward algorithm (12) for HMM will be used to compute $\mathrm{LIS}_i = \mathrm{Prob}(\theta_i = 0|z; \widehat{\Psi_{\mathrm{L}}})$.
7. Repeat the above steps 1–6 for each chromosome and have LIS statistics for SNPs from *all* chromosomes (see Note 7).
8. Combine and rank the LIS statistic from *all* chromosomes. Denote by $\mathrm{LIS}_{(1)}, \ldots, \mathrm{LIS}_{(p)}$ the ordered values, and $H_{(1)}, \ldots, H_{(p)}$ the corresponding SNPs. Find k such that $k = \max\left\{i : (1/i)\sum_{j=1}^{i} \mathrm{LIS}_{(j)} \le \alpha, 0\right\}$.

9. If $k > 0$, claim SNPs $H_{(1)}, \ldots, H_{(k)}$ as disease-associated, and the nominal FDR level is controlled at α; otherwise ($k = 0$) claim no SNPs are disease-associated under FDR level α.

4. Notes

1. HWE can also be tested using an exact test, described and implemented by Wigginton et al. (13), which is more accurate for rare genotypes.
2. We can also use Fisher's exact test (14) to generate association significance, which is more applicable when sample sizes are small.
3. We may have very small p-values, e.g., 1E − 20. It should be paid attention that without sufficient precision, it may be approximated as 0 and leads to infinites when transformed to z-values. One possible solution is to do all intermediate transformations in log-scale.
4. The initial value π_1 represents the proportion of disease-associated SNPs in a chromosome. The transition probabilities a_{00}anda_{11} represent the likelihood of SNP state changing from nondisease-associated to nondisease-associated ($0 \rightarrow 0$) and disease-associated to disease-associated ($1 \rightarrow 1$), respectively. We may use different proper values for different chromosomes as determined by related genetic domain knowledge. For example, chromosome 6 has a higher (expected) number of disease susceptibility loci then we can set π_1 to be a higher value. We represent positively and negatively associated SNPs by the signs of z-values, as bisected by odds ratio. Because of the (expected) existence of both susceptibility and protective loci, we include into the normal mixture a negative and a positive initial normal component with the initial μ values of -1.5 and 0.5. Other negative and positive pairs can also be tried.
5. Since EM algorithm does only local optimization, we may try different initial values and select the ones with the highest likelihood.
6. Based on our experience, a two- or three-component normal mixture model is sufficient in most situations for GWAS, i.e., $L = 2$ or 3. Occasionally we observe four-component normal mixture ($L = 4$) but rarely $L > 4$. If not considering computational cost, we may try as large L as we want, though not necessary.
7. The HMM fitting program is the most time-consuming part. It takes a few hours for analyzing one chromosome using a

computer equipped with Intel® Xeon® Processor 5160 3.00 GHz and memory 8 GB. But the program can be executed in parallel for different chromosomes so as to save time for genome-wide analysis (all chromosomes).

References

1. McCarthy MI, Abecasis GR, Cardon LR et al (2008) Genome-wide association studies for complex traits: consensus, uncertainty and challenges. Nat Rev Genet 9:356–69.
2. Sabatti C, Service S, Freimer N (2003) False discovery rate in linkage and association genome screens for complex disorders. Genetics 164:829–833.
3. Benjamini Y, Hochberg Y (1995) Controlling the False Discovery Rate: A Practical and Powerful Approach to Multiple Testing. Journal of the Royal Statistical Society. Series B (Methodological) 57:289–300.
4. Wei Z, Sun W, Wang K et al (2009) Multiple testing in genome-wide association studies via hidden Markov models. Bioinformatics 25:2802–2808.
5. Cardon LR, Bell JI (2001) Association study designs for complex diseases. Nat Rev Genet 2:91–9.
6. http://web.njit.edu/~zhiwei/hmm/
7. Ephraim Y, Merhav N (2002) Hidden Markov processes. IEEE transactions on Information Theory 48:1518–1569.
8. Sun W, Cai TT (2009) Large-scale multiple testing under dependence. Journal Of The Royal Statistical Society Series B 71:393–424.
9. Schwarz G (1978) Estimating the dimension of a model. Ann. Statist. 6:461–464.
10. Hardy GH (1908) Mendelian Proportions in a Mixed Population. Science 28:49–50.
11. Weinberg W (1908) Über den Nachweis der Vererbung beim Menschen. Jahresh Wuertt Ver vaterl Natkd 64:368–382.
12. Rabiner LR (1989) A tutorial on hidden markov models and selected applications in speech recognition. In *Proceedings of the IEEE*, p.257–286.
13. Wigginton JE, Cutler DJ, Abecasis GR (2005) A note on exact tests of Hardy-Weinberg equilibrium. Am J Hum Genet 76:887–893.
14. Fisher RA (1932) Statistical Methods for Research Workers. Oliver & Boyd, Edinburgh

Chapter 23

Employing Gene Set Top Scoring Pairs to Identify Deregulated Pathway-Signatures in Dilated Cardiomyopathy from Integrated Microarray Gene Expression Data

Aik Choon Tan

Abstract

It is well accepted that a set of genes must act in concert to drive various cellular processes. However, under different biological phenotypes, not all the members of a gene set will participate in a biological process. Hence, it is useful to construct a discriminative classifier by focusing on the core members (subset) of a highly informative gene set. Such analyses can reveal which of those subsets from the same gene set correspond to different biological phenotypes. In this study, we propose *G*ene *S*et *T*op *S*coring *P*airs (*GSTSP*) approach that exploits the simple yet powerful relative expression reversal concept at the gene set levels to achieve these goals. To illustrate the usefulness of *GSTSP*, we applied this method to five different human heart failure gene expression data sets. We take advantage of the direct data integration feature in the *GSTSP* approach to combine two data sets, identify a discriminative gene set from >190 predefined gene sets, and evaluate the predictive power of the *GSTSP* classifier derived from this informative gene set on three independent test sets (79.31% in test accuracy). The discriminative gene pairs identified in this study may provide new biological understanding on the disturbed pathways that are involved in the development of heart failure. *GSTSP* methodology is general in purpose and is applicable to a variety of phenotypic classification problems using gene expression data.

Key words: Gene set analysis, Top scoring pairs, Relative expression classifier, Microarray, Gene expression

1. Introduction

Functional genomics technologies such as expression profiling using microarrays provide a global approach to understanding cellular processes in different biological phenotypes. Microarray technologies have been applied to a wide range of biological problems and have yielded success in the identification of new biomarkers and disease subtypes for better disease treatments. Identifying and relating candidate genes and their relationships

Junbai Wang et al. (eds.), *Next Generation Microarray Bioinformatics: Methods and Protocols*,
Methods in Molecular Biology, vol. 802, DOI 10.1007/978-1-61779-400-1_23,

to each other in the biological context remains the challenges in the analysis of gene expression data. Much of the initial work has focused on the development of tools for identifying differentially expressed genes using a variety of statistical confidence. These analyses typically reveal large numbers of genes ranging from hundreds to thousands with altered expression. Mining through such large gene lists in order to identify "candidate genes" that participate in disease development and progression represents a challenging task in functional genomics. An expert is required to examine the gene list and select those genes that are correlated with a disease state or represent the activity of a known molecular mechanism (e.g., biological process), based on the availability of functional annotations and one's own knowledge (1, 2). While useful, these are ad hoc approaches, are subjective, and tend to exhibit bias in their analyses (2). Furthermore, the lists of "candidate genes" identified from various studies have little overlap between them (3), questioning their validity. This is partly due to ad hoc biases and limited sample sizes in gene expression studies (large *P* small *N* problem).

Recently, several computational methods have improved the ability to identify candidate genes that are correlated with a disease state by exploiting the idea that gene expression alterations might be revealed at the level of biological pathways or co-regulated gene sets, rather than at the level of individual genes (1, 4–6). Such approaches are more objective and robust in their ability to discover sets of coordinated differentially expressed genes among pathway members and their association to a specific biological phenotype. These analyses may provide new insights linking biological phenotypes to their underlying molecular mechanisms, as well as suggesting new hypotheses about pathway membership and connectivity. However, under different disease phenotypes, not all the members of a gene set will participate in a biological process. Hence, it is useful to construct a discriminative classifier by focusing on the core members (subset) of an informative gene set. Such analyses can reveal which of those subsets from the same gene set correspond to different biological phenotypes. Using this information, core gene members from a biological process that were systematically altered from one biological phenotype to another can be identified.

Here, we present a novel data-driven machine learning method, *G*ene *S*et *T*op *S*coring *P*airs (*GSTSP*), to achieve the above-mentioned goals. *GSTSP* relates the results in the biological context (e.g., pathways) about genes and their relationships to each other with respect to different biological phenotypes, based on the *relative expression reversals* of gene pairs. In this study, we apply *GSTSP* to the analysis of human heart failure gene expression profiles. Heart failure (HF) is a progressive and complex clinical syndrome that affects 4.9 million people in the USA, and 550,000 new cases are diagnosed each year (7). Dilated cardiomyopathy (DCM) is a common cause of this cardiac disease and is primarily characterized by the development and progression of left ventricular (LV) remodeling, specifically dilatation of the LV and dysfunction of the myocardium, leading to the inability of the cardiac pump to support the energy requirements of the body (8, 9). Several inherited and environmental factors can initiate dilatation of the LV by disrupting various cellular pathways, leading to the development of DCM. As a dynamic system, the heart initially responds to these perturbations by altering its gene expression pattern (compensated stage). The heart undergoes physiological "remodeling" during this period and the long-term effects of these changes prove to be harmful, triggering a different set of cellular processes which eventually lead to progression of the heart failure phenotype (8, 10, 11). It is necessary to improve our understanding of the disrupted molecular pathways that are involved in the development of heart failure, as the details of the molecular mechanisms that are involved remain unclear.

2. Methods

2.1. The Relative Expression Reversal *Learning Method*

In prior work, we have implemented the *relative expression reversal* learning method as a *T*op *S*coring *P*air (*TSP*) classifier (12) and a *k*-*T*op *S*coring disjoint *P*airs (*k-TSP*) classifier (13). The *k-TSP* classifier uses exactly *k* top disjoint gene pairs for classifying gene expression data. When $k = 1$, this algorithm, referred to simply as *TSP*, selects a unique pair of genes. We demonstrated that the *TSP* and *k-TSP* methods can generate simple and accurate decision rules by classifying 19 different sets of cancer gene expression profiling data (13). Furthermore, the performance of the *k-TSP* classifier is comparable to PAM (predictive analysis of microarray (14)) and support vector machines, and outperforms other classical machine learning methods (decision trees, naïve Bayes classifier, and *k*-nearest neighbor classifier) on these human cancer gene expression data sets (13). The *TSP* classifier and its variants are *rank-based*, meaning that the decision rules only depend on the relative ordering of the gene expression values within each profile. Due to the *rank-based* property, these methods can be applied directly to integrate data generated from different studies and to perform cross-platform analysis without performing any normalization of the underlying data (15, 16).

The *k-TSP* method is implemented as follows. Let the gene expression training data set *S* be a $P \times N$ matrix $\mathbf{X} = [\mathbf{x}_{p,n}]$, $p = 1, 2, \ldots, P$ and $n = 1, 2, \ldots, N$, where *P* is the number of genes in a profile and *N* is the number of samples (profiles).

Each sample has a class label of either C_1 (DCM) or C_2 (NF for mRNA isolated from nonfailing human heart). For simplicity, let nC_1 and nC_2 be the number of examples in C_1 and C_2, respectively. Expression values of the P genes are then ordered (most highly expressed, second most highly expressed, etc.) within each fixed profile. Let $R_{i,n}$ denote the rank of the ith gene in the nth array (profile). Replacing the expression values $x_{i,n}$ by their ranks $R_{i,n}$ results in a new data matrix $\mathbf{R}$ in which each column is a permutation of $\{1, \ldots, P\}$.

The learning strategy for the *k-TSP* classifier is to exploit discriminating information contained in the $\mathbf{R}$ matrix by focusing on marker gene pairs (i, j), for which there is a significant difference in the probability of the event $\{R_i < R_j\}$ across the N samples from class C_1 to C_2. For every pair of genes $i, j \in \{1, \ldots, P\}$, $i \neq j$, compute $p_{ij}(C_m) = \text{Prob}(R_i < R_j | Y = C_m)$, $m = \{1, 2\}$, i.e., the probabilities of observing $R_i < R_j$ (equivalently, $x_i < x_j$) in each class. These probabilities are estimated by the relative frequencies of occurrences of $R_i < R_j$ within profiles and over samples. Next, define Δ_{ij} as the "score" for each gene pair (i, j), where $\Delta_{ij} = -p_{ij}(C_1) - p_{ij}(C_2)|$, and identify pairs of genes with high scores Δ_{ij}. Such pairs are the most informative for classification. Define a "rank score" Γ_{ij} for each pair of genes (i, j) that incorporates a measure of the gene expression level inverted from one class to the other within a pair of genes (13). Sorting of each gene pair (i, j), first according to the score Δ_{ij} and then by the rank score Γ_{ij}, yields a set of k top scoring gene pairs. The prediction of the *k-TSP* classifier is based on the majority voting scheme of these k top scoring gene pairs. Details of the *k-TSP* algorithm can be found in ref. 13.

2.2. Overview of the GSTSP Approach

The *GSTSP* approach builds on the advantages of *k-TSP* strategy and performs learning at the gene set level. Given the gene expression profiles from two different biological states and a set of M a priori defined gene sets, the *GSTSP* performs the following steps:

Step 1. Calculation of the Gene Set enrichment score. For each gene set GS_m, $m = 1, \ldots, M$, a *k-TSP* classifier with $k = 1$ is constructed and the TSP score $(\Delta_{\max})_m$ is recorded. We defined the score $(\Delta_{\max})_m$ as the enrichment score for gene set m. This step generates a list of $(\Delta_{\max})_m$ scores. GS_m with the highest score $(\Delta_{\max})_m$ is selected as the most *enriched* gene set GS_{enriched} for the classification problem. If ties occur (i.e., if more than one single gene set is identified), then the gene set with the lowest cross-validation error rate is selected as the most *enriched* gene set.

Step 2. Construction of the GSTSP classifier. Given the most *enriched* gene set GS_{enriched}, the *GSTSP* classifier is constructed using the *k-TSP* algorithm (13) on this gene set. Let Q denotes the number of genes in the *enriched* gene set GS_{enriched}, where $Q \ll P$.

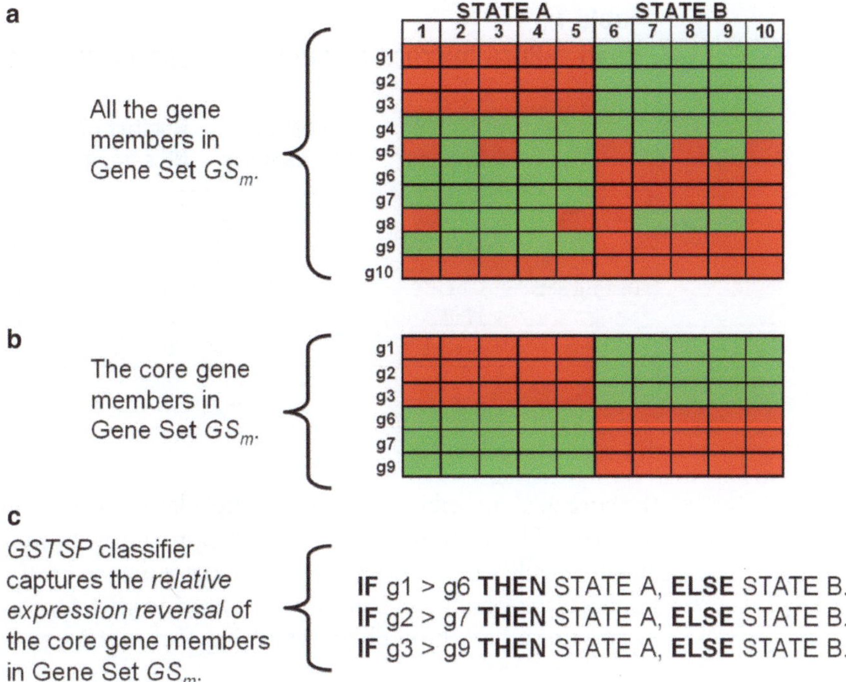

Fig. 1. Overview of the *GSTSP* approach. (**a**) Gene expression profiles of all the gene members (g1, g2, . . ., g10) in gene set GS_m under two different biological states (A and B). Each *row* corresponds to a gene and each *column* corresponds to a sample array. The expression level of each gene is represented by *red* (upregulation) or *green* (downregulation) in the sample array. (**b**) The core gene members in gene set GS_m showed the expression levels of the genes *reversed* from state A to state B. The core gene members are informative features in distinguishing state A from state B. (**c**) The goal of the *GSTSP* approach is to construct a classifier that automatically captures the core gene members from a list of predefined gene sets. The *GSTSP* classifier generates *IF–ELSE* rules in describing the relationships of the gene pair for each biological state.

The *k-TSP* algorithm returns the top *k* disjoint pairs as the *GSTSP* classifier for the *enriched* gene set $GS_{enriched}$.

The idea of the *GSTSP* approach is illustrated by the following example. Given five expression profiles from each of the two different biological states (A and B) and an informative gene set GS_m (Fig. 1a) of ten gene members (g1, g2, . . ., g10), the *GSTSP* approach is ideally suited for finding gene pairs where their relative expression levels are reversed to one another from State A to B in the gene set GS_m. Genes (g1, g2, g3, g6, g7, g9) represent the core gene members in gene set GS_m as their relative expression levels can be used as informative features in distinguishing state A from B (Fig. 1b). Genes that have little or no observed changes (g4 and g10), and those that are randomly expressed (g5 and g8) in all the states, are uninformative features in the classification problem. A gene set GS_m is considered *enriched* or informative if

Table 1
Heart failure microarray data sets used in this study

	Data set	Number of DCM samples	Number of NF samples	References
Training sets	Yung	12	10	(11)
	Harvard	27	14	(20)
Testing sets	Chen	7	0	(17)
	Hall	8	0	(18)
	Kittleson	8	6	(19)

the core gene members exhibited many relative expression reversal patterns. The output of the *GSTSP* approach is a classifier that automatically captures these "*relative expression reversal*" patterns between the core gene members of the gene set GS_m in discriminating these biological phenotypes (Fig. 1c) (see Note 1).

2.3. Other Information

2.3.1. Microarray Data

Gene expression profiles from human DCM and NF were collected from five different published data sets, where each used Affymetrix oligonucleotide microarray technology. Four of the data sets were generated from Affymetrix U133A array with 22,215 probe sets (11, 17–19), while the other one was collected from an Affymetrix U133 Plus 2.0 array consisting of 54,675 probe sets (20). Probe sets of the Affymetrix U133A array represent a subset of the Affymetrix U133 Plus 2.0 array probe sets. In this study, we focused on the analysis of the 22,215 probe sets common to both arrays. Table 1 summarizes the data sets used in this study.

2.3.2. Data Integration

Owing to the limited availability of human heart failure microarray data, it is very unlikely to generate a robust classifier due to the small size of the training sample set. In this study, we integrated the Yung and Harvard data sets to increase the training set sample size. The integrated data set (Yung–Harvard) consists of 63 samples (39 DCM and 24 NF). The direct data integration capability of the *TSP* and its variants allowed the integrated data set (Yung–Harvard) to be applied directly by our learning methods without any normalization procedure (15, 16).

2.3.3. Compilation of Pathway Gene Sets

We analyzed 193 gene sets consisting of pathways defined by public databases. First, we downloaded human pathway annotations from KEGG (Release 32.07m07) (21) and GenMAPP (Hs-Contributed-20041216 version, March 2005) (22) databases. We mapped the pathway annotations to Affymetrix HG-U133A probe sets using the gene symbols available from Affymetrix Web site (April 2005). Pathways that have less than five gene members in a set were removed from this analysis. We also manually combined gene sets that overlapped between KEGG and GenMAPP annotations, based on literature reviews. The final gene sets included 126 sets from KEGG, 61 sets from GenMAPP, and 6 from manually combined pathways.

2.3.4. Estimation of Classification Rate

We performed Leave-One-Out Cross-Validation (LOOCV) to estimate the classification rate of the training data listed in Table 1. In LOOCV, for each sample $\mathbf{x}_n$ in the training set S, we train a classifier based on the remaining $N - 1$ samples in S and use that classifier to predict the label of $\mathbf{x}_n$. The LOOCV estimate of the classification rate is the fraction of the N samples that are correctly classified.

2.3.5. Classification Measurements on Independent Test Sets

We trained the classifiers on the training set and evaluated their performance on the independent test sets. We measured the classifiers' accuracy ($\text{Acc} = (\text{TP} + \text{TN})/N$), sensitivity ($S_\text{n} = \text{TP}/nC_1$), specificity ($S_\text{p} = \text{TN}/nC_2$) and precision ($\text{Prec} = \text{TP}/(\text{TP} + \text{FP})$) on the independent test set, where TP, TN, and FP are the number of correctly classified samples from C_1, number of correctly classified samples from C_2, and the number of incorrectly classified samples from C_2, respectively. We also computed the F_1-measure (23) of each classifier that combines sensitivity and precision into a single efficiency measure, $F_1 = (2 \times S_\text{n} \times \text{Prec})/(S_\text{n} + \text{Prec})$. The F_1-measure represents the harmonic mean of the sensitivity and precision, and it has a value between 0 and 1, where a higher value (close to 1) represents a better classifier.

2.3.6. Significance Analysis of Microarray

Tusher et al. (24) introduced significance analysis of microarray (SAM) method that scores genes with statistically significant changes in expression by assimilating a set of gene-specific t tests. A score is assigned to each gene, based on its expression change relative to its standard deviation gene expression across experiments (profiles). Genes with scores greater than a threshold (based on false-discover rate, FDR, and q-value of permutation tests) are selected as potentially significant. SAM is currently the most popular method for analyzing differential gene expression (24).

2.3.7. Gene Set Enrichment Analysis

Gene set enrichment analysis (GSEA) (6) is a computational method that employs statistical significance tests to determine if a given gene set is enriched in a biological phenotype gene expression profile. The idea of GSEA is to evaluate microarray data at the level of gene sets (defined based on prior biological knowledge), coupled with a weighted Kolmogorov–Smirnov-like statistic to calculate its enrichment score (ES). GSEA employs phenotype-based permutation test to estimate the statistical significance

(P-value) of the ES, taking into account multiple hypothesis testing by calculating the false discovery rate (FDR). In this study, we used GSEA desktop application v1.0. We performed 1,000 permutation tests on the integrated data (Yung–Harvard) to assess the enrichment of these gene sets.

2.4. Effects of Data Integration on k-TSP Classifiers

The first experiment of this study is to investigate the effect of increased training sample size by direct data integration using the *k-TSP* method. In this experiment, we compared the classifiers generated from two single data sets in Table 1 (Yung and Harvard) and the combined data set (Yung–Harvard). We also generated 100 permutated data sets of the same size as the integrated data set (Random) by shuffling the actual class labels and maintaining the expression values. We trained *k-TSP* classifiers from these permutated data sets to obtain the null distribution of the classifier's performance on this increased sample size. The random results are presented as mean ± SD. We performed statistical analysis using the single-tailed Z-test where a P-value < 0.05 was accepted as statistically significant compared to the random classifiers. The results for this experiment are presented in Table 2.

From this experiment, we observed that the increased sample size in training data improved the classifiers' LOOCV accuracies (Table 2). In Table 2, the classifier trained on the integrated data set (Yung–Harvard) achieved the highest accuracy in both LOOCV (93.7%) and independent test set (72.41%). Furthermore, the Yung–Harvard classifier achieved the highest F_1-measure, outperforming classifiers induced from individual training set and random data sets. Although the *k-TSP* classifiers trained on Yung, Harvard and Yung–Harvard data sets are statistically significant in LOOCV accuracies, their prediction accuracies and F_1-measures on independent test set are not statistically significant when compared to the random classifiers (P-values > 0.05). These results suggest that it is more likely to overfit a classifier when training with a limited number of samples and a large number of features.

2.5. Effects of Incorporating Gene Sets Information on GSTSP Classifiers

The second experiment is to evaluate the effect of using gene sets to define a priori with the *GSTSP* method. We applied the *GSTSP* algorithm to individual training data sets in Table 1, the integrated data set (Yung–Harvard), and the permutation data sets (Random) as described previously. The Random results are presented as mean ± SD. We performed statistical analysis using single-tailed Z-test where a P-value < 0.05 was accepted as statistically significant compared to the random classifiers. Table 3 summarizes the results of this experiment.

By incorporating the gene set information to the training set, the classifiers' prediction accuracies of Yung, Harvard, and

Table 2
***k-TSP* classifiers' performance on using all genes**

	Sample size	LOOCV	Independent test sets				
		Acc (%)	Acc (%)	S_n (%)	S_p (%)	*Prec* (%)	F_1-measure
Yung	22	**86.40**	55.17	43.48	100.00	100.00	0.6061
Harvard	41	**90.20**	44.18	34.78	100.00	100.00	0.5161
Yung–Harvard	63	**93.70**	72.41	69.57	83.33	94.12	0.8000
Random (100×)	63	51.47 ± 17.76	54.83 ± 14.24	57.22 ± 24.67	46.57 ± 45.42	84.14 ± 13.13	0.6352 ± 0.1838

Results shown in bold are statistically significant than the random classifiers (*P*-value < 0.05)

Table 3
Results for *GSTSP* classifiers

		LOOCV	Independent test sets				
	Gene set #	Acc (%)	Acc (%)	S_n (%)	S_p (%)	*Prec* (%)	F_1-measure
Yung	190	**95.50**	72.41	65.22	100.00	100.00	0.7895
Harvard	103	82.90	75.86	95.65	0.00	78.57	0.8627
Yung–Harvard	127	**84.10**	**79.31**	86.96	50.00	86.96	**0.8696**
Random (100×)	Varies	70.20 ± 7.24	54.03 ± 15.27	57.09 ± 24.48	42.33 ± 45.59	81.76 ± 15.63	0.6308 ± 0.1930

Results shown in bold are statistically significant compared to the random classifiers (*P*-value < 0.05)

Yung–Harvard on the independent test set were improved, except for the random classifiers (Table 3). The Yung–Harvard classifier identified Gene Set #127 as the *enriched* set that performed statistically better than the random classifiers on LOOCV accuracy, test accuracy, and the F_1-measure (P-values < 0.05). Although the test accuracies for *GSTSP* classifiers generated from the Yung and Harvard data set were improved over the classifiers trained on all of the genes, the F_1-measures are not statistically significant when compared to the random *GSTSP* classifiers (P-values > 0.05). The *GSTSP* classifier generated from Yung–Harvard is more robust than the classifiers constructed from Yung or Harvard alone, as it achieved statistically significant prediction accuracy and F_1-measure on the independent test sets. This result shows that the gene set selected by the *GSTSP* approach is correlated to the biological phenotypes of DCM and NF. This result also confirms the findings in ref. 15 that an advantage of the *k-TSP* classifier is that it enables direct data integration across studies, thus providing a larger sample size from which to learn a more robust and accurate *relative expression reversal* classifier.

2.6. Statistical Significance of the Gene Set Identified bythe GSTSP Classifier

We next asked whether the gene set identified by the *GSTSP* classifier is statistically significant, as compared to any random gene sets. *GSTSP* classifier constructed from Yung–Harvard data has identified Gene Set #127 as the most enriched gene set in distinguishing DCM from NF samples (Table 3). Gene Set #127 represents the Cardiac-Ca^{2+}-cycling gene set, with 777 gene members involved in ATP generation and utilization regulated by Ca^{2+} in the cardiac myocyte. We performed the following permutation test to evaluate the statistical significance of this *enriched* gene set. First, we randomly grouped 777 (out of 22,215) genes from the training data to form a random gene set. Next, we constructed a *GSTSP* classifier from this random gene set, and assessed its prediction accuracy on the test set. We repeated this procedure 2,000 times to obtain the prediction accuracy enriched by these random gene sets (the null distribution). Finally, we performed statistical analysis using single-tailed Z-test where a P-value < 0.05 was accepted as statistically significant compared to the random gene sets. The results from this experiment show that the Cardiac-Ca^{2+}-cycling gene set identified by the *GSTSP* approach is significantly enriched in classifying DCM and NF samples (P-value < 0.05).

2.7. GSTSP Classifier for Distinguishing DCM from NF Samples

In this study, the *GSTSP* classifier constructed from the integrated data sets consists of seven pairs of genes derived from the Cardiac-Ca^{2+}-cycling gene set (Fig. 2). These 14 genes are regulated by intracellular Ca^{2+} cycling and they all involved in ATP generation and utilization in the cardiac myocyte. The *GSTSP* classifier can be easily translated into a simple set of IF–ELSE decision rules. For

a

IF ATP5I ≥ MYH6 **THEN** DCM, **ELSE** NF.
IF ATP6V1C1 ≥ ATP2A2 **THEN** DCM, **ELSE** NF.
IF CAMK2B ≥ TMP4 **THEN** DCM, **ELSE** NF.
IF PRKACA ≥ P2RX5 **THEN** DCM, **ELSE** NF.
IF ATPIF1 ≥ ATP1B3 **THEN** DCM, **ELSE** NF.
IF NDUFS1 ≥ OGDH **THEN** DCM, **ELSE** NF.
IF NDUFB5 ≥ ATP1B1 **THEN** DCM, **ELSE** NF.

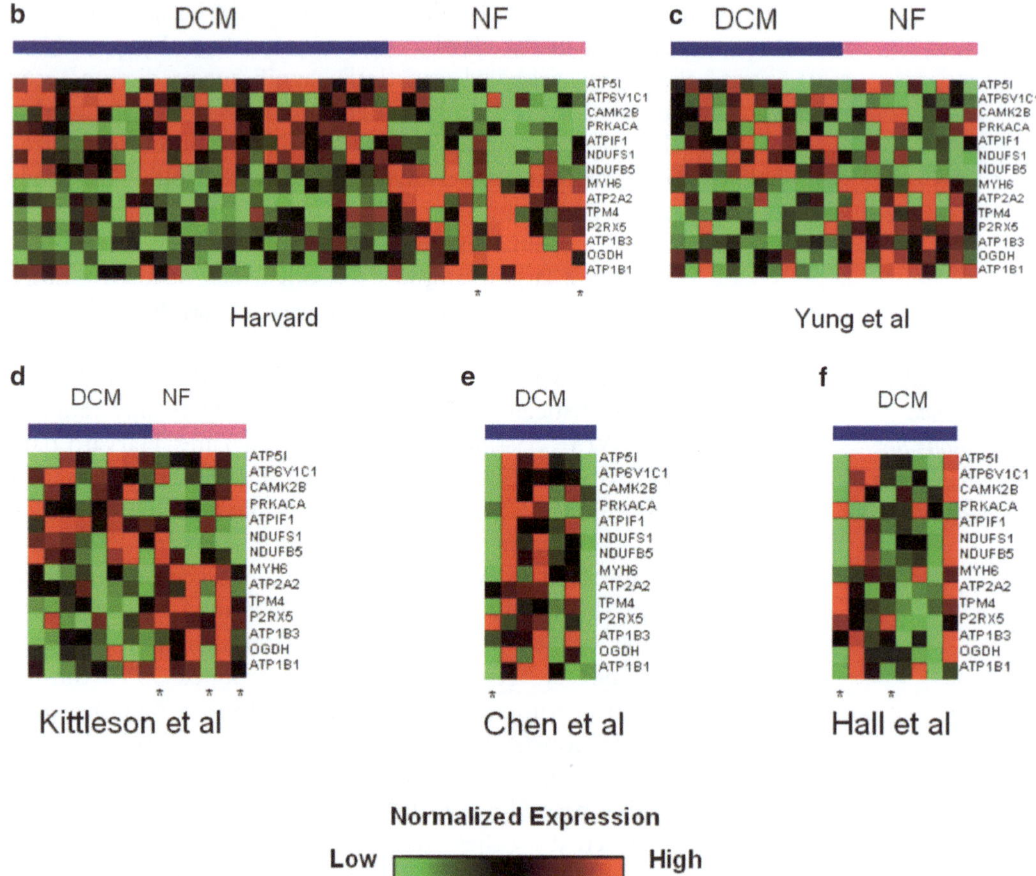

Fig. 2. The *GSTSP* classifier for distinguishing DCM from NF samples. (**a**) Decision rules for *GSTSP* classifier. Heat maps of genes that distinguish DCM from NF from the Cardiac-Ca^{2+}-cycling gene set for Harvard (**b**), Yung et al. (**c**), Kittleson et al. (**d**), Chen et al. (**e**) and Hall et al. (**f**) data sets. (**b–f**) The *blue* and *pink panels* denote the DCM and NF samples, respectively. *Row* and *columns* in the heatmap correspond to genes and samples, respectively. The expression level for each gene is normalized across the samples such that the mean is 0 and the standard deviation (SD) is 1. Genes with expression levels greater than the mean are colored in *red* and those below the mean are colored in *green*. The scale indicates the number of SDs above or below the mean. Columns labeled with an *asterisk* (*) were misclassified by the *GSTSP* classifier.

example, the corresponding decision rule for the first gene pair of the classifier (ATP5I, MYH6) is:

IF ATP5I ≥ MYH6 *THEN* DCM; *ELSE* NF.

In words: *if the expression of ATP5I is greater than or equal to MYH6, then the sample is classified as DCM, otherwise it is NF.* Since the *GSTSP* classifier contains more than one decision rule, the final prediction of the new sample is based on the majority votes from these seven rules. The order of the decision rules in the classifier is based on the consistency and differential magnitude between the gene pairs in the training samples. Figure 2 illustrates the heat map and the decision rules of these genes in training and testing data sets.

2.8. Biological Significance and Experimental Supports for the Genes Identified by the GSTSP *Classifier*

Here we provide the biological significance of the genes selected by the *GSTSP* classifier in discriminating DCM from NF samples. Genes identified by *GSTSP* classifier are from the cardiac calcium cycling gene set and they are involved in ATP utilization processes (myosin ATPase and ion channels/pumps), ATP generation pathways (tricarboxylic acid (TCA) cycle and oxidative phosphorylation), and β-adrenergic receptor signaling pathway. These pathways have direct influence on myocyte excitation–contraction–relaxation mechanisms, all of which are regulated by intracellular Ca^{2+} cycling. The expression changes of these genes are supported by published experimental results, suggesting that the alteration mechanism of the ATP generation and utilizing processes regulated by intracellular Ca^{2+} cycling have direct correlation to the development of human heart disease. In the DCM (heart failure) phenotype, the gene expression of major ATP consumers (myosin ATPase and ion channels/pumps) is downregulated, while the expression of several ATP synthase genes is upregulated. This may suggest that in heart failure, the heart is under an "energy starvation" state (lack of ATP), where the ATP generated from the mitochondria is insufficient to sustain the energy needs of the myocyte (9, 25) (see Note 2).

2.9. Validation of the List of Significant Differentially Expressed Genes

One of the limitations in analyzing human heart failure gene expression data is the difficulty in collecting heart tissue samples. The size of human heart samples is considered *small* when compared to the collection of human cancer samples. Hence, it is not surprising that most of the results reported from analysis of human heart expression data contain hundreds (11) or thousands (26) of significantly expressed genes. Here we applied SAM (24) to identify genes that are significantly expressed in each individual training data set in Table 1. Out of 22,215 gene probes, SAM identified 5,907 genes in the Yung data set and 7,266 genes in the Harvard data set that have more than 1.2-fold change in expression. The direct approach to assess the common differentially expressed genes between each set is to look for overlap in the corresponding data sets using a Venn diagram, as illustrated in Fig. 3. There are 2,127 genes that overlap between the two sets. In a conventional microarray analysis, sifting through this gene list (>2,000 genes) represents a daunting task for any biologist.

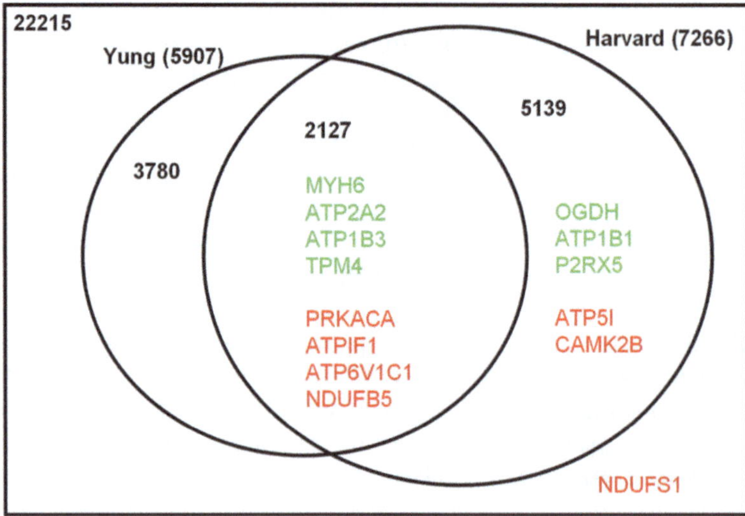

Fig. 3. SAM analysis of gene expression data with fold change ≥1.2. Gene names in the figure represent genes that have been identified by the *GSTSP* classifier. *Red* and *green* color represents upregulation and downregulation, respectively, for that gene under DCM condition.

By using the *GSTSP* classifier, trained from the integrated (Yung–Harvard) data sets, we have identified seven gene pairs for distinguishing DCM from NF samples. Thirteen of these genes have more than 1.2-fold change of expression, as identified by SAM; and eight of them overlap between the two training data sets (Fig. 3). This analysis indicates that gene pairs in which their *relative expressions* are *reversed* from DCM to NF states make up the *GSTSP* classifier. In addition, the decision rules generated by the classifier are simpler (14 genes) and easy to interpret when compared to the SAM outputs, facilitating follow-up study on these genes (see Note 4). We also compared the SAM outputs with a published data set (see Note 3).

2.10. Validation of the Core Gene Members by GSEA Analysis

To validate that the genes selected by the *GSTSP* method are the core members that contribute to the enrichment in distinguishing DCM from NF, we performed the GSEA analysis on this gene set against the compilation pathway gene sets. Based on the statistical analysis of GSEA, 90 gene sets had enrichment in DCM, only 33 of them are significant at a nominal *P*-value < 0.05 and only 1 gene set (*GSTSP-DCM*) is significant at FDR < 0.25. For the NF phenotype, there are 104 enrichment gene sets, 38 gene sets are significant at a nominal *P*-value < 0.05 and only one gene set (*GSTSP-NF*) is significant at FDR < 0.25. The *GSTSP-DCM* gene set is significantly enriched in the DCM phenotype (*P*-value = 0, FDR = 0.001) while the gene set *GSTSP-NF* is significantly enriched in the NF samples (*P*-value = 0, FDR = 0.093). Using

GSEA, we found that the genes identified by the *GSTSP* classifier were significantly enriched in DCM versus NF. The GSEA analysis provides additional support for the enrichment of the *GSTSP* in classifying the human heart failure microarray data (see Note 5).

3. Notes

1. *Concept of GSTSP classifier:* We present a computational method that is based on the concept of *relative expression reversal* coupled with gene set information to identifying discriminative and biological meaningful gene pairs from integrated data sets.
2. *Statistical and biological validation of the GSTSP classifier:* The *GSTSP* classifier is robust and accurate when tested on independent interstudy test sets. The classifier is also simple and easy to interpret. Furthermore, the identified gene pairs have been confirmed by published experimental results showing that they are significantly differentially expressed in DCM and NF phenotypes. The gene set that *enriched* the gene pairs classifier is involved in ATP generation and utilization in the myocyte regulated by intracellular Ca^{2+} cycling.
3. *Comparing differentially expressed gene list with published data:* Margulies et al. (26) performed a large-scale gene expression analysis on 199 human myocardial samples from nonfailing, failing, and LV assist device-supported human hearts using the Affymetrix microarray platform. To date, their study represents one of the largest microarray analyses on human heart samples. Unfortunately, their data is not publicly available, and therefore is not included in this study. The only way to crosscheck our results with theirs is by comparing the gene list provided in their online supplements. When we compared the genes identified by the *GSTSP* classifier with their 3,088 gene list (26), 12 genes were listed in their gene list with more than 1.2-fold differential expression. This result provides additional support that the *GSTSP* approach identifies genes with differential expression that differs significantly between DCM and NF states.
4. *Gene set analysis:* The *GSTSP* approach shares the same spirit with recent computational approaches using gene set concept (4–6, 27) in analyzing microarray data. The gene pairs are easy to interpret, involving a small number of core gene members of the *enriched* pathway. These results illustrate the value of analyzing complex processes in terms of higher-level gene modules and biological processes. This type of analysis increases our ability to identify the signal in microarray data

and provides results that are easier to interpret than gene lists. The *GSTSP* methodology is general in purpose and is applicable to a variety of phenotypic classification problems using gene expression data.

5. *Summary:* The results from these experiments are twofold: first, the gene set selected by the *GSTSP* approach in this study is correlated to the biological phenotypes of DCM and NF; and second, it highlights the importance of integrating multiple data sets to train a robust classifier.

References

1. Mootha VK, Lindgren CM, Eriksson K-F et al (2003) PGC-1alpha-responsive genes involved in oxidative phosphorylation are coordinately downregulated in human diabetes. Nature Genetics 34:267–273.
2. Winslow RL, Gao Z (2005) Candidate gene discovery in cardiovascular disease Circ Res 96:605–606.
3. Sharma UC, Pokharel S, Evelo CTA et al (2005) A systematic review of large scale and heterogeneous gene array data in heart failure. J Mol Cell Cardiol 38: 425–432.
4. Rhodes DR, Chinnaiyan AM (2005) Integrative analysis of the cancer transcriptome. Nature Genetics 37:S31-S37.
5. Segal E, Friedman N, Kaminski N et al (2005) From signatures to models: understanding cancer using microarrays. Nature Genetics 37:S38-S45.
6. Subramanian A, Tamayo P, Mootha VK et al (2005) Gene Set Enrichment Analysis: a knowledge-based approach for interpreting genome-wide expression profiles. Proc Natl Acad Sci U S A 102:15545–15550.
7. AHA. (2005) Heart Disease and Stroke Statistics - 2005 Update. American Heart Association.
8. Liew CC, Dzau VJ (2004) Molecular genetics and genomics of heart failure. Nature Reviews Genetics 5:811–825.
9. Ventura-Clapier R, Garnier A, Veksler V (2004) Energy metabolism in heart failure. Journal of Physiology 555:1–13.
10. Barrans JD, Allen PD, Stamatiou D et al (2002) Global gene expression profiling of end-stage dilated cardiomyopathy using a human cardiovascular-based cDNA microarray. American Journal of Pathology 160:2035–2043.
11. Yung CK, Halperin VL, Tomaselli GF et al (2004) Gene expression profiles in end-stage human idiopathic dilated cardiomyopathy: altered expression of apoptotic and cytoskeletal genes. Genomics 83:281–297.
12. Geman D, d'Avignon C, Naiman DQ et al (2004) Classifying gene expression profiles from pairwise mRNA comparisons. Statistical Applications in Genetics and Molecular Biology 3:Article 19.
13. Tan AC, Naiman DQ, Xu L et al (2005) Simple decision rules for classifying human cancers from gene expression profiles. Bioinformatics 21:3896–3904.
14. Tibshirani R, Hastie T, Narasimhan B et al (2003) Class prediction by nearest shrunken centroids, with applications to dna microarrays. Statistical Science 18:104–117.
15. Xu L, Tan AC, Naiman DQ et al (2005) Robust prostate cancer marker genes emerge from direct integration of inter-study microarray data. Bioinformatics 21: 3905–3911.
16. Xu L, Tan AC, Winslow RL et al (2008) Merging microarray data from separate breast cancer studies provides a robust prognostic test. BMC Bioinformatics 9:125.
17. Chen YJ, Park S, Li Y et al (2003) Alterations of gene expression in failing myocardium following left ventricular assist device support. Physiology Genomics 14:251–260.
18. Hall JL, Grindle S, Han X et al (2004) Genomic profiling of the human heart before and after mechanical support with a ventricular assist device reveals alterations in vascular signaling networks. Physiology Genomics 17:283–291.
19. Kittleson MM, Ye SQ, Irizarry RA et al (2004) Identification of a gene expression profile that differentiates between ischemic and nonischemic cardiomyopathy. Circulation 110:3444–3451.
20. Harvard. (2005) Genomics of Cardiovascular Development, Adaptation, and Remodeling. NHLBI Program for Genomic Applications,

Harvard Medical School. URL: http://www.cardiogenomics.org.

21. Kanehisa M, Goto S, Kawashima S et al (2004) The KEGG resource for deciphering the genome. Nucleic Acids Research 32:D277-D280.
22. Dahlquist KD, Salomonis N, Vranizan K et al (2002) GenMAPP: a new tool for viewing and analyzing microarray data on biological pathways. Nature Genetics 31:19–20.
23. van Rijsbergen CJ (1979) *Information Retrieval*, 2nd ed., Butterworths.
24. Tusher VG, Tibshirani R, Chu G (2001) Significance analysis of microarrays applied to the ionizing radiation response. PNAS 98:5116–5121.
25. Sanoudou D, Vafiadaki E, Arvanitis DA et al (2005) Array lessons from the heart: focus on the genome and transcriptome of cardiomyopathies. Phyisology Genomics 21:131–143.
26. Margulies KB, Matiwala S, Cornejo C et al (2005) Mixed messages: transcription patterns in failing and recovering human myocardium. Circ Res 96:592–599.
27. Rhodes DR, Kalyana-Sundaram S, Mahavisno V et al (2005) Mining for regulatory programs in the cancer transcriptome. Nature Genetics 37:579–583.

Chapter 24

JAMIE: A Software Tool for Jointly Analyzing Multiple ChIP-chip Experiments

Hao Wu and Hongkai Ji

Abstract

Chromatin immunoprecipitation followed by genome tiling array hybridization (ChIP-chip) is a powerful approach to map transcription factor binding sites (TFBSs). Similar to other high-throughput genomic technologies, ChIP-chip often produces noisy data. Distinguishing signals from noise in these data is challenging. ChIP-chip data in public databases are rapidly growing. It is becoming more and more common that scientists can find multiple data sets for the same transcription factor in different biological contexts or data for different transcription factors in the same biological context. When these related experiments are analyzed together, binding site detection can be improved by borrowing information across data sets. This chapter introduces a computational tool JAMIE for *J*ointly *A*nalyzing *M*ultiple Ch*I*P-chip *E*xperiments. JAMIE is based on a hierarchical mixture model, and it is implemented as an R package. Simulation and real data studies have shown that it can significantly increase sensitivity and specificity of TFBS detection compared to existing algorithms. The purpose of this chapter is to describe how the JAMIE package can be used to perform the integrative data analysis.

Key words: Tiling array, ChIP-chip, Transcription factor binding site, Data integration

1. Introduction

ChIP-chip is a powerful approach to study protein–DNA interactions (1). The technology has been widely used to create genome-wide transcription factor (TF) binding profiles (2, 3). Similar to other microarray technologies, ChIP-chip often produces noisy data. The low signal-to-noise ratio (SNR) can cause low sensitivity and specificity of transcription factor binding site (TFBS) detection. ChIP-chip data in public databases (e.g., the NCBI Gene Expression Omnibus (4)) are rapidly growing. With the enormous amounts of public data, scientists can now easily find multiple data sets for the same TF, possibly collected from different biological contexts, or data for different TFs but in the same biological context. When such multiple data sets are available, one

Junbai Wang et al. (eds.), *Next Generation Microarray Bioinformatics: Methods and Protocols*,
Methods in Molecular Biology, vol. 802, DOI 10.1007/978-1-61779-400-1_24,

can combine information across data sets to improve statistical inference. This is very useful if the data of primary interest is noisy and additional information from other experiments is required to distinguish signals from noise. For this reason, there is an increasing need for statistical and computational tools to support integrative analysis of multiple ChIP-chip experiments.

1.1. A Motivating Example

The advantage of integrative data analysis can be illustrated by Fig. 1a. The figure shows ChIP-chip data from four experiments (GEO accession no.: GSE11062 (5); GSE17682 (6)). The data were generated by two different laboratories to study transcription factors Gli1 and Gli3. Both TFs belong to the Gli family of transcription factors and recognize the same DNA motif TGGGTGGTC. Their binding sites were profiled using Affymetrix Mouse Promoter 1.0R arrays in three different cell types (Limb: developing limb; Med: medulloblastoma; GNP: granule neuron precursor). The plot displays log 2 ratios of normalized ChIP and control probe intensities for each data set in a genomic region on chromosome 6.

A visual examination suggests that the "Gli1_Limb" data set has a low SNR. This is likely due to an unoptimized ChIP protocol and use of a mixed cell population which dilutes the biological signal. Importantly, the figure also shows that "peaks" (i.e., binding sites) from different data sets are correlated, that is, they tend to occur at the same genomic loci. The observed similarities

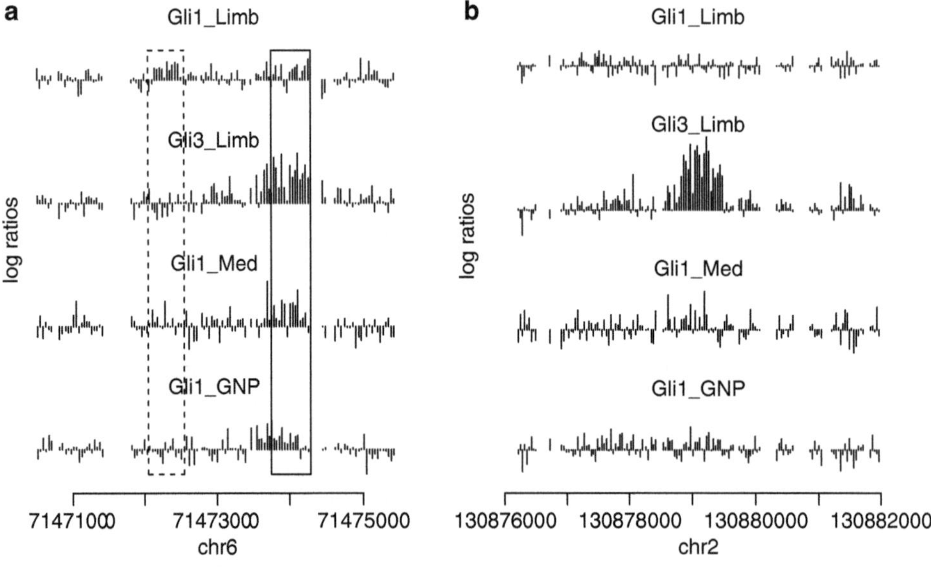

Fig. 1. Motivation of JAMIE. (**a**) Four Gli ChIP-chip datasets show co-occurrence of binding sites at the same genomic locus. This correlation may help distinguish real and false TFBSs. Each bar in the plot corresponds to a probe. Height of the bar is the log 2 ratio between IP and control intensities. (**b**) An example that shows context dependency of TF–DNA binding. The figure is reproduced from ref. 7 with permission from Oxford University Press.

among data sets can be utilized to improve peak detection. For instance, the small peak highlighted in the solid box in the "Gli1_Limb" data set cannot be easily distinguished from background if this data set is analyzed alone. However, when all data sets are analyzed together, presence of strong signals at the same location in the other three data sets strongly indicates that the weak peak in "Gli1_Limb" is a real binding site. In contrast, the peak in the dashed box has about the same magnitude in "Gli1_Limb," but it is less likely to be a real binding site since no binding signal is observed in the other data sets.

To conduct integrative data analysis, one should keep in mind that the protein–DNA interactions can be condition-dependent. In Fig. 1b, for instance, the signal in the "Gli3_Limb" data set is strong enough to be called as a binding site regardless of what happens in the other data sets. However, this peak is likely to be specific to "Gli3_Limb." One should avoid calling peaks from "Gli1_Limb" and "Gli1_GNP" only because there is a strong peak in "Gli3_Limb." Ideally, there should be a mechanism that automatically integrates and weighs different pieces of information, and ranks peaks according to the combined evidence. This cannot be easily achieved by analyzing each dataset separately and taking unions/intersections of the reported peaks.

In order to have a data integration tool that allows context-specific TF–DNA binding, we have proposed a hierarchical mixture model JAMIE for *J*ointly *A*nalyzing *M*ultiple related Ch*I*P-chip *E*xperiments (7). The algorithm is implemented as an add-on package for R which is a popular statistical programming language (8). Previously, a number of software tools have been developed for analyzing ChIP-chip data (e.g., Tiling Analysis Software (TAS) (9), MAT (10), TileMap (11), HGMM (12), Mpeak (13), Tilescope (14), Ringo (15), BAC (16), and DSAT (17), etc.). These tools, however, are all designed for analyzing one data set at a time. A recently developed HHMM approach (18) can be used to jointly analyzing one ChIP-chip data set with one related ChIP-seq data set. However, it is difficult to generalize this method to handle multiple data sets, since its parameter number grows exponentially with the number of data sets. Compared to these tools, JAMIE allows one to simultaneously handle multiple data sets and take full advantage of the data to improve the analysis. The number of parameters in JAMIE increases linearly with the data set number. As a result, the algorithm scales well with the increasing data set number. The model behind JAMIE can be generalized to analyzing multiple ChIP-seq data sets. This generalization is beyond the scope of this chapter and will not be discussed here. The statistical model used by JAMIE will be briefly reviewed in Subheading 2. Readers are referred to ref. 7 to learn the technical details of the model and its implementation. Subheading 3 briefly introduces the JAMIE software. The procedure in which the software is used to analyze data is described in Subheading 2.

1.2. JAMIE Model

JAMIE uses a hierarchical mixture model to capture the correlations among data sets (Fig. 2). The model is based on a concept called "potential binding region" (PBR). A PBR is a genomic region that can be potentially bound by the TFs of interest. Whether it is actually bound is dataset dependent. JAMIE assumes that protein–DNA binding can only occur within the PBRs. More precisely, it is assumed that any arbitrary L base pair (bp) long window has a prior probability π to become a PBR. Alternatively, it has probability $1 - \pi$ to become background. Let B_i (=1 or 0) indicate whether the ith window is a PBR or not. If window i is a PBR, then in data set d it can either become an active binding region with prior probability q_d, or remains silent (i.e., background) with probability $1 - q_d$. Let A_{id} (=1 or 0) indicate whether the window is actively bound by the TF in data set d or not. Conditional on $B_i = 1$, A_{id}s are assumed to be independent. The ChIP-chip probe intensities $\mathbf{Y}_i$ (normalized and $\log_2$ transformed) in a window are assumed to be generated according to the actual binding status of the window. If there is no active

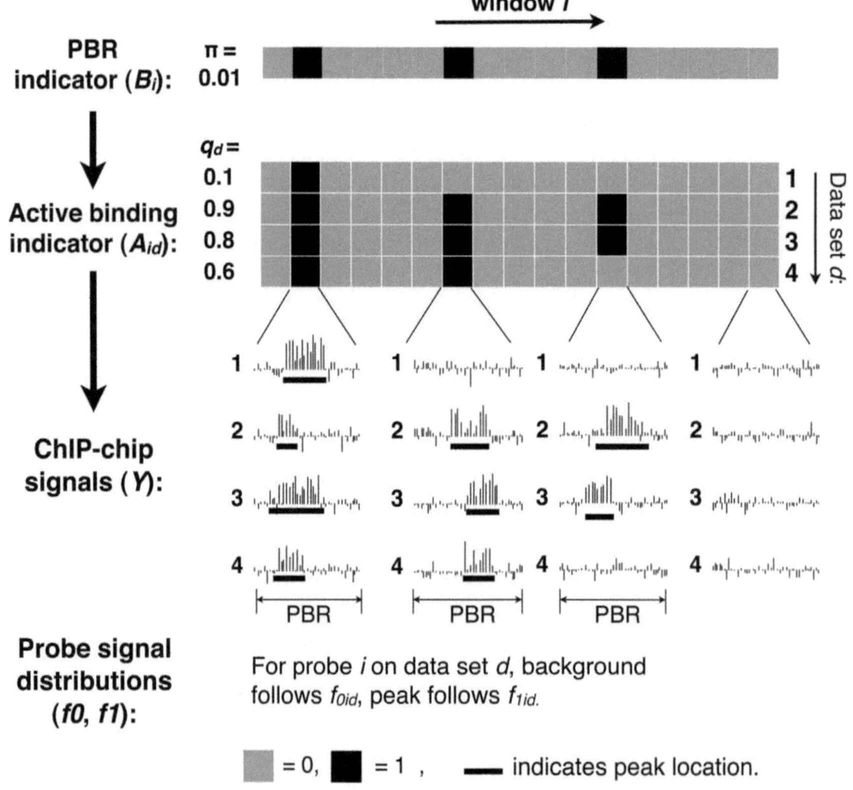

Fig. 2. An illustration of the JAMIE hierarchical mixture model. The figure is reproduced from ref. 7 with permission from Oxford University Press.

binding (i.e., $A_{id} = 0$), the intensities in window i and data set d are assumed to be independently drawn from a background distribution f_0. If there is active binding (i.e., $A_{id} = 1$), then the window will contain a peak (i.e., binding site). Instead of forcing the peak to occupy the whole window, JAMIE assumes that the peak can have several possible lengths and can start at any position within the window. The allowable peak lengths are denoted by $\mathbf{W}$ (e.g. {500, 600,..., 1,000} bps). The peak start and peak length have to satisfy the constraint that the peak is fully covered by the PBR. For a particular PBR and data set in which the PBR is active, all possible peak (start, length) configurations that meet this constraint can occur with an equal prior probability. This assumption allows one to model multiple TFs that bind to the same promoter or enhancer region but recognize different DNA motifs. The probe intensities within the peaks are assumed to be independently drawn from a distribution f_1. All the other probes, including those in background windows ($B_i = 0$), in PBRs but in a silent data set ($B_i = 1$ but $A_{id} = 0$), and in active PBRs ($B_i = 1$ and $A_{id} = 1$) but not covered by a peak, follow distribution f_0.

Let $\mathbf{A}_i$ denote the collection of all indictors A_{id} in window i. Let $\boldsymbol{\Theta}$ be the parameters including π, q_ds, L, $\mathbf{W}$, and parameters that specify f_0 and f_1. Given the parameters $\boldsymbol{\Theta}$, one can derive the joint probability of $\mathbf{Y}_i$, $\mathbf{A}_i$, and B_i, denoted by $P(\mathbf{Y}_i, \mathbf{A}_i, B_i|\boldsymbol{\Theta})$. In reality, only the probe intensities $\mathbf{Y}_i$ are observed. The parameters $\boldsymbol{\Theta}$ are unknown except for L and $\mathbf{W}$ which are configured by users. The problem of interest is to infer the true values of $\mathbf{A}_i$ and B_i which are also unknown. JAMIE employs a two-step algorithm to solve this problem. First, a fast algorithm tailored from TileMap (11) is used to analyze each data set separately to quickly identify potential TF binding regions. Using these candidate regions, an Expectation–Maximization (EM) algorithm (19) is developed to estimate $\boldsymbol{\Theta}$. Second, given $\boldsymbol{\Theta}$, JAMIE uses an L bp window to scan the genome. For each window, it first computes the posterior probability that the window is a PBR, $P(B_i = 1|\mathbf{Y}_i, \boldsymbol{\Theta})$, using the Bayes law. It then infers whether or not the PBR is active in data set d based on the posterior probability:

$$\begin{aligned} P(A_{id} = 1|\mathbf{Y}_i, \Theta) &= P(A_{id} = 1, B_i = 1|\mathbf{Y}_i, \Theta) \\ &= P(A_{id} = 1|B_i = 1, \mathbf{Y}_i, \Theta) \\ &\quad \times P(B_i = 1|\mathbf{Y}_i, \Theta). \end{aligned} \tag{1}$$

This probability has two components. The first component $P(A_{id} = 1|B_i = 1, \mathbf{Y}_i, \boldsymbol{\Theta})$ depends only on information in data set d due to the assumption that A_{id}s are independent conditional on $B_i = 1$. The second component $P(B_i = 1|\mathbf{Y}_i, \boldsymbol{\Theta})$ is the posterior probability that window i is a PBR given all the data, and it depends on information from all data sets. From this decomposition, it is clear that JAMIE uses information from other data sets to weigh information from dataset d in order to determine whether window i is actively bound by the TF in dataset d or not. For each data set, windows with $P(A_{id} = 1|\mathbf{Y}_i, \boldsymbol{\Theta})$ bigger than a user chosen cutoff will be selected, and overlapping windows will be merged. Peaks within the selected window will be identified and reported as the final binding regions.

Simulation and real data tests in ref. 7 have demonstrated that JAMIE performs either better than or comparable to MAT (10) and TileMap (11), two popular ChIP-chip analysis tools, in a variety of data sets. Both MAT and TileMap analyze individual data sets separately. Peaks reported by JAMIE usually have better ranking when benchmarked using the DNA motif enrichment or leave-one out consistency test (7). The results have also shown that the gain can be substantial in noisy data sets, consistent with the expectation that pooling information across data sets will help most when individual data sets have limited amounts of information. When using JAMIE, one should keep in mind that it is based on a number of model assumptions, and if the data dramatically violate these assumptions, the performance is not guaranteed to improve (see Note 1 for a discussion).

1.3. Software

JAMIE has been implemented as an add-on package for R (version 2.10) which is a freely available statistical programming language (8). The package has been tested on different operating systems including Red Hat Enterprise Linux Server release 5.4 (Tikanga), Windows XP/7, and Mac OS 10.6.3 (Snow leopard). It has been tested under R versions 2.8 or higher. Users might encounter problems in other operating systems or older versions of R. Compared to some existing methods, JAMIE requires more computation. However, as most of the engine functions are written in C, JAMIE provides reasonable computational performance. In a test run involving four data sets, each with 3 IP, 3 control, and 3.8 million probes, the whole process took around 15 min on a PC running Linux with 2.2 GHz CPU and 4 G RAM. The source codes and Windows binary package can be downloaded from ref. 20.

2. Methods

This section describes how to install and use JAMIE to analyze multiple related ChIP-chip experiments.

2.1. Installation

JAMIE shall be installed using the standard R package installation procedure. Briefly, one first installs R, perl, latex, and gcc (or g++) on the computer, and then edits the system's environment variable PATH to include the paths of the executable files of these programs (see Note 2). Download JAME (e.g., jamie_0.91.tar.gz), and enter the folder that contains the downloaded file. Type the following command will install JAMIE.

```
> R CMD INSTALL -l /path/to/library jamie_0.91.tar.gz
```

Here, "/path/to/library" is the folder name where the R packages are installed. To learn more about installing R packages, readers should refer to the R installation manual at (21).

JAMIE depends on two Bioconductor packages "affy" and "affyparser" to read and parse BPMAP and CEL files from Affymetrix arrays. These packages need to be installed in R if data are from Affymetrix platforms. To install these packages, type the following commands in the R environment:

```
> source("http://bioconductor.org/biocLite.R")
> biocLite()
> biocLite("affyparser")
```

Details of Bioconductor installation can be found at (22).

2.2. Data Preparation

JAMIE works for all types of tiling arrays. However, it requires that multiple data sets are from the same platform (i.e., probe locations are identical). For data from Affymetrix platforms, JAMIE requires BPMAP (which contains array platform designs) and CEL (for probe intensities) files. For data from other platforms, users need to prepare a single text file without column headers to include all data. In the text file, each row corresponds to one probe. The first two columns are chromosome and genomic coordinates of the probes. The rest of the columns contain probe intensities, or log ratios between IP and control channels in two-color arrays.

2.3. Configuration File

In addition to the data file(s), users need to prepare a plain text configuration file to provide necessary parameter information. Examples of configuration files can be found at (23). The file consists of several sections. Each section has a title which must be surrounded by square brackets. Each title occupies a line. Within each section, parameters are configured in the "parameter = value" format. Different array platforms and experimental designs require one to include different sections in the file.

2.3.1. Configuration Files for Non-Affymetrix Data

When data are from platforms other than Affymetrix, users need to provide a single text file containing both the array designs (chromosomes and locations for probes) and the probe data as described above. In this case, the configuration file should contain three sections *without* any particular order: "data," "Condition" and "peak finding."

Below is an example of the "data" section:

```
[data]
Title=project
Format=text
file=/directory/to/file/ChIP-chip.txt
WorkFolder=/directory/to/project/
```

Here, "Title" specifies the title of the project. Temporary files will be saved under this title (i.e., named as "project_*"). "Format" specifies the input data format. Valid options are "cel" if the data are from Affymetrix arrays, and "text" if the data are non-Affymetrix arrays and in text format. "file" provides the location of the data file (must be a single text file in this case). "WorkFolder" indicates the working directory. All temporary files and analysis results will be exported to this folder.

An example of the "Condition" section is shown below:

```
[Condition]
cond1=3 4
cond2=5 6
cond3=7 8
```

In this section, each row corresponds to a data set. Left-hand sides of the equal signs are user specified dataset names; in this example, they are "cond1," "cond2," and "cond3." The files storing final result will be called after these names, e.g., result for cond1 will be called "cond1-peak.txt," and so on. Right-hand sides of the equal signs specify the column ids of each data set in the input data file. These numbers need to be separated by white spaces in each row. In the example above, columns 3 and 4 in the data file are two replicate samples in the "cond1" data set, columns 5 and 6 are two replicate samples from the "cond2" data set, and so on. The numbers of replicates in different data sets do not need to be the same, and a single sample (no replicate) is allowed.

A sample "peak finding" section is shown below:

```
[peak finding]
candidateLength=1000
bumpLength=300 500 700 900
maxGap=300
MinProbe=6
FDRcutoff=0.2
computeFDR=0
```

Here, "candidateLength" specifies the length of PBRs L in bps. This number should be obtained by exploratory data analysis. The ideal PBR length should be bigger than most (95%) of the peaks. In most cases 1,000 bp is a good choice for TFBS detection. However, if the probes are sparse or DNA fragments are long after sonication, users should increase this number to increase the robustness of the results. A longer PBR length requires more computation. "bumpLength" specifies the allowable peak lengths **W** within a PBR. Again these numbers should be obtained by exploratory data analysis. Introducing more peak lengths will allow JAMIE to define peak boundaries more precisely, but it also increases computational burden. "maxGap" specifies the maximal gap (in bps) allowed between two adjacent probes within a peak. "MinProbe" specifies the minimal number of probes required in order to call a peak. "FDRcutoff" specifies the maximal false-discovery rate (FDR) for reporting peaks (see Note 3). Finally, "computeFDR" specifies the method for estimating FDR. The valid values are 0 or 1. 0 means that the FDRs are computed from the posterior probabilities. 1 means that the FDRs are estimated empirically from the data by swapping IP and control sample labels. After the label swap, JAMIE will be run on the label-swapped data using the model parameters estimated from the original data. The FDRs are then estimated using the ratio between the peak numbers from the label-swapped and non-swapped (original) data. Simulation results in ref. 7 have shown that these two estimates are fairly close when the model assumptions are reasonable. When the model assumptions are violated, however, the second method provides relatively more robust estimation. In practice, users are advised to specify "0" first for better computational efficiency. If the reported FDRs look suspicious, one can then specify "1" and use the empirical procedure instead.

2.3.2. Configuration Files for Affymetrix Data with Paired Samples

When data are from Affymetrix platforms, and if the IP and control arrays are paired, the following changes need to be made to the configuration file described above. First, in the "data" section, users need to specify "Format = cel." Two additional lines need to be provided to specify the location of BPMAP and CEL files. For example:

```
Bpmap=/dir/to/bpmap/Mm_PromPR_v02-1_NCBIv36.bpmap
CelFolder=/dir/to/CEL
```

A new parameter "Pair = 1" need to be provided to indicate that the arrays are paired.

The "Condition" section will be replaced by a new section "cel," with an example below:

```
[cel]
cond1=Cond1_IP1.CEL Cond1_CT1.CEL Cond1_IP2.CEL Cond1_CT2.CEL
cond2=Cond2_IP1.CEL Cond2_CT1.CEL Cond2_IP2.CEL Cond2_CT2.CEL
cond3=Cond3_IP1.CEL Cond3_CT1.CEL Cond3_IP2.CEL Cond3_CT2.CEL
cond4=Cond4_IP1.CEL Cond4_CT1.CEL Cond4_IP2.CEL Cond4_CT2.CEL
```

Here, each row corresponds to a data set. Again the left-hand sides of the equal signs are the user-specified dataset names. The right-hand sides are lists of CEL files for each data set. In the paired experiments, CEL files in each data set must be specified in the order of IP1, control1, IP2, control2, etc., based on the pairing relationship between IP and control samples.

The "peak finding" section and its format remain unchanged.

2.3.3. Configuration Files for Affymetrix Data with Nonpaired Samples

When data are from Affymetrix arrays, and if the IP and control samples are not paired, then the configuration file for the paired Affymetrix experiment should be changed as follows. First, in "data" section, users should specify "Pair = 0." The CEL files can now be listed in any order in the "cel" section. Second, a new section "Group" has to be provided to specify the identity (IP or control) of the CEL files. An example "Group" section is provided below:

```
[Group]
cond1=1100
cond2=1100
cond3=1100
cond4=1100
```

In this section, the number of lines must match those in the "cel" section. In each line, the left-hand sides of the equal signs are dataset names. These names must match the names provided in the "cel" section. The right-hand sides specify the IP/control identities. "0" represents control and "1" means IP. In this example, cond1 = 1100 means that for the "cond1" CEL files listed in the "cel" section, the first two files are IP samples and the last two files are control samples.

2.4. Run JAMIE

After the configuration file is set, the joint peak detection can be achieved by typing two lines of R commands. Assume that the configuration file is named as "config.txt," users can type:

```
> library(jamie)
> jamie("config.txt")
```

JAMIE will run the integrative data analysis. The results will contain a peak list for each data set. The peaks will be ranked according to the posterior probabilities. These results will be saved into tab-delimited text files in the user-specified working directory.

JAMIE saves several intermediate results in the working directory as rda files (binary files to save R objects). For example, if the project title in the configuration file is "project", after a full run of JAMIE, the following rda files will be generated:

- project-data.rda: saves normalized data and calculated probe level variances.
- project-candidate.rda: saves the calculated likelihood and estimated model parameters.

- project-postprob.rda: saves the posterior probabilities from the whole genome scan.

The purpose of saving these results is to speed up calculations. For instance, if one changes parameters in the "peak finding" section, the data reading and normalization steps do not have to be repeated again, and the normalized data can be read from the previously saved results. Users need to be cautions here: the rda files for saving the candidate regions and posterior probabilities need to be manually deleted if users want to change the configuration files to analyze new data. Otherwise JAMIE will merely read the saved results instead of redoing the calculation.

2.5. Downstream Analyses

With the peak lists produced by JAMIE, one can perform several subsequent analyses using the CisGenome software (24). For example, one can associate the peaks with neighboring genes, extract DNA sequences from the peaks, discover enriched DNA sequence motifs, and study the enrichment level of the motifs compared to negative control regions. Users are referred to (25) to learn more about CisGenome.

3. Notes

1. *Model assumptions.* JAMIE is developed based on a number of model assumptions. The model brings the statistical power. However, it is important to note that like all model-based approaches, the performance of JAMIE is highly dependent on how well the data fit the model assumptions. Based on the extensive simulation studies provided in the supplemental materials in ref. 7, JAMIE is fairly robust against violation of model assumptions and consistently outperforms MAT and TileMap. However, the simulation results have also shown that in cases of dramatic violation of the assumptions, the FDR estimates provided by JAMIE could be very biased. For this reason, in practice we recommend users to use JAMIE mainly as a tool to rank peaks, and use qPCR to obtain a more reliable FDR estimates whenever possible. It is also important to mention that the foundation of JAMIE is that multiple data sets are "related." Intuitively, when all q_ds are close to one, different data sets will share a large fraction of peaks, therefore data sets are highly correlated and borrowing information across data sets can significantly help peak detection. If the correlations among data sets are low, the gain will be minimal. For this reason, users are advised to use only related data sets in the analysis. For example, if one has a data set for one TF, he/she can go to public databases to find other data sets for the same TF and jointly analyze these

data sets together. Doing so will be more likely to obtain better results.

2. *Install R Packages.* In order to install an R package, one needs to have R, perl, latex and gcc (or g++) installed on the computer. R can be downloaded from ref. 26. Perl, latex, and gcc are installed in many Unix systems. For Windows, one can install perl and gcc by downloading Rtools from ref. 27, and install latex by downloading MiKTeX from ref. 28.

   In addition to installing these programs, one also needs to set an environment variable PATH to include the folders in which the executable files of these programs are installed. In Unix, this can be done by opening the user's shell profile file (e.g., .bash_profile), find the line in the file that sets the PATH variable, and edit the line. For example,

```
PATH=.:$PATH:$HOME/bin:$HOME/R/bin: $HOME/perl/bin: $HOME/latex/bin
```

   Save the file. Log out and then log in again. Check whether the system recognizes these programs by typing:

```
> R
> perl
> latex
> gcc
```

   If the PATH variable is set up correctly, typing the commands above will start the corresponding programs. If not, go back to edit PATH again.

   To set the PATH variable in windows, open "My Computer." Right click "Computer," choose "Properties," then choose "Advanced system settings." In the dialog that jumps out, click "Environment Variables." Choose "Path" in the "System variables," and click "Edit." Edit PATH and save it. To check whether the PATH variable is set up correctly, click "Start > Accessories > Command Prompt." In the command window that jumps out, type "R," "perl," "latex," "gcc" to check whether these programs are recognized by the system.

3. *FDR estimation.* Note that the FDR estimation could be biased if the model assumptions are dramatically violated. Users are advised to use a relaxed cutoff to obtain more peaks. The lowly ranked peaks can always be discarded in downstream analysis if needed.

Acknowledgments

The authors thank Drs. Eunice Lee, Matthew Scott, and Wing H. Wong for providing the Gli data, Dr. Rafael Irizarry for providing financial support, and Dr. Thomas A. Louis for insightful discussions. This work is partly supported by National Institute of Health R01GM083084 and T32GM074906.

References

1. Ren B, Robert F, Wyrick JJ et al (2000) Genome-wide location and function of DNA binding proteins. Science 290:2306–2309
2. Boyer LA, Lee TI, Cole MF et al (2005) Core transcriptional regulatory circuitry in human embryonic stem cells. Cell 122:947–956
3. Cawley S, Bekiranov S, Ng HH et al (2004) Unbiased mapping of transcription factor binding sites along human chromosomes 21 and 22 points to widespread regulation of noncoding RNAs. Cell 116:499–509
4. Barrett T, Troup DB, Wilhite SE et al (2009) NCBI GEO: archive for high-throughput functional genomic data. Nucleic Acids Res. 37:D885–890
5. Vokes SA, Ji H, Wong WH et al (2008) A genome-scale analysis of the cis-regulatory circuitry underlying sonic hedgehog-mediated patterning of the mammalian limb. Genes Dev. 22:2651–2663
6. Lee EY, Ji H, Ouyang Z et al (2010) Hedgehog pathway-regulated gene networks in cerebellum development and tumorigenesis. Proc. Natl. Acad. Sci. USA 107: 9736–9741
7. Wu H, Ji H (2010) JAMIE: joint analysis of multiple ChIP-chip experiments. Bioinformatics 26:1864–1870
8. The R Development Core Team (2010) R: A Language and Environment for Statistical Computing. http://cran.r-project.org/doc/manuals/refman.pdf
9. Kapranov P, Cawley SE, Drenkow J et al (2002) Large-scale transcriptional activity in chromosomes 21 and 22. Science 296:916–919
10. Johnson WE, Li W, Meyer CA et al (2006) Model-based analysis of tiling-arrays for ChIP-chip. Proc. Natl. Acad. Sci. USA 103:12457–12462
11. Ji H, Wong WH (2005) TileMap: create chromosomal map of tiling array hybridizations. Bioinformatics 21:3629–3636
12. Keles S (2007) Mixture modeling for genome-wide localization of transcription factors. Biometrics 63:10–21
13. Zheng M, Barrera LO, Ren B et al (2007) ChIP-chip: data, model, and analysis. Biometrics 63:787–796
14. Zhang ZD, Rozowsky J, Lam HY et al (2007) Tilescope: online analysis pipeline for high-density tiling microarray data. Genome Biol. 8:R81
15. Toedling J, Skylar O, Krueger T et al (2007) Ringo - an R/Bioconductor package for analyzing ChIP-chip readouts. BMC Bioinformatics 8:221
16. Gottardo R, Li W, Johnson WE et al (2008) A flexible and powerful bayesian hierarchical model for ChIP-Chip experiments. Biometrics 64:468–478
17. Johnson WE, Liu XS, Liu JS (2009) Doubly Stochastic Continuous-Time Hidden Markov Approach for Analyzing Genome Tiling Arrays. Ann. Appl. Stat 3:1183–1203
18. Choi H, Nesvizhskii AI, Ghosh D et al (2009) Hierarchical hidden Markov model with application to joint analysis of ChIP-chip and ChIP-seq data. Bioinformatics 25:1715–1721
19. Dempster AP, Laird NM, Rubin DB (1977) Maximum Likelihood from Incomplete Data Via Em Algorithm. J. Roy. Stat. Soc. B. 39:1–38
20. JAMIE download: http://www.biostat.jhsph.edu/~hji/jamie/
21. R installation manual: http://cran.r-project.org/doc/manuals/R-admin.html
22. Bioconductor manual: http://www.bioconductor.org/docs/install-howto.html
23. JAMIE configuration files: http://www.biostat.jhsph.edu/~hji/jamie/use.html
24. Ji H, Jiang H, Ma W et al (2008) An integrated software system for analyzing ChIP-chip and ChIP-seq data. Nat Biotechnol. 26:1293–1300
25. CisGenome website: http://www.biostat.jhsph.edu/~hji/cisgenome/
26. R download: http://www.r-project.org/
27. Rtools download: http://www.murdoch-sutherland.com/Rtools/
28. MiKTeX download: http://miktex.org/

Chapter 25

Epigenetic Analysis: ChIP-chip and ChIP-seq

Matteo Pellegrini and Roberto Ferrari

Abstract

The access of transcription factors and the replication machinery to DNA is regulated by the epigenetic state of chromatin. In eukaryotes, this complex layer of regulatory processes includes the direct methylation of DNA, as well as covalent modifications to histones. Using next-generation sequencers, it is now possible to obtain profiles of epigenetic modifications across a genome using chromatin immunoprecipitation followed by sequencing (ChIP-seq). This technique permits the detection of the binding of proteins to specific regions of the genome with high resolution. It can be used to determine the target sequences of transcription factors, as well as the positions of histones with specific modification of their N-terminal tails. Antibodies that selectively bind methylated DNA may also be used to determine the position of methylated cytosines. Here, we present a data analysis pipeline for processing ChIP-seq data, and discuss the limitations and idiosyncrasies of these approaches.

Key words: ChIP-seq, Chromatin immunoprecipitation, Transcription factor binding sites, Peak calling, Histone modification, DNA methylation, Next-generation sequencing, Poisson statistics

1. Introduction

The DNA sequence is the primary blueprint that controls cellular function. However, a complex layer of molecular modifications that are referred to as the epigenetic code affects the transcription and replication of DNA. Epigenetic modifications include the direct methylation of cytosines, as well as modifications to the structure of chromatin. In particular, the N-terminal tails of histones can be modified by a large number of enzymes that add or remove methyl, acetyl, phosphorous, or ubiquitin groups, among others (1). The characterization of the epigenetic state of chromatin is complicated by the fact that each cell type in an organism has a different epigenetic state. In fact, the epigenetic differences

Junbai Wang et al. (eds.), *Next Generation Microarray Bioinformatics: Methods and Protocols*,
Methods in Molecular Biology, vol. 802, DOI 10.1007/978-1-61779-400-1_25, © Springer Science+Business Media, LLC 2012

between cells are fundamental to the generation of diversity between cell types that all arise from a clonal population with identical DNA sequences.

The readout of epigenetic modification on a genome-wide scale can be carried out using chromatin immunoprecipitation techniques (2). In brief, these methods involve the crosslinking of DNA to protein using crosslinking agents as a first step, in order to freeze protein–DNA and protein–protein interactions. Subsequently, the chromatin is sonicated to yield fragments of protein-bound DNA that are typically a few hundred bases long. These fragments are then purified using antibodies that are specific to the particular modification that is being profiled (e.g., a specific modification of the histone tail, or cytosine methyl groups). The immunoprecipitated fraction is isolated, and the crosslinks are reversed to yield the DNA fragments bound to the protein of interest. These fragments are then either hybridized to a microarray (ChIP-chip) or sequenced using a high-throughput sequencing platform (ChIP-seq). The immunoprecipitated fragments are then compared to the fragments that were not selectively immunoprecipitated, often referred to as the input material, to identify sequences that enriched in the former with respect to the latter. These enriched regions correspond to the DNA sequences that are bound by the protein of interest.

Before the advent of next-generation sequencing, ChIP-chip was the standard technique for these types of assays (3). However, for many organisms it is not practical to generate genome-wide tiling arrays, and hence ChIP-chip data sets were often not genome-wide. Furthermore, the ability to detect a binding site in a ChIP-chip experiment is limited by the resolution of the probes on the array. Finally, the signal obtained by hybridization intensities on an array is analog, and it is often difficult to determine levels of enrichment that are statistically significant and hence indicative of true binding sites. Many of these limitations are overcome by using ChIP-seq (4). Since sequencing is not limited in any way by probes, and it is therefore a truly genome-wide approach. The only limitation is that it is impossible to definitively determine the position of a peak if it lies within a sequence that is repeated in the genome. For this reason, often ChIP-seq peaks are only called when they are associated with unique sequences that appear only once in the genome, and this can be a significant limitation since repetitive sequences are very abundant in large genomes such as that of humans. Nonetheless, Chip-seq technology is rapidly eclipsing the older ChiP-chip approach and we therefore present detailed protocols for the analysis of this latter data rather than the former.

2. Materials

In this chapter, we describe the computational protocols for analyzing ChIP-seq data. We will not discuss the experimental protocols for generating ChIP-seq libraries, as these have been published elsewhere.

2.1. Base Calls

From our standpoint, therefore, the material to carry out the analyses, we describe consist of the base calls that are output by the DNA sequencer. For the most common case of data generated by Illumina sequencers, this data consists of tens of millions of short reads that typically range from 36 to 76 bases in length (5). Several data standards have been developed for the encoding of these reads into flat files. The most common is the FASTQ standard which contains both the base calls at each position of the read as well as the quality scores that denote the confidence in the base calls (6) (see Note 1).

2.2. Alignment Software

The second essential material is an alignment tool to align the reads to a reference sequence. Over the past couple of years there has been a proliferation of new alignment tools that are specialized for the rapid alignment of millions of short reads to large reference genomes. These tools include Bowtie (7), Maq (8), and Soap (9) among others (see Note 2). Since the reads contain fragments of DNA from the genome, the alignments do not need to consider gaps (although some of these tools do permit the inclusion of small gaps). Similarly one only expects a few mismatches between the read sequence and reference genome due to base calling errors or polymorphisms in the genome sequence, and all these aligners allow for the inclusion of several mismatches in the alignment. Finally, most of the alignment tools do not explicitly consider base call quality scores when attempting to identify the optimal alignment for a read. However, some tools, such as Bowtie, do consider the quality scores after the alignment has been performed using only the base calls.

2.3. Genome Browser

The other critical tool to enable the analysis and interpretation of ChIP-seq data is a genome browser. This application allows one to zoom and pan to any position in the genome, and view the mapped reads. This is critical for both verifying the data analysis protocols and to generate detailed information for specific loci. Several tools are available for this purpose including the Integrated Genome Browser (10), and the UCSC genome browser (11) among many others (see Note 3). Typically, the data is uploaded in formats that depict either individual reads (e.g., bed format) or the accumulated counts associated

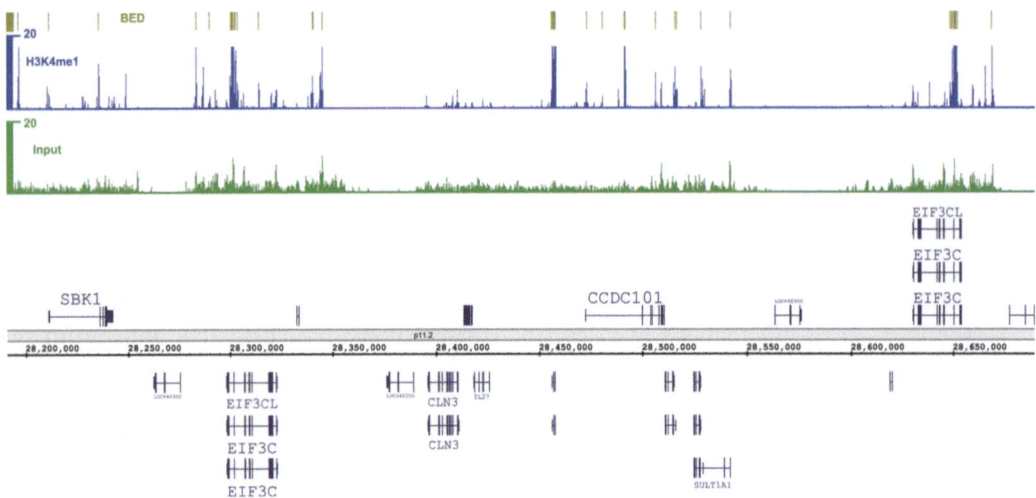

Fig. 1. A sample locus viewed using the UCSC genome browser. The first track from the *top* contains the windows that are found to be significantly enriched in the IP vs. input for H3K4me1, a histone mark. The second track, labeled H3K4me1, shows the counts for each 100 base window. The third track contains the input control. The tracks on the *bottom* contain the gene annotation which indicates the transcriptional start and end sites and the positions of introns for the two genes in this locus.

with reads that overlap a specific base (e.g., wiggle tracks). Examples of the output of these browsers may be seen in Fig. 1.

3. Methods

The methods that we describe will utilize the base calls described above, in conjunction with an alignment tool, to identify all the regions of the genome that containing significant peaks for the particular DNA binding protein that is being tested. Along with a description of the methods for data analysis, we also discuss software that has been developed to visualize the resulting data on the genome.

3.1. Read Alignment

The first step in the data analysis pipeline is to align the reads to a reference genome or other reference sequence of interest. Usually, alignments do not allow for gaps to be inserted between the reads and the reference sequence. For a 36-base reads it is customary to accept all alignments that generate no more than two mismatches between the reads and the reference sequence. The number of allowed mismatches can be adjusted to a higher level for longer reads, but it is difficult to come up with systematic approaches to determine what the optimal number of allowed mismatches should be, and thus this value is nearly always assigned based on ad hoc criteria. Finally, as we discussed above, reads that align with equal scores to multiple locations on the genome are most often thrown out, since they cannot be unambiguously assigned to a single peak. A variety of approaches have been developed to deal with multiple mapping problems. These include the probabilistic reassignment of reads based on the surrounding region (12) (which assumes that if a read maps to two locations, it is more likely to originate from the one that has more reads mapping in the immediate neighborhood), to the use of representations of the genome that explicitly account for the repeat structure of the sequence (13), to the simple addition of a weight to each read based on the multiplicity of its binding sites. While accounting for repeats is more critical in other applications (such as RNA-seq), in general people have found that it is less important in ChIP-seq applications, and generally none of these more sophisticated approaches are used.

Once the alignments have been completed the next step involves the evaluation of the alignment quality. This is measured using several criteria, the first and most significant of which is the fraction of reads that map to a unique location in the genome. In general, not all reads can map to unique locations because the reference sequence contains repetitive regions and because the sequencing process usually introduces random errors in the base calls. However, a well-prepared ChIP-seq library should yield unique alignments for somewhere around half of the reads. If the actual number is significantly lower (i.e., less than 30%) then this might indicate that there was a problem in the library preparation or the sequencing run. To attempt to optimize the number of reads that map to unique location on the reference sequence, it is common to attempt to trim the end of the reads as these often have lower base calling accuracy. As we see in Fig. 2 for a typical case, the number of mismatches tends to be high at the very start of the reads, low in the middle, and increases toward the end of the read. By trimming these locations it is possible to increase the number of reads that can be uniquely mapped to the genome.

One final consideration that is important for ChIP-seq libraries is that they are often plagued by low complexity. That is, the number of unique reads that are generated by the sequencer is often significantly smaller than the total number of reads, due to the resequencing of the same read multiple times. This phenomenon tends to be more common in ChIP-seq experiments because it is often difficult to produce large quantities of DNA using chromatin immunoprecipitation, due to the limits of the antibody affinity for its target, and potentially due to the limited number of sites where the target protein is bound (see Note 4). However, if we observe the same read multiple times, this does not necessarily imply that the target protein has higher affinity for the corresponding sequence, but could also be due to the fact that the particular read sequence is more efficiently amplified during the library preparation protocol. As a result, to minimize these

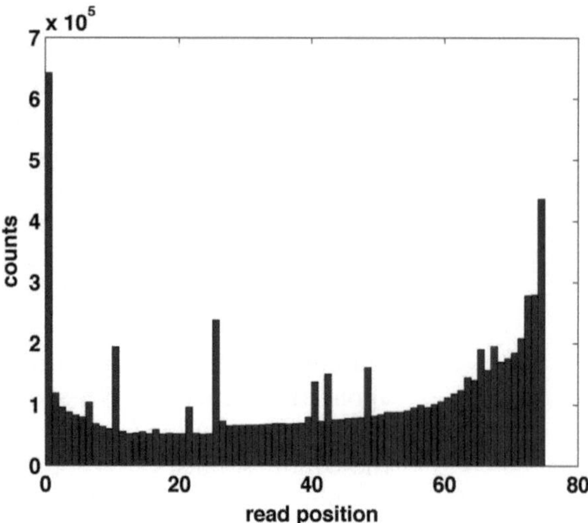

Fig. 2. Mismatch counts as a function of position in read. Reads were aligned to the genome using Bowtie. Up to two mismatches were allowed per alignment. The position of the mismatch along the read is indicated on the *x*-axis, and the total number of mismatches at this position is shown on the *y*-axis. The first base has a significant number of mismatches compared to the first 50 bases. The last ten bases show an increasing number of mismatches. A few positions in the middle of the read also show anomalously high mismatch counts, possibly due to some perturbation to the sequencing cycle during this run.

biases, we usually only align the unique reads in the library, and not the total reads. This may be accomplished by either sorting the reads in the library and selecting unique reads, or by combining reads that map to the same location into a single read that contributes only one count.

3.2. Peak Detection

Once the reads have been aligned to the genome, the binding sites of the target protein can be indentified. To accomplish this it is customary to first tile the genome using windows, within which we attempt to detect peaks. The size of the window is typically between 100 and a couple of hundred bases. This roughly corresponds to the size of the sonicated DNA fragments that are used to generate the ChIP-seq library. Due to the limited sequencing depth (currently 30–40 million reads are produced for each library), and the size of sonication fragments, it is usually not possible to detect peaks with more than 100 base resolution. The tiling can either be sequential, or interleaved.

The counts within each window are determined by computing both the number of reads whose alignment starts directly within the window, as well as reads that align outside, but near the edges of the window. If we assume that each read corresponds to a one to two hundred base DNA fragment, then even reads that align to a position 100 bases upstream of the window, overlap and contribute to the counts in the window. Each read can either contribute a fractional count to the window, measured by the fraction of the read that overlaps the window, or more simply any level of overlap can lead to a discrete increment of one count. It is also important to realize that reads that map to the negative DNA strand contribute to windows that are upstream of the start site, while reads that map to the positive strand contribute to windows that are downstream of the start site.

To determine whether the counts within a window are significant, it is necessary to compare these to a background level. The most simplistic model is that the background level of each window is simply the average counts for all the windows across the genome. However, it is more customary to sequence a control library, usually referred to as the input library, to estimate the background counts. The input library consists of all the DNA fragments that were not immunoprecipitated during the course of the chromatin immunoprecipitation protocol. It should certainly have a more uniform distribution across the genome than the immunoprecipitated (IP) library, however, recent studies have shown that sonication and DNA purification methods result in biases that often lead to additional peaks around transcription start sites (14). Therefore, comparing the IP libraries with the input can remove some false-positive peaks that are just due to sonication biases. However, in order or this comparison to be meaningful, the input library must first be normalized so that it contains the same total numbers of counts as the IP library (see Note 5).

Once the counts of the IP and input libraries in each window in the genome have been computed, the final step involves that determination of the statistical significance of the increase in IP over input, if any. It is assumed that the counts in each window are approximately distributed according to the Poisson distribution, as the generation of a sequence library fragments from a genome is essentially a Poisson process (15). Therefore, to estimate the probability of observing the IP counts we use the cumulative Poisson distribution with an expected value provided by the input counts. That is, we compute the probability of observing the IP counts, or a higher value, given the expected number provided by the input counts. This approach will be noisy when the input counts are low, or zero. If the input counts are zero we can set the expected distribution to the genome average. This method will generate a *P*-value for each window in the genome. The last step requires one to estimate false-discovery rates (FDRs) based on this *P*-value distribution. There are many statistical approaches for estimating FDRs from *P*-value distribution, and we will not discuss these in detail here other than to provide several references (16, 17).

3.3. Data Visualization

An important component of ChIP-seq data analysis is the visualization of the data on a genome browser. As discussed above there are various tools that can be used for this purpose. Here, we illustrate the use of the UCSC Genome Browser (18). We illustrate a sample locus in Fig. 1. We show tracks for the IP counts the input counts, as well as the regions that are deemed to be significantly enriched in IP vs. input. The data is generated using a variety or formats. The counts files are generated using the wiggle format that describes the chromosome, position, and counts in each window. The significant peaks are displayed using the bed format, which denotes that boundaries of the region with significant enrichment. It is critical to generate these types of files when analyzing ChIP-seq data, to determine whether the peak finding algorithm, and the particular parameters chosen by the user, are in fact yielding reasonable peaks. The tool also allows one to visualize the data in any region of interest in the genome, in order to answer specific question about loci of interest.

3.4. Downstream Analysis

There are a multitude of possible downstream analyses that can be conducted on ChIP-seq data and here we limit ourselves to describe only a small set. It is, for instance, customary to overlay the peaks identified in the ChIP-seq data with positions of transcriptional start sites (TSS), as these can be directly associated regulatory regions. In this regard, it is customary to generate "meta plots" that display the total number of peaks a certain distance from the TSS. For example, in Fig. 3 we show the total number of peaks around the TSS for a specific histone modification. We note right away the modification is enriched around the TSS but depleted right at the TSS. Similar analyses can be performed for any other genomic feature, such as transcription termination sites, intron–exon boundaries, or repeat boundaries.

A slightly different representation of the enrichment around features identifies the average trends along the entire length of the feature (e.g. (19)) (Fig. 3, bottom panel). That is each gene is rescaled so that it is covered by a fixed number of bins (typically 100 or so). The density of peaks in each bin is then computed (i.e., the number of peaks divided by the bin length). The values of the bins are averaged or summed over all the genes in the genome to generate the average trend of peaks across the genome. The same analysis is usually performed on the upstream and downstream regions of the genes, which can comprise 50% or so of the total gene length. The combination of the upstream, gene, and downstream region then generates a comprehensive view of the trends in the data around genes. Thus, unlike the previous plots, these provide a more global view of the peak trends across genes. As before, these types of analyses may be performed across any genomic feature, and not just genes. It may be of interest to generate the average trends across repetitive elements in the genome, or internal exons.

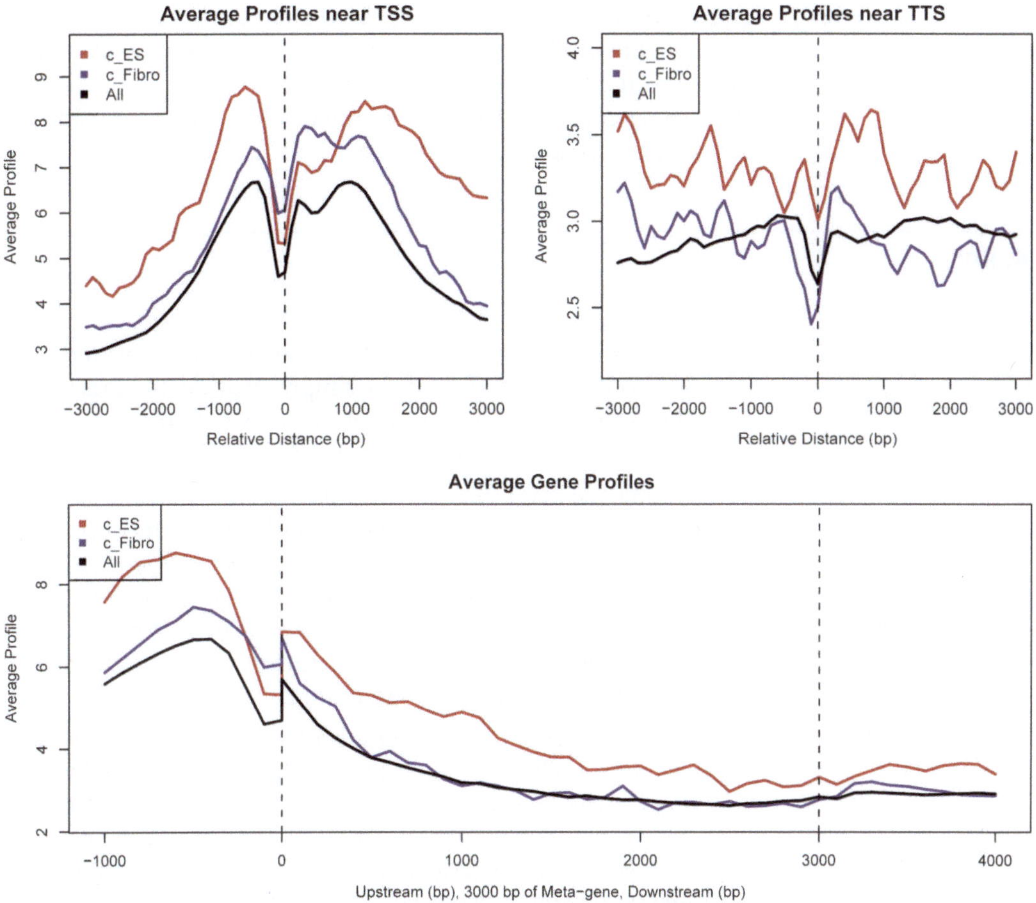

Fig. 3. Average levels of H3K4me1 acetylation at the start and end of genes. This meta-analysis computes the average levels of H3K4me1 in a 6-kb region surrounding the transcriptional start site (*top right*) and end site (*top left*). We see that H3K4me1 positive regions are preferentially located around, but not right over the start sites. In the bottom panel we show a scaled metagene analysis, where all genes have been aligned so that they start at 0 and end at 3,000. The average H3K4me1 levels 1 kb upstream and downstream of all genes are also shown. In all cases, genes are grouped into three groups. c_ES are genes that are differentially induced in embryonic stem cells and c_Fibro are those induced in fibroblasts (24), while All are all the genes.

Another common analysis attempts to summarize the locations of peaks throughout the genome. While the previous two procedures summarize the distribution of peaks around genes, a large fraction of the peaks may lie far from genes, and thus would not be considered in these analyses. To account for these, it is customary to generate a table that describe the fractions of peaks that are within genes, or a certain distance from genes. Such a table might include categories that correspond to regions that are, for example, tens of kilobases away from genes.

Of course the analyses described above are only a small sampling of all the possible downstream analyses that can be attempted on this data. It is also possible to analyze the sequence composition of peak regions, or search for specific sequence motifs. One might also consider the distribution of peaks across chromosomes to identify large-scale trends. However, a comprehensive description of all of these methodologies lies outside the scope of this chapter (see Note 6).

4. Notes

1. Many aligners do not use base call information and it is therefore often sufficient to simply provide the base calls. These files are sometimes referred to as raw formats and are significantly smaller in size than the FASTQ format.
2. Among the many alignment tools that have become available over the past few years, Bowtie is probably the most popular, as it tends to be one of the fastest, with an efficient indexing scheme that requires relatively small amounts of memory. For a typical mammalian genome the indices built from the reference sequence are around 4 gigabytes, and a single lane of data can be aligned in about an hour.
3. The UCSC genome browser is probably the most widely used browser. It allows users to upload data onto the UCSC site, where it can be compared to data that permanently resides on the server (such as annotation files). However, if the genome of interest is not preloaded in the browser, it is very difficult to upload it onto the browser. Nonetheless, various instances of the browser are maintained by other groups that contain additional genomes (e.g. (20)).
4. To increase the complexity of ChIP-seq libraries it is necessary to immunoprecipitate as much material as possible, which in typical circumstances may require performing multiple immunoprecipitations on batches of millions of cells.
5. Other popular peak calling approaches can be significantly more sophisticated, by taking into consideration the shape of the peak, the length of reads, and the posterior probabilities (21, 22).
6. An example of a suite of tools that may be applied for these types of analyses may be found at ref. 23.

Acknowledgments

The authors would like to thank Professor Bernard L. Mirkin for development of the drug-resistant models of human neuroblastoma cells and for his advice and encouragement, and Jesse Moya for technical assistance. This work was supported by Broad Stem Cell Research Center and Institute of Genomics and Proteomics at UCLA.

References

1. Jenuwein T, Allis CD (2001) Translating the histone code. Science 293:1074–1080.
2. Nelson JD, Denisenko O, Bomsztyk K (2006) Protocol for the fast chromatin immunoprecipitation (ChIP) method. Nat Protoc 1:179–185.
3. Buck MJ, Lieb JD (2004) ChIP-chip: considerations for the design, analysis, and application of genome-wide chromatin immunoprecipitation experiments. Genomics 83:349–360.
4. Valouev A, Johnson DS, Sundquist A et al (2008) Genome-wide analysis of transcription factor binding sites based on ChIP-Seq data. Nat Methods 5:829–834.
5. Mardis ER (2008) The impact of next-generation sequencing technology on genetics. Trends Genet 24:133–141.
6. Cock PJ, Fields CJ, Goto N et al (2010) The Sanger FASTQ file format for sequences with quality scores, and the Solexa/Illumina FASTQ variants. Nucleic Acids Res 38:1767–1771.
7. Langmead B, Trapnell C, Pop M et al (2009) Ultrafast and memory-efficient alignment of short DNA sequences to the human genome. Genome Biol 10:R25.
8. http://maq.sourceforge.net/.
9. Li R, Li Y, Kristiansen K et al (2008) SOAP: short oligonucleotide alignment program. Bioinformatics 24:713–714.
10. Nicol JW, Helt GA, Blanchard SG Jr et al (2009) The Integrated Genome Browser: free software for distribution and exploration of genome-scale datasets. Bioinformatics 25:2730–2731.
11. Rhead B, Karolchik D, Kuhn RM et al (2010) The UCSC Genome Browser database: update 2010. Nucleic Acids Res 38:D613–619.
12. Clement NL, Snell Q, Clement MJ et al (2010) The GNUMAP algorithm: unbiased probabilistic mapping of oligonucleotides from next-generation sequencing. Bioinformatics 26:38–45.
13. Pevzner PA, Tang H (2001) Fragment assembly with double-barreled data. Bioinformatics 17:S225–233.
14. Auerbach RK, Euskirchen G, Rozowsky J et al (2009) Mapping accessible chromatin regions using Sono-Seq. Proc Natl Acad Sci U S A 106:14926–14931.
15. Mikkelsen TS, Ku M, Jaffe DB et al (2007) Genome-wide maps of chromatin state in pluripotent and lineage-committed cells. Nature 448:553–560.
16. Benjamini Y, Drai D, Elmer G et al (2001) Controlling the false discovery rate in behavior genetics research. Behav Brain Res 125:279–284.
17. Muir WM, Rosa GJ, Pittendrigh BR et al (2009) A mixture model approach for the analysis of small exploratory microarray experiments. Comput Stat Data Anal 53:1566–1576.
18. http://genome.ucsc.edu/.
19. Cokus SJ, Feng S, Zhang X et al (2008) Shotgun bisulphite sequencing of the Arabidopsis genome reveals DNA methylation patterning. Nature 452:215–219.
20. http://genomes.mcdb.ucla.edu.
21. Zhang Y, Liu T, Meyer CA et al (2008) Model-based analysis of ChIP-Seq (MACS). Genome Biol 9:R137.
22. Spyrou C, Stark R, Lynch AG et al (2009) BayesPeak: Bayesian analysis of ChIP-seq data. BMC Bioinformatics 10:299.
23. http://liulab.dfci.harvard.edu/CEAS/.
24. Chin MH, Mason MJ, Xie W et al (2009) Induced pluripotent stem cells and embryonic stem cells are distinguished by gene expression signatures. Cell Stem Cell 5:111–123.

Chapter 26

BiNGS!SL-seq: A Bioinformatics Pipeline for the Analysis and Interpretation of Deep Sequencing Genome-Wide Synthetic Lethal Screen

Jihye Kim and Aik Choon Tan

Abstract

While targeted therapies have shown clinical promise, these therapies are rarely curative for advanced cancers. The discovery of pathways for drug compounds can help to reveal novel therapeutic targets as rational combination therapy in cancer treatment. With a genome-wide shRNA screen using high-throughput genomic sequencing technology, we have identified gene products whose inhibition synergizes with their target drug to eliminate lung cancer cells. In this chapter, we described BiNGS!SL-seq, an efficient bioinformatics workflow to manage, analyze, and interpret the massive synthetic lethal screen data for finding statistically significant gene products. With our pipeline, we identified a number of druggable gene products and potential pathways for the screen in an example of lung cancer cells.

Key words: Next generation sequencing, shRNA, Synthetic lethal screen

1. Introduction

RNA interference (RNAi)-based synthetic lethal (SL) screens have potential for the identification of pathways that cancer cell viability in the face of targeted therapies (1–4). With a genome-wide short hairpin (sh)RNA interference-based screen using high-throughput genomic sequencing (Next Generation Sequencing, NGS) technology, we have identified gene products whose inhibition synergizes with their target drug to eliminate cancer cells.

In the SL screen experiment, cells are infected with lentiviral carrying individual shRNAs. After lentiviral infection, the cells are separated into vehicle and drug treatment groups. RNA is then harvested from the cells, reverse-transcribed, and PCR amplified. PCR products are then deep-sequenced using a next-generation sequencing machine. The experiment is generally repeated in duplicate or triplicate. Sequences obtained from the sequencer

Junbai Wang et al. (eds.), *Next Generation Microarray Bioinformatics: Methods and Protocols*,
Methods in Molecular Biology, vol. 802, DOI 10.1007/978-1-61779-400-1_26,

are then analyzed. ShRNAs that are enriched and depleted in treated samples represent "resistant hits" and "synthetic lethal hits" (SL hits) for investigational drug, respectively. We are more interested in the "SL hits," as these genes can be used as the targets for the drug tested. The main bottleneck in SL screening processes is data analysis and interpretation, similar to other NGS applications. Therefore, in this work, we developed an efficient computational analysis pipeline to manage, analyze, and interpret the massive data for finding statistically significant gene products that are SL with the drug.

2. Materials

2.1. shRNA Library

Cells were infected with lentiviral carrying individual shRNAs from the GeneNet™ Human 50K shRNA library (SBI, Mountain View, CA). The SBI genome-wide shRNA library contains 213,562 unique shRNA sequences (27 bp). The rules for selecting shRNA sequences that are likely to effectively silence target genes of interest are similar to rules used to select short-probe sequences that are effective for microarray hybridization. On average, every gene was targeted by four shRNAs. To build the reference shRNA library, we mapped 213,562 unique shRNA sequences against the latest human genome (GRCh37) using Bowtie (5). From this mapping, 111,849 shRNAs can be mapped to 18,106 known gene regions with maximum of two mismatches, while the other shRNAs were mapped to contig regions. We build BWT (Burrows–Wheeler Transformation) index (6) on this reference shRNA library for mapping the sequences.

2.2. Synthetic Lethal Screen Using Next Generation Sequencing

To identify gene targets whose inhibition will cooperate with tested drugs to more effectively eliminate cancer cells, we designed a genome-wide RNAi-based loss-of-function screen (Fig. 1a). In our screen, we utilized a lentiviral-expressed genome-wide human shRNA library from SBI. Cancer cells were infected with the lentiviral shRNA library to obtain a pure population of shRNA expressing cells. Some period of growth also allowed for the elimination of shRNAs that target ("knockdown") essential genes. Cell line was then divided into two groups: one is untreated, and the other is treated with the drug, followed by a couple of days of culture without drug. Generally, each group is repeated in triplicate. RNA was then harvested from the cells and the shRNA sequences reverse transcribed using a primer specific to the vector. The cDNA was amplified by nested PCR. The primers for the second amplification include adapter sequences

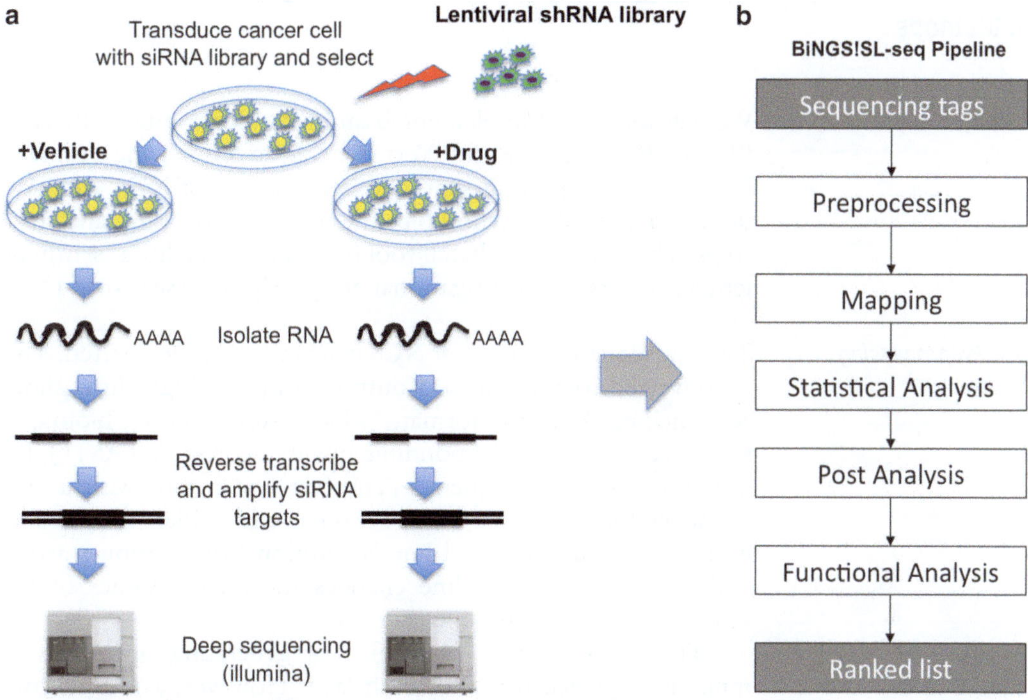

Fig. 1. Genome-wide RNAi-based loss-of-function screen. (**a**) Experimental approach. (**b**) Computational approach. The output of deep sequencing from (**a**) is the input for the BiNGS!SL-seq analysis pipeline (**b**).

specific for the Illumina Genome Analyzer$_{IIx}$. After the second amplification the cDNA includes only the 27-bp of the shRNAs followed by the short vector sequence. These PCR products were sequenced using the Genome Analyzer, which uses reversibly, fluorescence tagged bases and laser captured to perform massively parallel sequencing by synthesis. These sequences were then identified and the number of clusters for each shRNA sequence was quantified. In our example experiment, to identify synthetic lethal partners for the epidermal growth factor receptor (EGFR) inhibitor in lung cancer, we performed the genome-wide synthetic lethal screen by deep sequencing on two nonsmall cell lung cancer cell lines that exhibit intermediate and sensitive to this inhibitor (7). Over six million shRNAs were sequenced per sample by the NGS machine, representing more than 55,000 unique shRNAs. Candidate shRNA sequences underrepresented in the treated samples target genes whose inhibition sensitizes the cells to the drug. Conversely, those samples that are over-represented in the treated samples represent genes, the products of which are required for the cytotoxicity of the drug. Figure 1 describes the overall experimental and computational strategies.

3. Methods

We developed and implemented an innovative solution, BiNGS! SL-seq (*Bi*oinformatics for *N*ext *G*eneration *S*equencing), for analyzing and interpreting synthetic lethal screen of NGS data. We devised a general analytical pipeline that consists of five analytical steps. The pipeline is a batch tool to find the gene list as synthetic lethal partners for investigational drugs (Fig. 1b) (see Note 1).

3.1. Preprocessing

The raw sequence output of NGS machine is scarf formatted. This is converted to the standard output format of high-throughput sequencing, FASTQ format, which stores both biological sequence and its corresponding quality scores. A FASTQ file uses four lines per sequence. The first line begins with a "@" character followed by a unique sequence identifier. Second line is the raw sequence, third line is additional description starting with "+," and the last line encodes the quality values of the sequence in the second line (Fig. 2).

The NGS machine is capable of generating upto tens of millions sequence reads for each lane. However, as a trade-off, these speeds suffered from higher sequencing error rate. As an effort to avoid sequencing error, sometimes a barcode is used. In our preprocessing module, we filter out erroneous and low quality reads and converted the quality score from the sequencer to the quality score. Also, if the sequences were bar-coded, we use the barcode as reference for quality check and to filter out reads without barcode (Fig. 2). In this example, we used the 9-bp vector sequence as the barcode in this filtering step. As illustrated in Fig. 3, sequences contain a barcode, TTTTTGAAT, will be retained for further analysis while the last three sequences without the barcode will be discarded. Therefore, they are not converted to FASTQ formatted sequences and will not be mapped to the reference library, either.

The quality value of each sequence is calculated by two methods, the Sanger method known as Phred quality score and the Solexa method (see Note 2). The example of Fig. 2 is encode by Phred score + 64 (Illumina 1.3+). For raw reads, the range of scores is dependent on the technology and is generally up to 40. Generally, quality value decreases near the 3′ end of the reads (Fig. 3).

3.2. Mapping

Next, we mapped these reads against the shRNA reference library that we built based on the SBI shRNA sequences. The output from this step is a $P \times N$ matrix, where P and N represents the shRNA counts and samples, respectively. We use Bowtie (5), as the basis of the alignment and mapping component in the analysis

a

```
ILLUMINA-545855_0012:6:1:8865:1153#0/1:TTTGTCTTAAACAGATAGAGAGCATCATTTTTGAATTCGN:bbbbbbbbbbbbbbbbbbbbbbbbbbbbbbbbbbbbbbbB
ILLUMINA-545855_0012:6:1:8896:1155#0/1:CCATTTGAATACAGCGATGGATGGTCATTTTTGAATTCAN:]Y]KZ`_]___``\`]W``TOOOOJWY]]]````BBBBBB
ILLUMINA-545855_0012:6:1:9007:1160#0/1:CATCTTAGATATGGGAACTTAAGGGGATTTTTGAATTCGN:bbbbbbbbbbbbbbbbbbbb[W[[[bbbbbbbbbbaaa_B
ILLUMINA-545855_0012:6:1:9166:1162#0/1:CGGCTAGCTGAGACTCTTGGCCTGGAGTTTTTGAATTCGN:bbbbbbbbbabbbbab^aabbbbab_`_bbbbb`b]^^BB
ILLUMINA-545855_0012:6:1:9240:1158#0/1:TAACTAAACAATTGAAAGATACATTTCTTTTTGAATTCAN:bbbbbbbbbbbbbbbbbbbbbbbbbbbbbaaaabbbbbbB
ILLUMINA-545855_0012:6:1:1101:4281#0/1:AAAAAAAAAAAAAAAAAAAAAAAAAAAAANNNNNNNNNNN:`\Y`__BBBBBBBBBBBBBBBBBBBBBBBBBBBBBBBBBB
ILLUMINA-545855_0012:6:1:12940:1102#0/1:AGATCTGAAAGCATTTTTTGAATTCATGACCAATTCTTCN:bbbbbbbbbbbbbbbbbbbbbbbbbbbbbbbbbbbbbbbB
ILLUMINA-545855_0012:6:1:14732:1075#0/1:TTTAAGCTGTTTGGTAGGAANTCAAANATAAATTTAGTAN:bbbbbbbbbbbbbbBBBBBBBBBBBBBBBBBBBBBBBBBB
```

Raw read sequences (scarf format)

b

```
@ILLUMINA-545855_0012_6_1_8865_1153#0/1
TTTGTCTTAAACAGATAGAGAGCATCA
+
34 34 34 34 34 34 34 34 34 34 34 34 34 34 34 34 34 34 34 34 34 34 34 34 34 34 34
@ILLUMINA-545855_0012_6_1_8896_1155#0/1
CCATTTGAATACAGCGATGGATGGTCA
+
29 25 29 11 26 32 31 29 31 31 31 32 32 28 32 29 23 32 32 20 15 15 15 15 10 23 25
@ILLUMINA-545855_0012_6_1_9007_1160#0/1
CATCTTAGATATGGGAACTTAAGGGGA
+
34 34 34 34 34 34 34 34 34 34 34 34 34 34 34 34 34 34 34 34 27 23 27 27 27 34 34
@ILLUMINA-545855_0012_6_1_9166_1162#0/1
CGGCTAGCTGAGACTCTTGGCCTGGAG
+
34 34 34 34 34 34 34 34 34 33 34 34 34 34 33 34 30 33 33 34 34 34 34 33 34 31 32
@ILLUMINA-545855_0012_6_1_9240_1158#0/1
TAACTAAACAATTGAAAGATACATTTC
+
34 34 34 34 34 34 34 34 34 34 34 34 34 34 34 34 34 34 34 34 34 34 34 34 34 34 34
```

FASTQ formatted sequences

Fig. 2. An example of read sequences. (**a**) Scarf formatted sequences. These contain a 9-bp barcode, TTTTTGAAT at the 3′ end of the sequences. (**b**) FASTQ formatted sequences. The last three read sequences are not converted to FASTQ sequences because of barcode errors. FASTQ formatted sequences will be input of mapping programs.

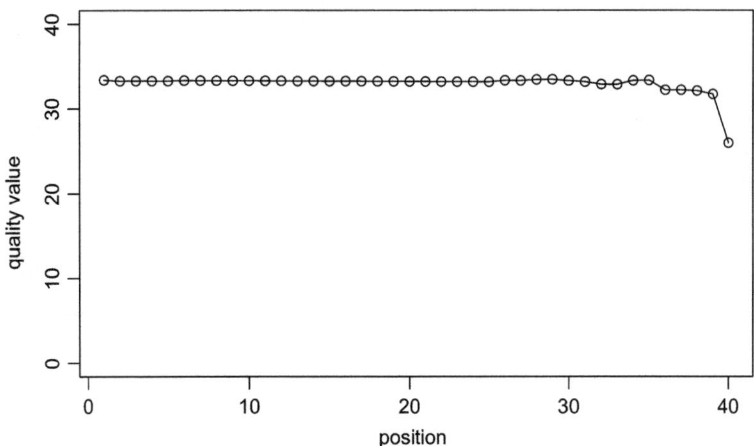

Fig. 3. Relationship between quality value and the position of reads. Sequencing qualities drop near the 3′ end of the reads.

pipeline. Bowtie employs the Burrows–Wheeler Transformation (BWT) indexing strategy (6), which allows large sets of sequences to be searched efficiently in a small memory footprint and performs faster as compared to the hash-based indexing methods, with equal or greater sensitivity. We allowed unique mapping with two mismatches. From our experience with more than ten synthetic lethal screen analyses, 60–70% of the raw reads are mapped

Table 1
Summary of synthetic lethal screen data of EGFR tyrosine kinase inhibitor experiment in two non-small cell lung cancer cell lines

Data			Number of sequence tags	Number of sequence tags passed filtering	Number of tags mapped to shRNA library (213,562 shRNAs)	Number of tags mapped to shRNA library represents to gene (111,849 shRNAs)
Cell line #1	Control group	C1	7,397,899	6,497,236 (87%)	4,530,246 (61%)	3,365,202 (45%)
		C2	7,189,768	6,286,679 (87%)	4,386,199 (61%)	3,257,177 (45%)
		C3	6,682,685	5,843,273 (87%)	4,081,528 (61%)	3,041,599 (46%)
	Treatment group	T1	6,019,739	5,117,651 (85%)	3,544,787 (59%)	2,625,611 (44%)
		T2	6,647,530	5,758,762 (87%)	3,994,710 (60%)	2,964,899 (45%)
		T3	6,630,475	5,733,016 (86%)	3,977,493 (60%)	2,960,004 (45%)
Cell line #2	Control group	C1	7,976,052	7,266,004 (91%)	4,791,506 (60%)	3,495,683 (44%)
		C2	8,084,137	7,382,139 (91%)	4,849,828 (60%)	3,538,347 (44%)
		C3	7,957,330	7,251,081 (91%)	4,770,303 (60%)	3,496,462 (44%)
	Treatment group	T1	7,925,668	7,233,517 (91%)	4,769,845 (60%)	3,473,641 (44%)
		T2	6,638,274	6,013,615 (91%)	3,968,719 (60%)	2,899,982 (44%)
		T3	6,470,612	5,883,321 (91%)	3,897,280 (60%)	2,850,055 (44%)

to the reference library. However, when we consider only shRNAs representing known genes, about 45% of raw reads are mapped. In our lung cancer examples (Table 1), all samples have 6–8 millions of 40 bp long reads. On average, 60% of the reads were mapped to the shRNA reference library from the two lung cancer cell experiments (Table 1).

3.3. Statistical Analysis

Before we performed the statistical test, we filtered out shRNAs where the median raw count in the control group is greater than the maximum raw count in the treatment group if the shRNA is enriched in the control group, and vice versa. This filtering step decreases the number of false positives, and gives us more confidence in detecting the real biological signals. After this filtering step, we employ Negative Binomial (NB) to model the read counts data. The Poisson distribution is commonly used to model count data. However, due to biological and genetic variations, for sequencing data the variance of a read is often much greater than the mean value. That is, the data are over dispersed in this case. From our preliminary study (8), we have identified that a NB distribution best models the count data generated by NGS. Here, we implemented NB as the statistical model in our pipeline to model the count distribution in the NGS data using edgeR (9). We also compute the q-value of FDR (false discovery rate) for multiple comparisons for these shRNAs.

3.4. Postanalysis

As a gene can be targeted by multiple shRNAs, we performed meta-analysis by combining p-values of all the shRNAs representing the same gene using weighted Z-transformation method. Fisher's combined probability test (10) is commonly used in meta-analysis (combining independent p-values). This method is based on the product of adjusted p-values, which follows a chi-square distribution with $2k$ degrees of freedom (where k = number of p-values). Variations of Fisher's combined probability test were introduced in the literatures, notably weighted Fisher's method (11). Alternative to Fisher's approach is to employ the inverse normal transformation (or Z-transformation) of the adjusted p-values and combined the Z-scores or the weighted Z-score method (12, 13). In (13), it was demonstrated that to test for a common null hypothesis, the Z-transformation approach is better than the Fisher's approach. As a procedure that combines the Z-transformation method, we adopted weighted Z-transformation (13) that puts more weight to the small adjusted p-value shRNA (see Note 3). Using this weighted Z-transformation method, we can collapse multiple shRNAs into genes, with an associated p-value ($P(wZ)$). We use $P(wZ)$ to sort the list for identifying synthetic lethal (SL) hits. Also, with another example of Leukemia cell line experiment, we noticed that from the distributions of p-values, the p-value distribution of combined genes by weighted Z-transformation method looks a mixture of distributions of null hypothesis and alternative hypothesis (Fig. 4).

From the BiNGS!SL-seq analysis, using $P(wZ) < 0.05$ as the cut-off, 1,237 and 758 genes were enriched in the EGFR inhibitor treatment group for cell line #1 and #2, respectively. We found 106 overlapping genes from both cell lines. These genes represent the SL hits for EGFR inhibitor in the lung cancer. These overlapping genes are statistically significant based on 10,000 simulations on randomly selected genes ($p < 0.0001$).

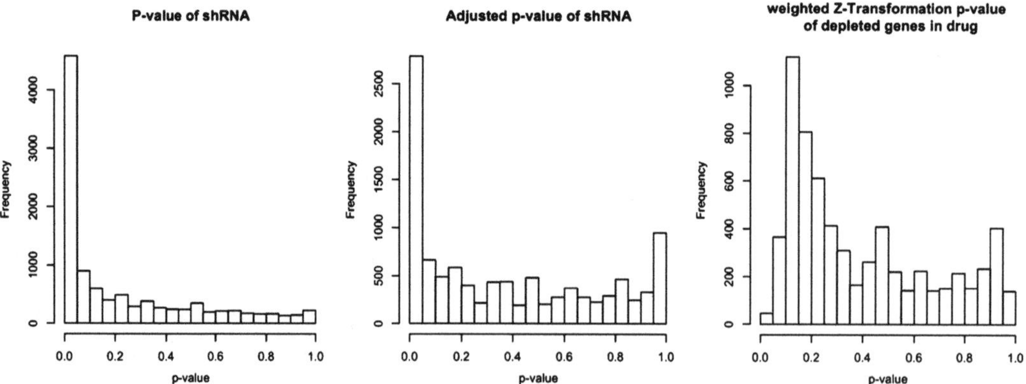

Fig. 4. Distributions of p-value, adjusted p-value by multiple correction, and p-value of weighted Z-transformation.

3.5. Functional Analysis

To delineate the functionality of the SL hits, we performed enrichment analysis on the final gene list using the NIH DAVID functional analysis tool (14, 15). In our lung cancer experiment, to identify synthetic lethal pathways to the EGFR inhibitor, we performed enrichment analysis on the 106 common SL hits using NIH DAVID. From the KEGG pathway results, we found several pathways enriched with multiple SL hits. The top two enriched pathways were "colorectal cancer pathway (hsa05210)" ($p = 0.02$) and "Wnt signaling pathway (hsa04310)" ($p = 0.02$). Both pathways were interconnected, and the enriched SL genes were involved in the canonical Wnt signaling pathway (16). Using the enriched pathway as the seed, we then extended the search in individual hits generated from both cell lines to identify additional SL partners in this pathway that are not defined by KEGG pathway.

4. Notes

1. *BiNGS!SL-seq:* We have developed BiNGS!SL-seq to analyze and interpret genome-wide synthetic lethal screen by deep sequencing. The BiNGS!SL-seq consists of five analytical steps: Preprocessing, Mapping, Statistical Analysis, Postanalysis, and Functional Analysis.
2. *Quality Score*: The following two equations represent both methods:

$$Q_{\text{sanger}} = -10\log_{10}(p) \quad (1)$$

$$Q_{\text{solexa}} = -10\log_{10}(p/(1-p)), \quad (2)$$

where p is the probability that the corresponding base call is incorrect. Both methods are asymptotically identical at higher values, approximately $p < 0.05$ is equivalent to $Q > 13$.

Alternatively, ASCII encoding can be applied for interpreting the quality score of the reads.

3. *Weighted Z-transformation method*: Let k shRNAs representing the gene g, we will use the weighted Z-transformation method to collapse these shRNAs to obtain an estimated p-value for gene g. The equation for weighted Z-transformation method:

$$Zw(g) = \frac{\sum_{i=1}^{k} w_i Z_i}{\sqrt{\sum_{i=1}^{k} w_i^2}}, \tag{3}$$

where $w_i = (1 - p_i)$, p_i is the adjusted p-value of ith shRNA calculated from exact test based on negative binomial model. Using this weighted Z-transformation method, we can collapse multiple shRNAs into genes, with an associated p-value ($P(wZ)$) for each gene.

4. *Summary*: Using this computational approach, we identified multiple pathways important for NSCLC survival following EGFR inhibition, and inhibition of these pathways has the potential to potentiate anti-EGFR therapies for NSCLC. We believe that the BiNGS!SL-seq can be applied to analyze and interpret different synthetic lethal screens using next generation sequencing in revealing novel therapeutic targets for various cancer types.

Acknowledgments

The authors unreservedly acknowledge the experimental and computational expertise of the BiNGS! Team – James DeGregori, Christopher Porter, Joaquin Espinosa, S. Gail Eckhart, John Tentler, Todd Pitts, Mark Gregory, Matias Casa, Tzu Lip Phang, Dexiang Gao, Hyunmin Kim, Tiejun Tong, and Heather Selby.

References

1. Gregory MA, Phang TL, Neviani P et al (2010) Wnt/Ca2+/NFAT signaling maintains survival of Ph + leukemia cells upon inhibition of Bcr-Abl. Cancer Cell 18: 74–87.
2. Luo J, Emanuele MJ, Li D et al (2009) A genome-wide RNAi screen identifies multiple synthetic lethal interactions with the Ras oncogene. Cell 137: 835–848.
3. Azorsa DO, Gonzales IM, Basu GD et al (2009) Synthetic lethal RNAi screening identifies sensitizing targets for gemcitabine therapy in pancreatic cancer. J Transl Med 7:43.
4. Whitehurst AW, Bodemann BO, Cardenas J et al (2007) Synthetic lethal screen identification of chemosensitizer loci in cancer cells. Nature 446:815–819.

5. Langmead B, Trapnell C, Pop M et al (2009) Ultrafast and memory-efficient alignment of short DNA sequences to the human genome. Genome Biol 10:R25.
6. Burrows M, Wheeler DJ. (1994) A block-sorting lossless data compression algorithm. HP Labs Technical Reports SRC-RR-124.
7. Helfrich BA, Raben D, Varella-Garcia M et al (2006) Antitumor activity of the epidermal growth factor receptor (EGFR) tyrosine kinase inhibitor gefitinib (ZD1839, Iressa) in non-small cell lung cancer cell lines correlates with gene copy number and EGFR mutations but not EGFR protein levels. Clin Cancer Res 12:7117–7125.
8. Gao D, Kim J, Kim H et al (2010) A survey of statistical software for analyzing RNA-seq data. Human Genomics 5:56–60.
9. Robinson MD, McCarthy DJ, Smyth GK (2010) edgeR: a Bioconductor package for differential expression analysis of digital gene expression data. Bioinformatics 26:139–140.
10. Fisher S (1932) Statistical methods for research workers. Genesis Publishing Pvt Ltd.
11. Goods I (1955) On the weighted combination of significance tests. Journal of the Royal Statistical Society. Series B (Methodological) 17:264–265.
12. Wilkinson B (1951) A statistical consideration in psychological research. Psychological Bulletin 48:156–158.
13. Whitlock MC (2005) Combining probability from independent tests: the weighted Z-method is superior to Fisher's approach. J Evol Biol 18:1368–1373.
14. Huang DW, Sherman B, Lempicki RA (2008) Systematic and integrative analysis of large gene lists using DAVID bioinformatics resources. Nature Protocols 4:44–57.
15. Dennis G Jr, Sherman BT, Hosack DA et al (2003) DAVID: Database for Annotation, Visualization, and Integrated Discovery. Genome Biol 4:P3.
16. Klaus A, Birchmeier W (2008) Wnt signalling and its impact on development and cancer. Nat Rev Cancer 8:387–98.

INDEX

Junbai Wang et al. (eds.), *Next Generation Microarray Bioinformatics: Methods and Protocols*,
Methods in Molecular Biology, vol. 802, DOI 10.1007/978-1-61779-400-1, © Springer Science+Business Media, LLC 2012

T

MIX
Papier aus verantwortungsvollen Quellen
Paper from responsible sources
FSC® C105338

If you have any concerns about our products,
you can contact us on
ProductSafety@springernature.com

In case Publisher is established outside the EU,
the EU authorized representative is:
Springer Nature Customer Service Center GmbH
Europaplatz 3, 69115 Heidelberg, Germany

Printed by Libri Plureos GmbH
in Hamburg, Germany